Graduate Texts in Physics

Graduate Texts in Physics

Graduate Texts in Physics publishes core learning/teaching material for graduate- and advanced-level undergraduate courses on topics of current and emerging fields within physics, both pure and applied. These textbooks serve students at the MS- or PhD-level and their instructors as comprehensive sources of principles, definitions, derivations, experiments and applications (as relevant) for their mastery and teaching, respectively. International in scope and relevance, the textbooks correspond to course syllabi sufficiently to serve as required reading. Their didactic style, comprehensiveness and coverage of fundamental material also make them suitable as introductions or references for scientists entering, or requiring timely knowledge of, a research field.

More information about this series at http://www.springer.com/series/8431

Frank Grossmann

Theoretical Femtosecond Physics

Atoms and Molecules in Strong Laser Fields

Third Edition

Springer

Frank Grossmann
Institut für Theoretische Physik
Technische Universität Dresden
Dresden, Saxony
Germany

ISSN 1868-4513 ISSN 1868-4521 (electronic)
Graduate Texts in Physics
ISBN 978-3-030-09016-6 ISBN 978-3-319-74542-8 (eBook)
https://doi.org/10.1007/978-3-319-74542-8

Softcover re-print of the Hardcover 3rd edition 2018

Printed on acid-free paper

This Springer imprint is published by the registered company Springer Nature Switzerland AG
The registered company address is: Gewerbestrasse 11, 6330 Cham, Switzerland

To the memory of my father

Hans Grossmann

Preface to the Third Edition

I would like to thank Claus Ascheron for suggesting a third edition of this book. This has allowed me to add a considerable amount of new material, mainly in the applications part, that is rounding off the presentation.

In Chap. 3, the discussion of the Volkov solution for the laser-driven free particle has been extended to 3D and plane waves. In the atomic physics chapter, a discussion of the two-electron Helium atom and of the knee structure in its double ionization, as well as a review of ATI rings and of low energy structures, that can be observed in the photo-electron spectra of rare gas atoms, have been added. Furthermore, dominant interaction Hamiltonians are discussed and used in the context of high-order harmonic generation. In addition, the section on the Keldysh parameter has been extended. In the molecular physics part, the sections on the Born-Oppenheimer approximation and on pump-probe spectrocopy have been expanded by a more explicit discussion of nonadiabatic dynamics and the inclusion of 2D IR spectroscopy in the frequency domain. For those readers who have caught fire and would like to dig deeper into the literature, I have now more clearly structured the "Notes and Further Reading" sections. Finally, some more exercises together with their solutions have been added, mainly to the chapters on atomic and molecular physics.

I am grateful for inspiring discussions and scientific as well as teaching collaborations with Jan-Michael Rost and Ulf Saalmann. Furthermore, Carlos Zagoya has produced exciting new results on high-order harmonic generation during his Ph.D. thesis, that are the basis for the new discussion of dominant interaction Hamiltonians in the present edition. Valuable input in terms of graphs and/or fruitful discussions from John Briggs, Thomas Fennel, Stefanie Gräfe, Jost Henkel, Manfred Lein, Max Möller, Yoshitaka Tanimura and Sandro Wimberger is gratefully acknowledged. For help with technical issues I thank Sandra Grossmann, and for uncountable day-to-day discussions on quantum dynamics in general, I am grateful to Max Buchholz, Tobias Fiedlschuster, Jan Handt, and Michael Werther.

Finally, I would once more like to express my pleasure and gratitude for the opportunity to interact with highly motivated students and to participate in the very inspiring scientific environment in atomic and molecular physics in Dresden.

Dresden, Germany
November 2017

Frank Grossmann

Preface to the Second Edition

The main motivation for this second edition of the book was the inclusion of more exercises and, most importantly, also a collection of solutions of these exercises at the end of the book. Some type errors and more severe flaws of the first edition that have come to my attention have been corrected. The book is, however, intended as a comprehensive introduction to a topic of current interest and not as a complete review of the field. I have therefore not added more (advanced) material into the main text. Furthermore, I have reservedly updated the literature list by including a few more research papers, although a lot more work has been done in the last five years and a related new field, the field of attosecond physics, has emerged. Some more textbook references and reviews have also been added where appropriate.

Again, I would like to express my thanks for enlightening discussions on the topics touched herein to more generations of students that attended my related lectures at TU Dresden and also to Larry Schulman, Jan Handt, Werner Koch, Christoph-Marian Goletz, Alexander Kästner, Michael Fischer, Carlos Zagoya, Sebastian Krause, Niklas Rohling, Max Buchholz, and Tobias Fiedlschuster for ongoing exchange of ideas and to Alexander Kästner for providing numerical results.

For making the book available as part of the series "Graduate Texts in Physics", and for support during the production stage I would like to thank Claus Ascheron and his team from Springer Verlag.

Dresden, Germany
March 2013

Frank Grossmann

Preface to the First Edition

The development of modern pulsed lasers with intensities larger than $10^{16}\,\frac{\mathrm{W}}{\mathrm{cm}^2}$ and with very short pulse duration in the femtosecond regime enables experimentalists to study elementary processes such as chemical reactions and excitation mechanisms in different areas in physics in the time domain. In parallel to the experimental investigations, analytical and numerical studies of laser driven atoms and molecules with a limited number of degrees of freedom are performed. These theoretical investigations have led to the prediction and/or the explanation of a large variety of partly counter-intuitive phenomena. Among those are the generation of high harmonics using laser excited atoms or molecules, the ionization of atoms above the continuum threshold, the stabilization of atoms against ionization in very strong fields, counter-intuitive pulse sequences to selectively populate vibrational states in molecules and, last but not least, the control of chemical reactions by specially tailored laser pulses.

This book originated out of a course, that I have given on a regular basis since 2000 for advanced undergraduate and graduate students at Technische Universität Dresden. It offers a theoretically oriented approach to the field of laser driven atomic and molecular systems and requires some knowledge of basic classical and quantum mechanics courses as well as of classical electrodynamics. The book has two introductory chapters in part I that pave the way for the applications in part II. Part I and also Chap. 3 of Part II contain a selection of textbook material that is needed to understand the rest of the book. The material presented in the last two chapters is close to the recent literature. I have chosen only such works, however, that deal with fundamental concepts and are based on simple model calculations. A biased and incomplete list a references is given at the end of the chapters, preceded by some notes and hints for further reading. For those readers who are interested in some computational details, these are given in the appendices at the end of the corresponding chapters. Furthermore, at several places throughout the text, exercises are placed, whose independent solution allows a deeper understanding of the material presented.

In Chap. 1, we start with a short introduction into the foundations of the laser. We will especially concentrate on those aspects of pulsed lasers that will be important for the theoretical investigations in part II of the book.

The next fundamental chapter is devoted to the non-relativistic time-dependent Schrödinger equation. In the case of lasers of up to atomic field strengths, this equation allows the theoretical description of the phenomena we want to investigate in part II. Analytical as well as numerical methods to solve the time-dependent Schrödinger equation are thus in the focus of Chap. 2. Throughout the whole presentation, in order to keep the approach as simple as possible, we touch the topic of correlated many particle dynamics only where necessary and concentrate on the description of electronic as well as nuclear dynamics with the help of models with as few degrees of freedom as possible. The contents and the presentation of Chap. 2 are inspired by the excellent new textbook by David Tannor, *Introduction to Quantum Mechanics: A Time-Dependent Perspective*, which hopefully will start a "revolution" in the way quantum mechanics is taught in the future.

The second part of the book, starting with Chap. 3, contains a collection of equivalent ways to couple a charged particle to a classical electromagnetic field. As the basic postulate we use the principle of minimal coupling. By using unitary transformations, one can then either derive the length form or the Kramers-Henneberger form of the coupling. As first examples of laser-matter interaction, we study the dynamics of (structure-less) two-level systems in laser fields. Phenomena like Rabi oscillations of the occupation probability, occurring there, will be encountered off an on in the remainder of the book. Furthermore, also the fundamental so-called rotating wave approximation will be discussed for the first time in this context.

Selected examples of laser-matter interaction in atomic physics are reviewed in Chap. 4. Here, we concentrate on the phenomena of ionization and high harmonic generation of a single electron in a Coulomb potential of possibly reduced dimensionality. It turns out that a perturbation theoretic approach would not be suited to understand most of the phenomena presented in this chapter. Thus the numerical wavepacket methods that were in the focus of Chap. 2 will find their first application.

The next step in the direction of higher complexity of the dynamics will be taken in Chap. 5. Here we deal with laser-driven systems in molecular and chemical physics. The simplest molecule, the hydrogen molecular ion, H_2^+, will serve as a vehicle to understand some of the basic concepts of molecular physics such as electronic potential surfaces. In the following, the full numerical solution of the coupled electron nuclear problem of H_2^+ in a monochromatic laser field will be reviewed. After discussing the fundamental Born-Oppenheimer approximation, for the rest of the chapter, we then assume that the solution of the electronic many body problem is at our disposal in the form of analytically or numerically given potential energy surfaces. After a short digression on nuclear motion on a single electronic surface, and the discussion of a simple two coupled surfaces problem, we then review some modern applications in the fields of femtosecond spectroscopy, optimal control theory, and quantum information processing under the foregoing assumption.

At this point I thank the students at TU Dresden who have attended my lectures. They have inspired me enormously, through their intense collaboration, during the lectures, as well as during the exercise classes. This has motivated me to rethink the material presented again and again and the students have thus contributed substantially to the improvement of the manuscript. Also the hospitality of the Max-Planck-Institute for the Physics of Complex Systems, that offered me the opportunity to attend and run several conferences in the field was very important to shape my understanding presented here. Furthermore, I express my deep gratitude to Jan-Michael Rost and Rüdiger Schmidt for their continuous availability for discussions and for long-term collaboration. Moreover, I am grateful to Peter Hänggi for the introduction to the field of driven quantum systems during my Ph.D. work with him, and to Eric Heller for opening the world of time-dependent semiclassics to me. In addition, I have benefitted from uncountable discussions with and valuable advice of former members of the Theoretical Quantum Dynamics Group in Freiburg, especially Gernot Alber, Richard Dehnen, Volker Engel, Christoph Meier, Gerd van de Sand, and Gerhard Stock. Furthermore, former and present members of the Theoretical Atomic and Molecular Physics Group at the Institute of Theoretical Physics of TU Dresden and the Finite Systems Department at the MPIPKS in Dresden have helped shape my understanding. Among many others these are Andreas Becker, Agapi Emmanouilidou, Celal Harabati, Anatole Kenfack, Thomas Kunert, Ulf Saalmann, and Mathias Uhlmann. For helping me by answering specific questions or supplying information and valuable graphs, I would like to thank Wolfgang Schleich, Jan Werschnik, Matthias Wollenhaupt and Shuhei Yoshida. For advice and help with respect to graphics issues, I thank Arnd Bäcker and Werner Koch. Finally, I am indebted to David Tannor, who supplied me with preliminary versions of his book at a very early stage and thus helped shape the presentation here to a substantial degree. The focus of David's book on a time-dependent view of quantum phenomena is an absolute necessity if one wants to study laser-driven systems.

Dresden, Germany Frank Grossmann
May 2008

Contents

Part I
Prerequisites

Chapter 1
A Short Introduction to Laser Physics

To study the influence of light on the dynamics of an atom or a molecule experimentally, laser-light sources are used most frequently. This is due to the fact that laser light has well-defined and often tunable properties. The theory of the laser dates back to the 1950s and 1960s and by now is textbook material. In this introductory chapter, we start by recapitulating some basic notions of laser theory, which will be needed later-on.

More recently, experimentalists have been focusing on pulsed mode operation of lasers with pulse lengths on the order of femtoseconds, allowing for time-resolved measurements. At the end of this chapter, we therefore put together some properties of pulsed lasers that are important for their application to atomic and molecular systems. The characterization of laser fields by windowed Fourier transforms concludes this chapter.

1.1 The Einstein Coefficients

Laser activity may occur in the case of nonequilibrium, as we will see below. Before dealing with this situation, let us start by considering the case of equilibrium between the radiation field and the atoms of the walls of a cavity. This will lead to the Einstein derivation of Planck's radiation law.

The atoms will be described in the framework of Bohr's model of the atom, allowing the electron to occupy only discrete energy levels. For the derivation of the radiation law, the consideration of just two of those levels is appropriate. They shall be indexed by 1 and 2 and shall be populated such that for the total number of atoms

$$N = N_1 + N_2 \tag{1.1}$$

F. Grossmann, *Theoretical Femtosecond Physics*, Graduate Texts in Physics,
https://doi.org/10.1007/978-3-319-74542-8_1

holds. This means that N_2 of the atoms are in the excited state with energy E_2 and N_1 atoms are in the ground state with energy E_1. Transitions between the states shall be possible by emission or absorption of photons of the appropriate energy. The following processes can be distinguished:

- Absorption of light, leading to a transition rate

$$\left.\frac{dN_2}{dt}\right|_{abs} = \rho N_1 B_{12} \tag{1.2}$$

  from the ground to the excited state
- Induced (or stimulated) emission of light, leading to a transition rate

$$\left.\frac{dN_1}{dt}\right|_{emin} = \rho N_2 B_{21} \tag{1.3}$$

  for the population change of the ground state
- Spontaneous emission of light, leading to a rate

$$\left.\frac{dN_1}{dt}\right|_{emsp} = N_2 A, \tag{1.4}$$

  which amounts to a further increase of the ground state population

The first two processes are proportional to the energy density ρ of the radiation field times the population of the initial state with the constants B_{12}, respectively B_{21}. The process of spontaneous emission does not depend on the external field and is proportional to A. These coefficients are called Einstein's A- and B-coefficients.

In thermal equilibrium, the rate of transition from level 1 to 2 has to equal that from 2 to 1, leading to the stationarity condition

$$N_1 B_{12} \rho = N_2 B_{21} \rho + N_2 A. \tag{1.5}$$

This equation can be resolved for the energy density ρ leading to

$$\rho = (N_1 B_{12}/(N_2 B_{21}) - 1)^{-1} A/B_{21}. \tag{1.6}$$

Furthermore, in thermal equilibrium the ratio of populations is given by the Boltzmann factor according to

$$N_1/N_2 = \exp\left\{-\frac{E_1 - E_2}{kT}\right\}, \tag{1.7}$$

with the temperature T and the Boltzmann constant k.

As $T \to \infty$ also $\rho \to \infty$, and the B-coefficients have to be identical $B_{12} = B_{21} = B$. Using Bohr's postulate

$$E_2 - E_1 = h\nu, \tag{1.8}$$

where ν is the frequency of the light, we can conclude from (1.6) that

$$\rho = \left(\exp\left\{\frac{h\nu}{kT}\right\} - 1\right)^{-1} A/B. \tag{1.9}$$

holds. The ratio of Einstein coefficients A/B can now be determined by comparing the formula above with the experimentally confirmed Rayleigh-Jeans law

$$\rho(\nu) = \left(8\pi/c^3\right)\nu^2 kT, \tag{1.10}$$

which is valid in the case of low frequencies (see also Fig. 1.1) and where c is the speed of light in vacuum. One then arrives at

$$A/B = \left(8\pi/c^3\right)h\nu^3 =: D(\nu)h\nu \tag{1.11}$$

for the ratio, where $D(\nu)\mathrm{d}\nu = 8\pi\nu^2/c^3\mathrm{d}\nu$ is the number of possible waves in the frequency interval from ν to $\nu + \mathrm{d}\nu$ in a cavity of unit volume [1]. Inserting this result into (1.9) yields Planck's radiation law

$$\rho(\nu)\mathrm{d}\nu = (D(\nu)\mathrm{d}\nu)h\nu\left(\exp\left\{\frac{h\nu}{kT}\right\} - 1\right)^{-1}. \tag{1.12}$$

The last factor in this expression is the number of photons with which a certain wave is occupied. As a function of the wavelength,[1]

$$\lambda = \frac{c}{\nu}, \tag{1.13}$$

Figure 1.1 shows a comparison of Planck's law with the two laws valid in the limits of either long or short wavelength. These are the Rayleigh-Jeans and Wien's law (which emerges from Planck's law by omitting the constant term in the denominator), respectively.

In the case of nonequilibrium an extension of the formalism just reviewed leads to the fundamentals of laser theory, as we will see in the following. The explicit calculation of the B-coefficient shall be postponed until Chap. 3.

[1]Due to $\mathrm{d}\nu = -c/\lambda^2\mathrm{d}\lambda$, we get $\rho(\lambda)\mathrm{d}\lambda = 8\pi hc/\lambda^5\mathrm{d}\lambda\left(\exp\left\{\frac{hc}{kT\lambda}\right\} - 1\right)^{-1}$.

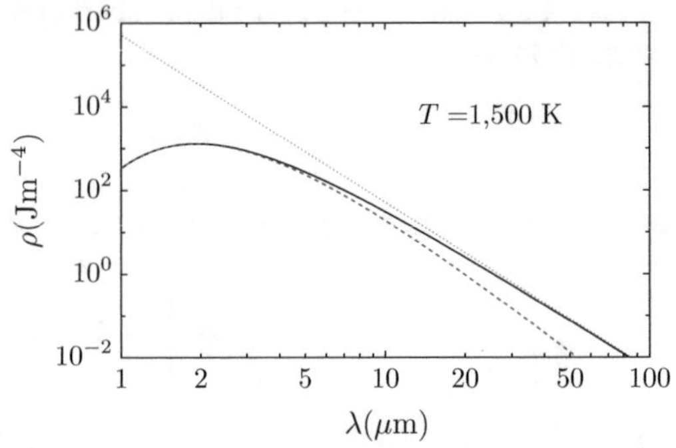

Fig. 1.1 Energy density (per wavelength interval) plotted double logarithmically as a function of wavelength for different radiation laws (Planck (*solid black*), Wien (*dashed blue*), Rayleigh-Jeans (*dotted red*)) at a temperature of T = 1,500 K

1.2 Fundamentals of the Laser

The derivation of the conditions for laser activity can be done in a crude way by again considering the populations of two levels between which the laser transition occurs. They shall be driven out of equilibrium by pumping and are interacting with a light field with photon number n (considered to be a continuous variable in the following) in a resonator [2].

First, we consider the processes leading to a change in the populations. In addition to the ones introduced in Sect. 1.1, these are pump (or gain) and loss processes. We concentrate on laser activity and therefore spontaneous emission can be neglected for the time being. Secondly, the realization of the laser process, requiring more than a bare two-level system will be discussed.

1.2.1 Elementary Laser Theory

In close analogy to the Einstein coefficients for the induced transition rates, coefficients can be defined that fulfill $W_{ij} = W_{ji} = W$, leading to an induced emission rate of $(N_2 - N_1)Wn$.[2] Including the gain and loss processes depicted in Fig. 1.2 the rate equations

$$\frac{\mathrm{d}N_1(t)}{\mathrm{d}t} = \gamma_{12}N_2 - \Gamma N_1 + (N_2 - N_1)Wn, \tag{1.14}$$

$$\frac{\mathrm{d}N_2(t)}{\mathrm{d}t} = \Gamma N_1 - \gamma_{12}N_2 - (N_2 - N_1)Wn \tag{1.15}$$

[2]Note that in the previous section the rate was proportional to ρ and here it is proportional to the dimensionless variable n; we therefore have to use a different symbol for the coefficients.

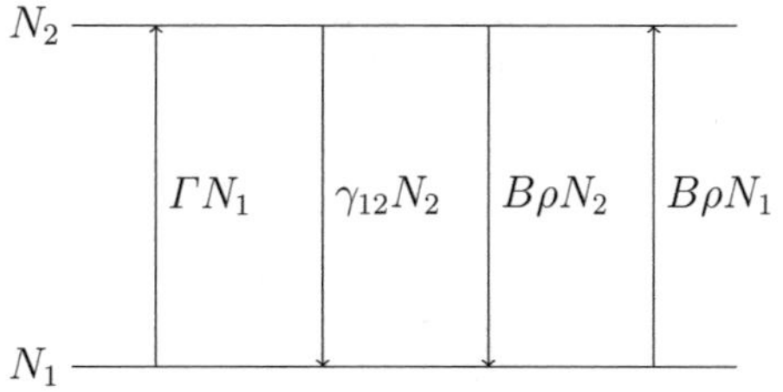

Fig. 1.2 Two-level system with elementary transitions (from the left to the right): Pump process, loss processes (e.g. by radiationless transitions), induced emission and absorption; spontaneous emission is not considered; adapted from [2]

emerge. Subtracting the first from the second equation leads to a rate equation for the difference $D = N_2 - N_1$, which is also referred to as population inversion

$$\frac{\mathrm{d}D}{\mathrm{d}t} = -2WnD - \frac{1}{T_1}(D - D_0). \tag{1.16}$$

Here the definitions of the unsaturated inversion $D_0 = N(\Gamma - \gamma_{12})/(\Gamma + \gamma_{12})$, which will become clear below, and the relaxation time $T_1 = (\Gamma + \gamma_{12})^{-1}$ have been introduced. Including loss effects of the optical cavity via a parameter t_{cav}, the rate equation for the photon number

$$\frac{\mathrm{d}n}{\mathrm{d}t} = WnD - \frac{n}{t_{\mathrm{cav}}} \tag{1.17}$$

follows, where the first term is due to the increase of radiation due to stimulated processes and the effect of spontaneous emission has been neglected. Equation (1.16) for the inversion together with (1.17) for the photon number are a simplified version of the full quantum mechanical laser equations, allowing one to understand some basic laser properties [2, 3].

For an amplification of the light field to occur by starting from a low initial photon number n_0 with unsaturated inversion D_0, the right hand side of (1.17) has to be larger than zero. For reasons of simplicity, let us here just consider the steady state defined by

$$\frac{\mathrm{d}n}{\mathrm{d}t} = 0 \quad \frac{\mathrm{d}D}{\mathrm{d}t} = 0, \tag{1.18}$$

however. For the inversion we get from (1.16)

$$D = D_0/(1 + 2T_1Wn), \tag{1.19}$$

i.e., a reduction for a finite photon number as compared to the unsaturated value D_0. The photon number in the steady state is

$$n\left(\frac{WD_0}{1 + 2T_1Wn} - \frac{1}{t_{\mathrm{cav}}}\right) = 0, \tag{1.20}$$

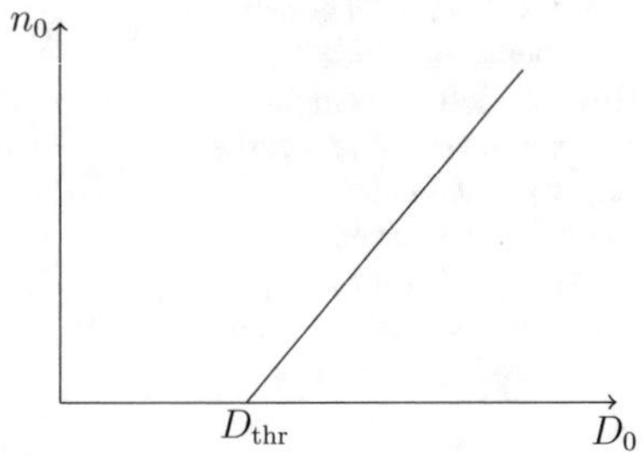

Fig. 1.3 Steady state photon number versus unsaturated inversion

leading to two different solutions:

(1) $n_0 = 0$
(2) $n_0 = (D_0 - D_{\text{thr}})\frac{t_{\text{cav}}}{2T_1}$.

In order for the nontrivial solution to be larger than zero, the inversion has to be larger than a threshold value $D_{\text{thr}} = 1/(Wt_{\text{cav}})$. As a function of D_0, the transition from the trivial solution to the one with a finite number of photons is depicted in Fig. 1.3.

In principle, laser theory has to be formulated quantum theoretically. This is done in, e.g., in [2]. There the transition from a standard light source to a laser above threshold is explained in a consistent framework. For large photon numbers one finds the phenomenon of anti-bunching, i.e. the photons leave the cavity equidistantly. The corresponding laser light has a constant amplitude. Therefore in the applications part of this book, we will assume that the field can be described classically by using a sinusoidal oscillation.

1.2.2 *Realization of the Laser Principle*

As we have just seen, nonequilibrium, characterized by population inversion, is crucial for operating a laser. Since the invention of the first maser,[3] it has been shown that inversion can be achieved in many different ways. A small collection of possibilities (including also the microwave case) will now be discussed.

1.2.2.1 The Ammonia Maser

In the NH_3-maser [4], the umbrella mode (see Fig. 1.4) leads to a double-well potential and thus quantum mechanically tunneling is possible. A corresponding doublet of levels in the double well exists, which is used for the maser process. Inversion is created by separating the molecules in the upper level from the ones in the lower level by using the quadratic Stark effect in an inhomogeneous electric field.

[3]Maser stands for "Microwave amplification by stimulated emission of radiation".

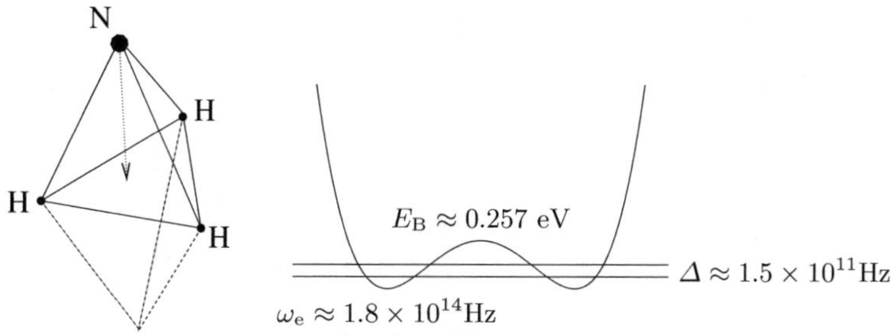

Fig. 1.4 Umbrella mode of NH_3 indicated by the arrow and schematic double well (frequency of oscillation around the minima ω_e, tunneling frequency Δ, and barrier height E_B are indicated) with the two levels (their separation is vastly exaggerated for reasons of better visibility) used for the maser process

This principle cannot be applied in the optical (laser) case, however, since typically $h\nu \gg kT$ at optical frequencies and therefore $N_2 \ll N_1$. Increasing the number of atoms in the upper level via pumping is therefore necessary.

1.2.2.2 The Ruby Laser

In order to achieve inversion in a laser, more than two levels are needed. Solid state lasers like the three level ruby laser [5] are pumped optically. Lasing is then done out of the metastable level E_2, shown in Fig. 1.5. By considering just the pumping and the loss terms in the stationary rate equations for the 3 level system one can show that

$$\Gamma > \gamma_{12}\left(1 + \frac{\gamma_{13}}{\gamma_{23}}\right) \tag{1.21}$$

has to hold for $N_2 > N_1$, which can be fulfilled with moderate pumping under the conditions $\gamma_{12} \ll \gamma_{13}$ and $\gamma_{23} \gg \gamma_{13}$ [3].

1.1. *In the stationary case, consider an extension of the rate equations to the three level case and neglect the induced terms. Under which condition for the pumping rate Γ can population inversion between the second and first level be achieved? Discuss the final result.*

1.2.2.3 Other Types of Lasers

Other types of lasers are gas lasers, in which the laser active medium is pumped by collisions with electrons or atoms and the transitions can be either electronic (He-Ne laser) or ro-vibronic ones (CO_2 laser).

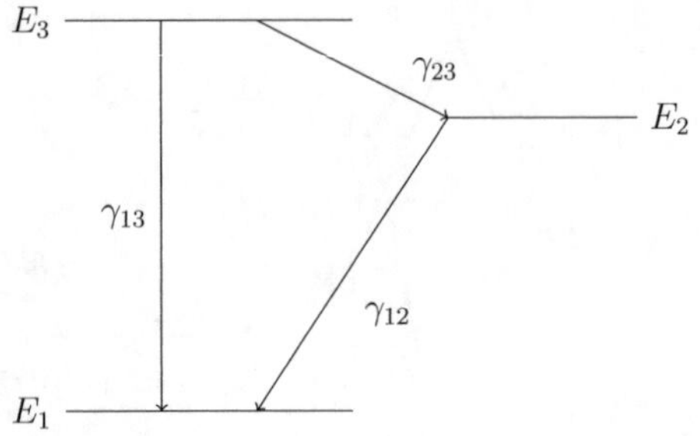

Fig. 1.5 3 level system of the ruby laser with the metastable level E_2

In addition, there are semiconductor based lasers, dye lasers, excimer lasers, to name but a few. Their working principles are described in some detail in [2, 6]. A common principle in the experimental setup of all lasers is the fact that spontaneous emission (being a form of isotropic noise) should be suppressed. This is a difficult task, especially for high frequencies, however, due to the fact that $A \sim B\nu^3$, see (1.11), holds for the Einstein coefficients. Details of the experimental setup, like the quality factor of the cavity, have to be considered in order to understand how temporal fluctuation tend to get washed out [7].

A special laser type is the free electron laser (FEL), where a high speed electron beam is accelerated in a spatially modulated magnetic field and thereby emits coherent light. The role of the laser medium as well as that of the energy pump are both played by the electron beam [8]. The FEL principle has been and will be realized in several large scale facilities around the globe.

1.3 Pulsed Lasers

Experimentally, lasers have led to a revolution in the way spectroscopy is performed. This is due to the fact that lasers are light sources with well defined properties. They can be operated in a single mode continuous wave (cw) modus with a fixed or a tunable frequency or in a multi-mode modus [6]. Most importantly for the remainder of this book is the possibility to run lasers in a pulsed mode. There the laser only oscillates for a short time span (some femtoseconds) with the central frequency of the atomic transition that is used.

1.3.1 Frequency Comb

Experimentally, ultrashort laser pulses can be created by using the principle of mode locking, explained in detail in [6]. We will shortly discuss the superposition of a central mode with side bands, underlying that principle below. The net result is shown

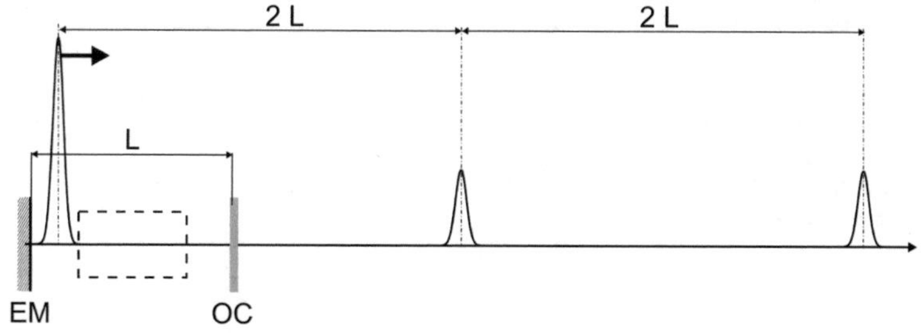

Fig. 1.6 Laser resonator with end mirror (EM), output coupler (OC) and a pulse, propagating between EM and OC and being partially transmitted, from [10]

in Fig. 1.6, where a train of femtosecond pulses coupled out of a cavity is depicted. Among other possible applications to be discussed in detail in later chapters, a pulse train can be used to measure frequencies very precisely [9].

How are the side bands obtained experimentally, and why does their superposition together with the central frequency ν lead to a train of pulses? The first question can be answered by considering the periodic modulation of the inversion with the frequency

$$\delta\nu = c/(2L) = 1/T_{\mathrm{RT}}, \tag{1.22}$$

corresponding to the round trip time of the light in a resonator of length. With the modulator placed at some position inside the cavity, the possible resonator modes with the angular frequencies

$$\omega + 2\pi n\,\delta\nu, \tag{1.23}$$

with $\omega = 2\pi\nu$ and $n = 0, \pm 1, \pm 2, \pm 3, \ldots$ are amplified [6]. Peaks at these equidistantly spaced frequencies are called the frequency comb.

To answer the second question, the amplitude of the electric field at a fixed point in space,

$$\mathcal{E}(t) = \sum_{n=-p}^{p} \mathcal{E}_n \cos[(\omega + 2\pi n\,\delta\nu)t + \varphi_n], \tag{1.24}$$

has to be considered, where $\varphi_n = n\alpha$ are the locked phases. A total of $2p + 1$ modes shall have a gain above the threshold value. In the case of constant $\mathcal{E}_n = \mathcal{E}_0$ and for $\alpha = 0$ this leads to an intensity of

$$I(t) \sim \mathcal{E}_0^2 \left| \frac{\sin[(2p+1)\pi\,\delta\nu\,t]}{\sin(\pi\,\delta\nu\,t)} \right|^2 \cos^2(\omega t). \tag{1.25}$$

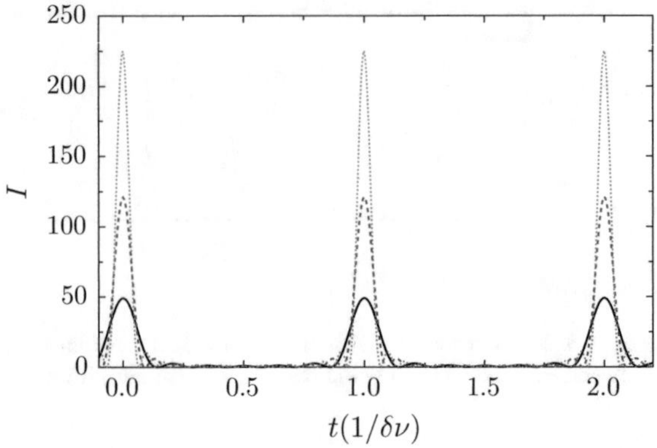

Fig. 1.7 Envelope of the intensity (arbitrary units) of a pulse train as a function of time for the superposition of 7 (*solid black line*) 11 (*dashed blue line*) and 15 modes (*dotted red line*)

1.2. *Derive a closed expression for the electric field and the corresponding intensity in the case of the mode locked laser by using the geometric series.*

The intensity of (1.25) contains a term describing a fast oscillation with the central frequency and an envelope function leading to peaks separated by the round trip time $T_{\mathrm{RT}} = 1/\delta\nu$. Furthermore, the pulse length[4] is $T_p \approx 1/\Delta\nu$, with the inverse width parameter $\Delta\nu = (2p+1)\delta\nu$, increasing linearly with the number of participating modes. The intensity as a function of time for three different total numbers of contributing modes is displayed in Fig. 1.7. The peak intensity increases proportional to $(2p+1)^2$, whereas the pulse length decreases with $1/(2p+1)$.

The effect of the pulse generation can also be understood in the photon picture. Those photons passing through the modulator at times where its transmission has a maximum will experience a minimum loss and the corresponding light will be maximally amplified. Enormously high intensities on the order of $10^{16}\,\mathrm{W\,cm^{-2}}$ can be generated using the principle of passive mode locking [6]. They prevail only for short times on the order of several femtoseconds, however. Pulses with 6 fs length are nowadays generated with Ti:sapphire lasers with Kerr lens mode locking and operate at a center wavelength of 800 nm [11]. Only around 2 oscillations of the field are contained in such a short pulse at those wavelengths. The light is therefore extremely polychromatic. Many further details regarding experimental parameters can be found in Chap. 3 of [10].

1.3.2 Carrier Envelope Phase

Let us look at the electric field of the last section in a bit more detail. It consists of an oscillation with the central frequency ω under an envelope and is plotted for a certain choice of parameters in Fig. 1.8.

[4] Defined as the full width at half maximum (FWHM) of the intensity curve.

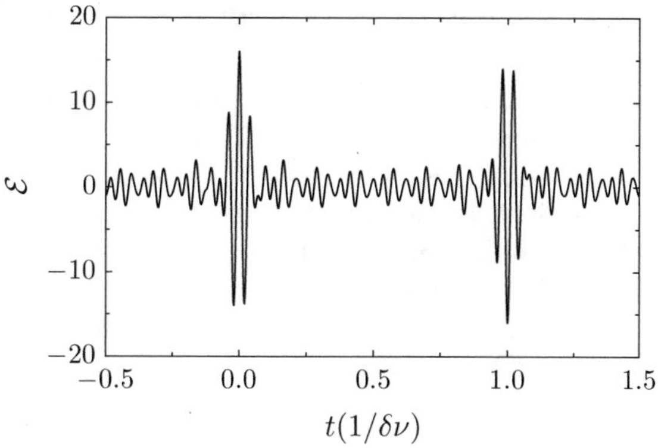

Fig. 1.8 Laser field (in units of $\mathcal{E}_0$) consisting of the superposition of 17 modes of maximum amplitude $\mathcal{E}_0$ with central frequency $\nu = 4$, $L = 3.0625$, and $\alpha = 0$ (all quantities in arbitrary units) as a function of time

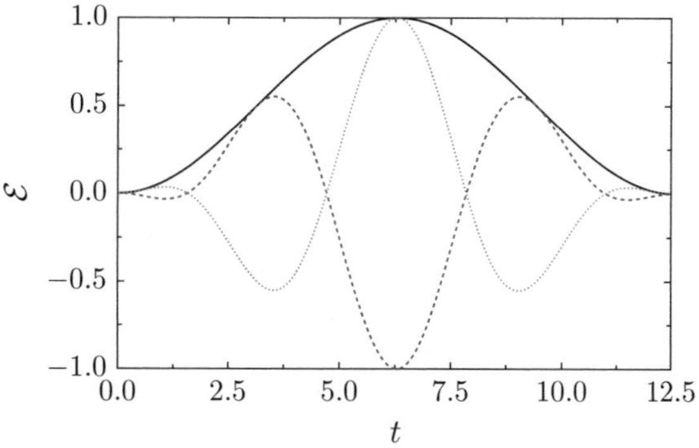

Fig. 1.9 Schematic (normalized) laser field oscillation for unit frequency under a *single* pulse envelope for two different values of the carrier envelope phase (*dotted red line*: $\Delta\varphi = 0$, *dashed blue line*: $\Delta\varphi = \pi$) as a function of time in arbitrary units, analogous to the two different pulses depicted in Fig. 1.8

The parameters in Fig. 1.8 have been chosen such that the separation of the main peaks does coincide with an odd integer multiple of half the period of the fundamental oscillation. This results in the fact that the phase of the fundamental oscillation is different by π, whenever the envelope has reached its next maximum. In general this phase difference is the so-called carrier envelope phase (CEP) $\Delta\varphi$, and later on we will adopt the form

$$\mathcal{E}(t) = \mathcal{E}_0 f(t) \cos\left(\int_0^t \omega(t')\mathrm{d}t' + \Delta\varphi\right) \tag{1.26}$$

for the amplitude of the laser field with an envelope function $f(t)$, which is chosen from a large variety of suitable analytic functions. In addition the frequency ω may be time-dependent and the carrier envelope phase can be varied, leading to tremendous effects as we will see later. In Fig. 1.9 a *single* pulse with fixed frequency is shown for two different values of $\Delta\varphi$ (0, respectively π).

1.3.3 Husimi Representation of Laser Pulses

As mentioned in Sect. 1.3.2, we will model a laser pulse by an oscillation times a freely chosen pulse envelope. This is reasonable due to the fact that arbitrarily formed laser pulses can be generated experimentally by so-called pulse shapers [10]. Use will be made of this fact in Chap. 5 in connection with the control of chemical reactions. In the following, the representation of the frequency content of general laser fields will be discussed.

A very intuitive way to characterize a laser pulse is given by a "windowed" Fourier transformation (or Husimi transformation)

$$F(\tau, \Omega) = \left| \int_{-\infty}^{\infty} \mathrm{d}t\, g(t-\tau)\mathcal{E}(t)\mathrm{e}^{-\mathrm{i}\Omega(t-\tau)} \right|^2, \tag{1.27}$$

with the window function

$$g(t) = \exp[-t^2/(2\sigma^2)]/\sqrt{2\pi\sigma^2}. \tag{1.28}$$

The function $F(\tau, \Omega)$ depending on a time-like variable, as well as on a frequency is also referred to as a spectrogram. It tells us at which time τ a certain frequency Ω is present in the original signal $\mathcal{E}(t)$. The term frequency resolved optical gating (FROG) is used for a measurement technique of a pulse which is designed by using (1.27) [10, 12]. In the field of molecular spectra the term vibrogram [13] is used for a quantity which is constructed in a similar way from a time-signal called auto-correlation function, to be defined in the next chapter.

The case of two pulses which are temporally delayed with respect to each other will occur frequently later-on. For such a so-called pump-dump pulse, with slightly different central frequency of the pump versus the dump pulse, a spectrogram is shown in Fig. 1.10. The frequency change and also the temporal delay is clearly visible in the spectrogram. Also the case of a single pulse with a so-called "up chirp" (central frequency increasing as a function of time) or a "down chirp" (central frequency decreasing) are very obvious in a corresponding Husimi plot. In order to verify this for a simple Gaussian pulse envelope, a Gaussian integral has to be performed (see Exercise 1.3). This is by far not the last one that appears in this book and for convenience some Gaussian integrals are collected in Appendix 1.A.

1.3. *For the case of a linearly chirped frequency,*

$$\omega(t) = \omega_0 \pm \lambda t,$$

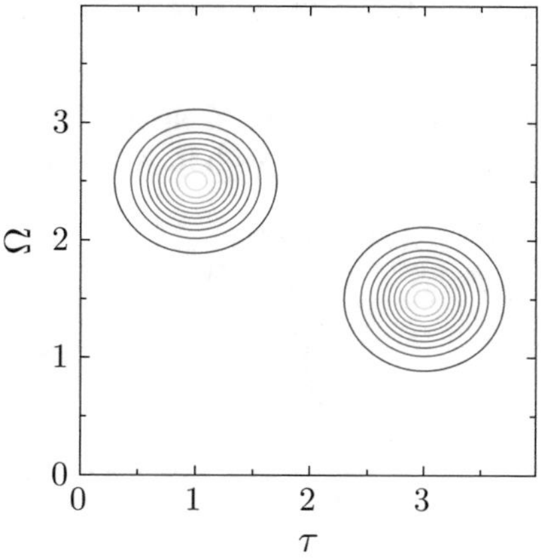

Fig. 1.10 Husimi transform of a pump-dump pulse as a function of τ and Ω in arbitrary units

calculate and schematically depict the Husimi transform of the pulsed field

$$\mathcal{E}(t) = \mathcal{E}_0 \exp\left[-\frac{(t-t_0)^2}{2\sigma^2} + \mathrm{i}\omega_0(t-t_0) \pm \mathrm{i}\frac{\lambda}{2}(t-t_0)^2\right].$$

1.4 Notes and Further Reading

Laser Theory

The theory of the laser is treated on the level of the rate equations as well as in its full quantum version in the book by Haken [2] (the first book of the series [1, 2] contains the derivation of Plancks's law that we have followed) and by Shimoda [3]. In these books one can find also a more detailed discussion of the rate equations beyond the steady state solution, especially concerning the build-up of the oscillation. A comment on the probabilistic nature of Einstein's derivation of Planck's law (as opposed to the concept of superposition of probability amplitudes that we will discuss in depth in Sect. 2.2.1) can be found in [14].

The concept of a photon, describing the particle aspect of electromagnetic radiation will be used many times in this book. It will become especially fruitful in the discussion of (above threshold) ionization in Chap. 4. The breakthrough in Einstein's famous 1905 work [15] is put into historical context in [16].

Short Laser Pulses

A lot of information on the experimental aspects of lasers and about mode locking are contained in the book by Demtröder [6]. The handbook article by Wollenhaupt et al. deals with the properties, the creation via mode locking, and the measurement of femtosecond laser pulses [10]. It also contains a long list of additional references. The characterization of short pulses by using FROGs is the topic of [10, 12].

1.A Some Gaussian Integrals

Throughout this book, Gaussian integrals will be encountered. For complex valued parameters a and b with $\mathrm{Re}\, a \geq 0$, the following formulae hold:

$$\int_{-\infty}^{\infty} \mathrm{d}x \, \exp\{-ax^2\} = \sqrt{\frac{\pi}{a}}, \tag{1.29}$$

$$\int_{-\infty}^{\infty} \mathrm{d}x \, x \exp\{-ax^2\} = 0, \tag{1.30}$$

$$\int_{-\infty}^{\infty} \mathrm{d}x \, x^{2k} \exp\{-ax^2\} = 1 \cdot 3 \cdots (2k-1) \left(\frac{1}{2a}\right)^k \sqrt{\frac{\pi}{a}}, \tag{1.31}$$

$$\int_{-\infty}^{\infty} \mathrm{d}x \, \exp\{-ax^2 + bx\} = \sqrt{\frac{\pi}{a}} \exp\left\{\frac{b^2}{4a}\right\}, \tag{1.32}$$

$$\int_{-\infty}^{\infty} \mathrm{d}x \, x \exp\{-ax^2 + bx\} = \left(\frac{b}{2a}\right) \sqrt{\frac{\pi}{a}} \exp\left\{\frac{b^2}{4a}\right\}, \tag{1.33}$$

$$\int_{-\infty}^{\infty} \mathrm{d}x \, x^2 \exp\{-ax^2 + bx\} = \left(\frac{1}{2a}\right) \left(1 + \frac{b^2}{2a}\right) \sqrt{\frac{\pi}{a}} \exp\left\{\frac{b^2}{4a}\right\}. \tag{1.34}$$

A generalization of one of the expressions above to the case of a d-dimensional integral that is helpful is

$$\int \mathrm{d}^d x \, \exp\{-\mathbf{x} \cdot \mathbf{A}\mathbf{x} + \mathbf{b} \cdot \mathbf{x}\} = \sqrt{\frac{\pi^d}{\det \mathbf{A}}} \exp\left\{\frac{1}{4} \mathbf{b} \cdot \mathbf{A}^{-1} \mathbf{b}\right\}, \tag{1.35}$$

valid for positive definite symmetric matrices $\mathbf{A}$. Like in the 1D case, it can be proven by using a "completion of the square" argument. Furthermore, the convention that non-indication of the boundaries implies integration over the whole range of the independent variables has been used.

References

1. H. Haken, *Licht und Materie Bd. 1: Elemente der Quantenoptik* (BI Wissenschaftsverlag, Mannheim, 1989)
2. H. Haken, *Licht und Materie Bd. 2: Laser* (BI Wissenschaftsverlag, Mannheim, 1994)
3. K. Shimoda, *Introduction to Laser Physics* (Springer, Berlin, 1984)
4. J.P. Gordon, H.J. Zeiger, C.H. Townes, Phys. Rev. **95**, 282 (1954)
5. T.H. Maiman, Phys. Rev. Lett. **4**, 564 (1960)
6. W. Demtröder, *Laser Spectroscopy* (Springer, Berlin, 1996)
7. F. Westermann, *Laser* (Teubner, 1976)
8. P. Luchini, H. Motz, *Undulators and free-electron lasers* (Clarendon Press, 1990)
9. T. Udem, J. Reichert, R. Holzwarth, T.W. Hänsch, Opt. Lett. **24**, 881 (1999)
10. M. Wollenhaupt, A. Assion, T. Baumert, in *Springer Handbook of Lasers and Optics*, ed. by F. Träger (Springer, Berlin, 2007), chap. 12, pp. 937–983

11. D.H. Sutter, G. Steinmeyer, L. Gallmann, N. Matuschek, F. Morier-Genoud, U. Keller, V. Scheuer, G. Angelow, T. Tschudi, Opt. Lett. **24**, 631 (1999)
12. R. Trebino, *Frequency Resolved Optical Gating: The Measurement of Ultrashort Laser Pulses* (Kluwer Academic Press, Boston, 2000)
13. K. Hirai, E.J. Heller, P. Gaspard, J. Chem. Phys. **103**, 5970 (1995)
14. C.W. Gardiner, P. Zoller, *Quantum Noise*, 3rd edn. (Springer, Berlin, 2004)
15. A. Einstein, Ann. Phys. (Leipzig) **17**, 132 (1905)
16. K. Hentschel, *Physics and Philosophy* (2006)

Chapter 2
Time-Dependent Quantum Theory

The focus of our interest will be the coupling of atomic and molecular systems to laser fields, whose maximal strength is of the order of the field experienced by an electron in the ground state of the hydrogen atom.[1] This restriction allows us to describe the field matter interaction non-relativistically by using the time-dependent Schrödinger equation (TDSE) [1]. Analytical solutions of this linear partial differential equation are scarce, however, even in the case without external driving.

In this chapter, we continue laying the foundations for the later chapters by reviewing some basic properties of the time-dependent Schrödinger equation and the corresponding time-evolution operator. After the discussion of two analytically solvable cases, we will consider various ways to rewrite and/or solve the time-dependent Schrödinger equation. Formulating the solution with the help of the Feynman path integral will allow us to consider an intriguing approximate, so-called semiclassical approach to the solution of the time-dependent Schrödinger equation by using classical trajectories. The last part of this chapter is dealing with numerical solution techniques for the time-dependent Schrödinger equation that will be referred to in later chapters.

2.1 The Time-Dependent Schrödinger Equation

In the heyday of quantum theory, Schrödinger postulated a differential equation for the wavefunction of a quantum particle. The properties of this partial differential equation of first order in time and the interpretation of the complex valued wavefunction are in the focus of this section. The importance of Gaussian wavepackets

[1]This is the atomic unit for the electric field, given by $\mathcal{E} = 5.14 \times 10^9$ V/cm.

The original version of this chapter was revised: Belated correction in figure has been incorporated. The erratum to this chapter is available at https://doi.org/10.1007/978-3-319-74542-8_7

F. Grossmann, *Theoretical Femtosecond Physics*, Graduate Texts in Physics,
https://doi.org/10.1007/978-3-319-74542-8_2

as (approximate) analytical solutions of the Schrödinger equation will show up for the first time by considering the so-called Gaussian wavepacket dynamics

2.1.1 Introduction

In position representation, the time-dependent Schrödinger equation for the wavefunction of a single particle of mass m, moving in three spatial dimensions (3D) with $\boldsymbol{r} = (x, y, z)$ is the linear partial differential equation

$$\mathrm{i}\hbar\dot{\Psi}(\boldsymbol{r}, t) = \hat{H}(\boldsymbol{r}, t)\Psi(\boldsymbol{r}, t). \tag{2.1}$$

The Hamilton operator

$$\hat{H}(\boldsymbol{r}, t) = \hat{T}_{\mathrm{k}} + V(\boldsymbol{r}, t) = \frac{\hat{\boldsymbol{p}}^2}{2m} + V(\boldsymbol{r}, t) = -\frac{\hbar^2}{2m}\Delta + V(\boldsymbol{r}, t) \tag{2.2}$$

is the sum of kinetic and potential energy, where

$$\Delta \equiv \frac{\partial^2}{\partial x^2} + \frac{\partial^2}{\partial y^2} + \frac{\partial^2}{\partial z^2} \tag{2.3}$$

is the Laplace operator in cartesian coordinates and the potential energy may (and in the cases to be considered later will) be time-dependent.

In order to gain a physical interpretation of the wavefunction, one multiplies (2.1) by $\Psi^*(\boldsymbol{r}, t)$ and from the resulting expression subtracts its complex conjugate. In the case of real valued potentials, $V(\boldsymbol{r}, t) = V^*(\boldsymbol{r}, t)$, this procedure yields the equation of continuity

$$\dot{\rho}(\boldsymbol{r}, t) = -\boldsymbol{\nabla} \cdot \boldsymbol{j}(\boldsymbol{r}, t), \tag{2.4}$$

with the so-called probability density

$$\rho(\boldsymbol{r}, t) = |\Psi(\boldsymbol{r}, t)|^2 \tag{2.5}$$

and the probability current density

$$\boldsymbol{j}(\boldsymbol{r}, t) = \frac{\hbar}{m}\mathrm{Im}\left\{\Psi^*(\boldsymbol{r}, t)\boldsymbol{\nabla}\Psi(\boldsymbol{r}, t)\right\} = \frac{1}{m}\mathrm{Re}\left\{\Psi^*(\boldsymbol{r}, t)\hat{\boldsymbol{p}}\,\Psi(\boldsymbol{r}, t)\right\}, \tag{2.6}$$

and where

$$\boldsymbol{\nabla} \equiv \boldsymbol{e}_x\frac{\partial}{\partial x} + \boldsymbol{e}_y\frac{\partial}{\partial y} + \boldsymbol{e}_z\frac{\partial}{\partial z} \tag{2.7}$$

is the nabla operator. We thus conclude that $|\Psi(\boldsymbol{r},t)|^2\mathrm{d}^3r$ is the probability to find the particle at time t in the volume element d^3r around $\boldsymbol{r}$. It may change by probability density flowing in or out, which is expressed via $\boldsymbol{j}$. Integration of (2.4) over all space yields that if Ψ is normalized at $t=t_0$ it will be normalized at all times, i.e.

$$\int_{-\infty}^{\infty} \mathrm{d}^3r|\Psi(\boldsymbol{r},t)|^2 = 1 \qquad \forall t \tag{2.8}$$

holds in the case of real potential functions and provided that the current density falls to zero faster than $1/r^2$ for $r\to\infty$.

2.1. *Derive the equation of continuity and prove that the norm is conserved in case that* $\boldsymbol{j}$ *falls to zero faster than* $1/r^2$ *for* $r\to\infty$.

For time-independent (autonomous) potentials $V(\boldsymbol{r},t)=V(\boldsymbol{r})$ equation (2.1) can be solved by separation of variables using the product Ansatz

$$\Psi(\boldsymbol{r},t)=\psi(\boldsymbol{r})\varphi(t). \tag{2.9}$$

After insertion into the time-dependent Schrödinger equation, we get

$$\mathrm{i}\hbar\frac{\dot{\varphi}(t)}{\varphi(t)} = \frac{\left\{-\frac{\hbar^2}{2m}\Delta + V(\boldsymbol{r})\right\}\psi(\boldsymbol{r})}{\psi(\boldsymbol{r})}. \tag{2.10}$$

Due to the fact that they depend on different variables, both LHS and RHS of this equation must be equal to a constant, which we name E. We thus arrive at the two equations

$$\mathrm{i}\hbar\dot{\varphi}(t) = E\varphi(t), \tag{2.11}$$

$$\hat{H}(\boldsymbol{r})\psi_E(\boldsymbol{r}) = E\psi_E(\boldsymbol{r}). \tag{2.12}$$

The first of these equations can be solved immediately by

$$\varphi(t)=\varphi_0\mathrm{e}^{-\mathrm{i}Et/\hbar}. \tag{2.13}$$

The second equation is the so-called time-independent Schrödinger equation [2]. It can be solved after specification of the potential $V(\boldsymbol{r})$. An exact analytical solution, however, is possible only in special cases. The energies E and the corresponding wavefunctions $\psi_E(\boldsymbol{r})$ are the eigenvalues and eigenfunctions of the problem. In their terms a particular solution of the time-dependent Schrödinger equation is given by

$$\Psi(\boldsymbol{r},t)=\psi_E(\boldsymbol{r})\varphi_0\mathrm{e}^{-\mathrm{i}Et/\hbar}, \tag{2.14}$$

where the constant φ_0 later on will be absorbed in ψ_E.

Due to the linearity of the time-dependent Schrödinger equation, its most general solution in the case of a discrete spectrum of energies, $\{E_n\}$, is a linear combination of eigenfunctions given by

$$\Psi(\boldsymbol{r},t)=\sum_{n=0}^{\infty}a_n\psi_n(\boldsymbol{r})\mathrm{e}^{-\mathrm{i}E_nt/\hbar}, \tag{2.15}$$

with

$$a_n=\int \mathrm{d}^3r\,\psi_n^*(\boldsymbol{r})\Psi(\boldsymbol{r},0)=\langle\psi_n|\Psi(0)\rangle \tag{2.16}$$

and where we have used that the eigenfunctions ψ_n corresponding to the discrete energies form a complete orthonormal set, i.e.,

$$\sum_n|\psi_n\rangle\langle\psi_n|=\hat{1}, \tag{2.17}$$

$$\langle\psi_n|\psi_m\rangle=\delta_{nm}, \tag{2.18}$$

formulated in Dirac's bra-ket notation.

In the case of a purely continuous spectrum, the general solution is

$$\Psi(\boldsymbol{r},t)=\int_0^{\infty}\mathrm{d}E\,a(E)\psi_E(\boldsymbol{r})\mathrm{e}^{-\mathrm{i}Et/\hbar}. \tag{2.19}$$

For spectra that consist of discrete as well as continuous parts, we get a corresponding sum of the two expressions given above.

For a Hamiltonian with a discrete spectrum of energy levels that can be arranged in ascending order, $E_0\le E_1\le E_2\le\ldots$, the expectation value of the energy with a trial wavefunction $|\phi\rangle$ is

$$\begin{aligned}\langle\phi|\hat{H}|\phi\rangle&=\sum_{n,m}\langle\phi|\psi_n\rangle\langle\psi_n|\hat{H}|\psi_m\rangle\langle\psi_m|\phi\rangle\\&=\sum_n E_n|\langle\psi_n|\phi\rangle|^2\\&\ge E_0\sum_n|\langle\psi_n|\phi\rangle|^2=E_0\langle\phi|\phi\rangle.\end{aligned} \tag{2.20}$$

For any trial wavefunction one therefore gets

$$\frac{\langle\phi|\hat{H}|\phi\rangle}{\langle\phi|\phi\rangle}\ge E_0, \tag{2.21}$$

the so-called Rayleigh-Ritz variational principle. In the variational method of Ritz, one is choosing the trial state as an explicit function of a parameter α and determines a minimum of the LHS of the equation above as a function of that parameter. In this way, an upper bound for the ground state energy can be determined.

In the derivation above, we started from the time-dependent Schrödinger equation in order to derive the time-independent Schrödinger equation. Schrödinger, however, published them in reverse order. Furthermore, one can derive the time-dependent from the time-independent version, if one considers a composite system of many degrees of freedom and treats the "environmental" degrees of freedom classically, leading to the "emergence of time" for the subsystem [3].

2.1.2 Time-Evolution Operator

Conservation of the norm of the wavefunction can also be proven on a more abstract level by using the unitarity of the so-called time-evolution operator. This operator allows for the formal solution of the time-dependent Schrödinger equation without going to a specific representation, and we therefore continue the discussion using the bra-ket notation. Furthermore, we consider time-independent Hamiltonians to start with. The time-dependent Schrödinger equation then reads

$$\mathrm{i}\hbar|\dot{\Psi}(t)\rangle = \hat{H}|\Psi(t)\rangle. \tag{2.22}$$

A formal solution of this equation is given by

$$|\Psi(t)\rangle = \mathrm{e}^{-\mathrm{i}\hat{H}(t-t_0)/\hbar}|\Psi(t_0)\rangle =: \hat{U}(t,t_0)|\Psi(t_0)\rangle, \tag{2.23}$$

where we have defined the time-evolution operator $\hat{U}(t,t_0)$ which "evolves" the wavefunction from time t_0 to time t. The solution above can be easily verified by inserting it into (2.22). However, we should be careful in differentiating an exponentiated operator, see also Exercise 2.3.

With the help of the formal solution of the time-dependent Schrödinger equation it can be shown that the normalization integral

$$\begin{aligned}\int_{-\infty}^{\infty} \mathrm{d}^3r|\Psi(\boldsymbol{r},t)|^2 &= \langle\Psi(t)|\Psi(t)\rangle \\ &= \langle\Psi(t_0)|\mathrm{e}^{\mathrm{i}\hat{H}^\dagger(t-t_0)/\hbar}\mathrm{e}^{-\mathrm{i}\hat{H}(t-t_0)/\hbar}|\Psi(t_0)\rangle \\ &= \langle\Psi(t_0)|\Psi(t_0)\rangle\end{aligned} \tag{2.24}$$

is equal to unity for all times, if it was unity at the initial time t_0. As in the previous subsection, this is true if the Hamiltonian is Hermitian

$$\hat{H}^\dagger = \hat{H}, \tag{2.25}$$

which is equivalent to the time-evolution operator being unitary

$$\hat{U}^\dagger(t, t_0) = \hat{U}^{-1}(t, t_0), \tag{2.26}$$

as can be inferred from the definition given in (2.23). Also the composition property of the time-evolution operator

$$\hat{U}(t, t_0) = \hat{U}(t, t')\hat{U}(t', t_0), \tag{2.27}$$

where $t' \in [t_0, t]$ can be deduced from its definition.

Things become much more involved as soon as the Hamiltonian is explicitly time-dependent. To investigate this case, it is very convenient to re-express the time-dependent Schrödinger equation in terms of an integral equation. It can be shown by insertion into (2.22) that

$$|\Psi(t)\rangle = |\Psi(t_0)\rangle - \frac{\mathrm{i}}{\hbar}\int_{t_0}^{t} \mathrm{d}t' \hat{H}(t')|\Psi(t')\rangle \tag{2.28}$$

is a formal solution of the time-dependent Schrödinger equation.[2] The wavefunction and thus the sought for solution also appears under the integral on the RHS, however. The equation above therefore is an implicit, so-called integral equation. It can be solved iteratively, starting with the zeroth iteration

$$|\Psi^0(t)\rangle = |\Psi(t_0)\rangle \tag{2.29}$$

of a constant wavefunction. For the first iteration, one uses the zeroth iteration for $|\Psi(t')\rangle$ on the RHS and finds

$$|\Psi^1(t)\rangle = |\Psi(t_0)\rangle - \frac{\mathrm{i}}{\hbar}\int_{t_0}^{t} \mathrm{d}t' \hat{H}(t')|\Psi(t_0)\rangle. \tag{2.30}$$

After infinitely many iterations, the full solution for the time-evolution operator follows to be

$$\hat{U}(t, t_0) = \hat{1} + \sum_{n=1}^{\infty}\left(\frac{-\mathrm{i}}{\hbar}\right)^n \int_{t_0}^{t} \mathrm{d}t_n \int_{t_0}^{t_n} \mathrm{d}t_{n-1} \cdots \int_{t_0}^{t_2} \mathrm{d}t_1 \hat{H}(t_n)\hat{H}(t_{n-1}) \cdots \hat{H}(t_1), \tag{2.31}$$

[2]We stress that the integral form given in (2.28) is equivalent to the differential form of the time-dependent Schrödinger equation (as can be shown by differentiation) and in addition it has the initial condition "built in".

where the integration variables in the nested integrals fulfill the inequalities $t_n \geq t_{n-1} \geq \cdots \geq t_1 \geq t_0$ and in general, the order of the Hamiltonians taken at different times may not be interchanged.

One can confirm this Dyson series solution [4] for the time-evolution operator by inserting it into (2.23) and finally into the time-dependent Schrödinger equation. Alternatively, one could have derived (2.31) also by "time-slicing" the interval $[t_0, t]$ into N equal parts of length Δt and by successively applying the infinitesimal time-evolution operator

$$\hat{U}(t_\nu + \Delta t, t_\nu) = \exp\left\{-\frac{\mathrm{i}}{\hbar}\hat{H}(t_\nu)\Delta t\right\} \overset{\lim \Delta t \to 0}{=} \hat{1} - \frac{\mathrm{i}}{\hbar}\hat{H}(t_\nu)\Delta t, \tag{2.32}$$

with constant Hamiltonians at the beginning of each time step.[3] It is rewarding to explicitly check the equivalence of this procedure with the general formula above by working through Exercise 2.2. The fact that unitarity (and thus norm conservation) and the composition property of the time-evolution operator also hold in the time-dependent case can be most easily shown by using the decomposition in terms of infinitesimal evolution operators.

2.2. *Verify the first three terms in the series for the time-evolution operator by collecting terms up to Δt^2 in the time-sliced expression*

$$\hat{U}(t, t_0) = [\hat{1} - \frac{\mathrm{i}}{\hbar}\hat{H}(t_{N-1})\Delta t][\hat{1} - \frac{\mathrm{i}}{\hbar}\hat{H}(t_{N-2})\Delta t]\cdots[\hat{1} - \frac{\mathrm{i}}{\hbar}\hat{H}(t_0)\Delta t]$$

and taking the limit $N \to \infty$, $\Delta t \to 0$, such that $N\Delta t = t - t_0$. Furthermore, show that $\hat{U}(t, t_0)$ is unitary, using the expression above.

Although (2.31) gives the time-evolution operator in terms of a series, this expression is the most convenient one to work with. For reasons of completeness it shall be mentioned that a closed form expression is possible. With the definition

$$\hat{T}[\hat{A}(t_1)\hat{B}(t_2)] \equiv \begin{cases} \hat{B}(t_2)\hat{A}(t_1) \text{ if } & t_2 > t_1 \\ \hat{A}(t_1)\hat{B}(t_2) \text{ if } & t_1 > t_2 \end{cases} \tag{2.33}$$

of the time-ordering operator,[4] it can be shown that

$$\hat{U}(t, t_0) = \hat{T}\mathrm{e}^{-\mathrm{i}/\hbar \int_{t_0}^{t} \mathrm{d}t' \hat{H}(t')} \tag{2.34}$$

[3] This is how Dyson proceeded in [4].

[4] At $t_1 = t_2$ and for $\hat{A} \neq \hat{B}$ additional assumptions on ordering would have to be made.

holds for the time-evolution operator in the general case of time-dependent Hamiltonians. Furthermore, by working through Exercise 2.3, or even more easily from (2.31), it can be shown that the time-evolution operator fulfills

$$\frac{\mathrm{d}}{\mathrm{d}t}\hat{U}(t) = -\frac{\mathrm{i}}{\hbar}\hat{H}\hat{U}(t), \tag{2.35}$$

i.e., the appropriate time-dependent Schrödinger equation.

2.3. *For the verification of the formal solution of the TDSE, we need the time derivative of an operator of the form*

$$\hat{U}(t) = \exp[\hat{B}(t)].$$

(a) *Calculate* $\frac{\mathrm{d}\hat{U}}{\mathrm{d}t}$ *by using Taylor expansion of the exponential function (keep in mind that, in general, an operator does not commute with its time derivative).*
(b) *Consider the special case* $\hat{B}(t) \equiv -\frac{\mathrm{i}}{\hbar}\hat{H}_0 t$ *and give a closed form solution for* $\frac{\mathrm{d}\hat{U}}{\mathrm{d}t}$.
(c) *Consider the special case* $\hat{B}(t) \equiv -\frac{\mathrm{i}}{\hbar}\int_0^t \mathrm{d}t'\hat{H}(t')$ *and convince yourself that a simple closed form expression for* $\frac{\mathrm{d}\hat{U}}{\mathrm{d}t}$ *can not be given!*
(d) *Show that the construction* $\hat{U}(t) = \hat{T}\exp[\hat{B}(t)]$ *with the time ordering operator and the operator* $\hat{B}$ *from part* (c) *allows for a closed form solution of the time-evolution operator as well as of its time derivative by proving that the relation*

$$\hat{T}\hat{B}^n = n!\left(\frac{-\mathrm{i}}{\hbar}\right)^n \int_0^t \mathrm{d}t_n \int_0^{t_n} \mathrm{d}t_{n-1} \cdots \int_0^{t_2} \mathrm{d}t_1 \hat{H}(t_n)\hat{H}(t_{n-1}) \cdots \hat{H}(t_1)$$

holds.

In order to study some further properties of the time-evolution operator, we go into position representation again by multiplication of (2.23) from the left with $\langle \boldsymbol{r}|$ and insertion of unity in terms of position states. Setting $t_0 = 0$, we find for the propagated wavefunction

$$\Psi(\boldsymbol{r}, t) = \int \mathrm{d}^3 r' \langle \boldsymbol{r}|\hat{U}(t, 0)|\boldsymbol{r}'\rangle \Psi(\boldsymbol{r}', 0). \tag{2.36}$$

The position matrix element of the time-evolution operator

$$K(\boldsymbol{r}, t; \boldsymbol{r}', 0) := \langle \boldsymbol{r}|\hat{U}(t, 0)|\boldsymbol{r}'\rangle \tag{2.37}$$

is frequently referred to as the propagator. As can be shown by differentiation of (2.36) with respect to time, $K(\boldsymbol{r}, t; \boldsymbol{r}', 0)$ itself is a solution of the time-dependent Schrödinger equation with initial condition $K(\boldsymbol{r}, 0; \boldsymbol{r}', 0) = \delta(\boldsymbol{r} - \boldsymbol{r}')$. For this reason, and under the assumption that $t \geq 0$, K is also termed *time-dependent* (retarded) Green's function of the Schrödinger equation.

Again starting from (2.37) another important property of the propagator can be shown. Due to the fact that the propagator itself is a special wavefunction, the closure relation[5]

$$K(\boldsymbol{r},t;\boldsymbol{r}',0) = \int \mathrm{d}^3r'' K(\boldsymbol{r},t;\boldsymbol{r}'',t'')K(\boldsymbol{r}'',t'';\boldsymbol{r}',0), \tag{2.38}$$

follows. It could have also been derived directly from (2.27) by going into position representation and inserting an additional unit operator in terms of position eigenstates.

2.1.3 Spectral Information

In the applications to be discussed in the following chapters, the initial state frequently is assumed to be the ground state of the undriven problem. In this section we will see how spectral information can, in general, be extracted form the propagator via Fourier transformation.

To extract spectral information of autonomous systems from time-series, we start from (2.37) in the case of a time-independent Hamiltonian and insert unity in terms of a complete set of orthonormal energy eigenstates $|\psi_n\rangle = |n\rangle$ twice, in order to arrive at the spectral representation

$$K(\boldsymbol{r},t;\boldsymbol{r}',0) = \sum_{n=0}^{\infty} \psi_n^*(\boldsymbol{r}')\psi_n(\boldsymbol{r}) \exp\left\{-\frac{\mathrm{i}}{\hbar}E_n t\right\} \tag{2.39}$$

of the propagator. Taking the trace (let $\boldsymbol{r}' = \boldsymbol{r}$ and integrate over position) of this expression one arrives at

$$\begin{aligned} G(t,0) &:= \int \mathrm{d}^3r\, K(\boldsymbol{r},t;\boldsymbol{r},0) \\ &= \int \mathrm{d}^3r \sum_{n=0}^{\infty} |\psi_n(\boldsymbol{r})|^2 \exp\left\{-\frac{\mathrm{i}}{\hbar}E_n t\right\} \\ &= \sum_{n=0}^{\infty} \exp\left\{-\frac{\mathrm{i}}{\hbar}E_n t\right\}. \end{aligned} \tag{2.40}$$

For large negative imaginary times only the ground state contribution to the sum above survives. This observation leads to the so-called Feynman-Kac formula

$$E_0 = -\lim_{\mathcal{T}\to\infty} \frac{1}{\mathcal{T}} \ln G(-\mathrm{i}\hbar\mathcal{T},0). \tag{2.41}$$

[5]In cases where no inverse group element exists this is called semi-group property.

If one performs a Laplace transform on $G(t, 0)$,[6] then the *energy-dependent* Green's function

$$
\begin{aligned}
G(z) &= \frac{\mathrm{i}}{\hbar}\int_0^\infty \mathrm{d}t\, G(t,0)\exp\left\{\frac{\mathrm{i}}{\hbar}zt\right\} \\
&= \sum_{n=0}^{\infty}\frac{1}{E_n - z}
\end{aligned}
\tag{2.42}
$$

emerges. This function has poles at the energy levels of the underlying eigenvalue problem.

For numerical purposes it is often helpful to study the time-evolution of wavepackets. By considering the auto-correlation function of an initial wavefunction

$$
|\Psi_\alpha\rangle = \sum_n |n\rangle\langle n|\Psi_\alpha\rangle = \sum_n c_n^\alpha |n\rangle, \tag{2.43}
$$

which is defined according to

$$
c_{\alpha\alpha}(t) := \langle\Psi_\alpha|\mathrm{e}^{-\mathrm{i}\hat{H}t/\hbar}|\Psi_\alpha\rangle = \sum_n |c_n^\alpha|^2 \mathrm{e}^{-\mathrm{i}E_n t/\hbar}, \tag{2.44}
$$

one gains the local spectrum by Fourier transformation

$$
\begin{aligned}
S(\omega) &= \frac{1}{2\pi\hbar}\int \mathrm{d}t\, \mathrm{e}^{\mathrm{i}\omega t} c_{\alpha\alpha}(t) \\
&= \sum_{n=0}^{\infty} |c_n^\alpha|^2 \delta(E_n - \hbar\omega).
\end{aligned}
\tag{2.45}
$$

This result is a series of peaks at the eigenvalues of the problem that are weighted with the absolute square of the overlap of the initial state with the corresponding eigenstate $|n\rangle$. A recent development in this area is the use of so-called harmonic inversion techniques instead of Fourier transformation. This is a nonlinear procedure which allows for the use of rather short time series to extract spectral information [5].

Not only the spectrum but also the eigenfunctions can be determined from time-series. To this end one considers the Fourier transform of the wavefunction at one of the energies just determined

$$
\begin{aligned}
\lim_{T\to\infty}\frac{1}{2T}\int_{-T}^{T}\mathrm{d}t\, \mathrm{e}^{\mathrm{i}E_m t/\hbar}|\Psi_\alpha(t)\rangle &= \sum_{n=0}^{\infty} c_n^\alpha \lim_{T\to\infty}\frac{1}{2T}\int_{-T}^{T}\mathrm{d}t\, \mathrm{e}^{-\mathrm{i}(E_n - E_m)t/\hbar}|n\rangle \\
&= c_m^\alpha |m\rangle.
\end{aligned}
\tag{2.46}
$$

[6] In order to ensure convergence of the integral, one adds a small positive imaginary part to the energy, $z = E + \mathrm{i}\epsilon$.

This procedure filters out an eigenfunction if the overlap with the initial state is sufficiently high.

Alternatively, the concept of imaginary time propagation is again helpful if one wants to determine the ground state. To this end the time evolution of (2.43)

$$|\Psi_\alpha(-\mathrm{i}\hbar\tau)\rangle = \sum_n c_n^\alpha |n\rangle \mathrm{e}^{-\tau E_n} \tag{2.47}$$

is considered for large τ, when only the ground state contribution survives. At the end of the calculation the ground state has to be renormalized and can be subtracted from the initial wavefunction. Repeating the procedure with the modified initial state, the next highest energy state can in principle be determined.

2.4. *Use the propagator of the 1D harmonic oscillator with $V(x) = \frac{1}{2}m\omega_\mathrm{e}^2 x^2$, given by [6]*

$$K(x,t;x',0) = \sqrt{\frac{m\omega_\mathrm{e}}{2\pi\mathrm{i}\hbar\sin(\omega_\mathrm{e}t)}}\exp\left\{\frac{\mathrm{i}m\omega_\mathrm{e}}{2\hbar\sin(\omega_\mathrm{e}t)}\left[(x^2+x'^2)\cos(\omega_\mathrm{e}t)-2xx'\right]\right\}$$

to derive the spectrum.
Hint: Use the geometric series.

2.1.4 *Analytical Solutions for Wavepackets*

To conclude this introductory section, we review an Ansatz for the solution of the time-dependent Schrödinger equation with the help of a Gaussian wavepacket and will derive equations of motion for the parameters of this wavepacket. We then continue with a review of the dynamics of a wavepacket in the box potential. Due to the specific nature of the spectrum in this case, insightful analytic predictions for the probability density as a function of time can be made.

2.1.4.1 Gaussian Wavepacket Dynamics (GWD)

Already in 1926, Schrödinger has stressed the central importance of Gaussian wavepackets in the transition from "micro-" to "macro-mechanics" [7]. For this reason, we will now consider a wavefunction in the form of a Gaussian as the solution of the time-dependent Schrödinger equation.

In Heller's GWD [8], one uses a complex-valued Gaussian in position representation (for reasons of simplicity here in 1D),

$$\Psi(x,t) = \left(\frac{2\alpha_0}{\pi}\right)^{1/4}\exp\left\{-\alpha_t(x-q_t)^2+\frac{\mathrm{i}}{\hbar}p_t(x-q_t)+\frac{\mathrm{i}}{\hbar}\delta_t\right\}, \tag{2.48}$$

as an Ansatz for the solution of (2.1). The expression above contains *real valued* parameters $q_t, p_t \in \mathbf{R}$ and *complex valued* ones $\alpha_t, \delta_t \in \mathbf{C}$ that are undetermined up to now. The initial parameters $q_0, p_0, \alpha_0 = \alpha_{t=0}$ shall be given and $\delta_0 = 0$. Equations of motion for the four time-dependent parameters can be gained by a Taylor expansion of the potential around $x = q_t$, according to

$$V(x,t) \approx V(q_t,t) + V'(q_t,t)(x-q_t) + \frac{1}{2!}V''(q_t,t)(x-q_t)^2 \tag{2.49}$$

and by using the time-dependent Schrödinger equation.

After insertion of the time- and position-derivatives of the wavefunction

$$\dot{\Psi}(x,t) = \left\{-\dot{\alpha}_t(x-q_t)^2 + 2\alpha_t\dot{q}_t(x-q_t) + \frac{\mathrm{i}}{\hbar}\dot{p}_t(x-q_t) - \frac{\mathrm{i}}{\hbar}p_t\dot{q}_t + \frac{\mathrm{i}}{\hbar}\dot{\delta}_t\right\} \Psi(x,t), \tag{2.50}$$

$$\Psi'(x,t) = \left[-2\alpha_t(x-q_t) + \frac{\mathrm{i}}{\hbar}p_t\right]\Psi(x,t), \tag{2.51}$$

$$\Psi''(x,t) = \left\{-2\alpha_t + \left[-2\alpha_t(x-q_t) + \frac{\mathrm{i}}{\hbar}p_t\right]^2\right\}\Psi(x,t), \tag{2.52}$$

the coefficients of equal powers of $(x - q_t)$ can be compared, as the time-dependent Schrödinger equation has to be fulfilled at every x. This leads to the following set of equations

$$(x-q_t)^2: \quad -\mathrm{i}\hbar\dot{\alpha}_t = -\frac{2\hbar^2}{m}\alpha_t^2 + \frac{1}{2}V''(q_t,t), \tag{2.53}$$

$$(x-q_t)^1: \quad \mathrm{i}\hbar 2\alpha_t\dot{q}_t - \dot{p}_t = \mathrm{i}\hbar 2\alpha_t\frac{p_t}{m} + V'(q_t,t), \tag{2.54}$$

$$(x-q_t)^0: \quad p_t\dot{q}_t - \dot{\delta}_t = \frac{\hbar^2}{m}\alpha_t + \frac{p_t^2}{2m} + V(q_t,t). \tag{2.55}$$

From (2.54), after separation of real and imaginary part, we find

$$\dot{q}_t = \frac{p_t}{m}, \tag{2.56}$$

$$\dot{p}_t = -V'(q_t,t). \tag{2.57}$$

The real-valued parameters therefore fulfill Hamilton equations with initial conditions $q_{t=0} = q_0$ and $p_{t=0} = p_0$. From (2.55) and with the definition of the classical Lagrangian

$$L = T_{\mathrm{k}} - V, \tag{2.58}$$

where

$$T_{\mathrm{k}} = \frac{p_t^2}{2m} \tag{2.59}$$

is the classical kinetic energy, the differential equation

$$\dot{\delta}_t = L - \frac{\hbar^2}{m}\alpha_t \tag{2.60}$$

can be derived. The remaining equation for the inverse width parameter α_t follows from (2.53). It is the nonlinear Riccati differential equation

$$\dot{\alpha}_t = -\frac{2\mathrm{i}\hbar}{m}\alpha_t^2 + \frac{\mathrm{i}}{2\hbar}V''(q_t, t). \tag{2.61}$$

The width of the Gaussian is a function of time and in contrast to an approach, that will be discussed later in this chapter, where the width of the Gaussians is fixed (i.e. "frozen"), the approach reviewed here is also referred to as the "thawed" GWD.

In the cases of the free particle and of the harmonic oscillator, all equations of motion can be solved exactly analytically and, because the Taylor expansion is also exact in these cases, the procedure above leads to an exact analytic solution of the time-dependent Schrödinger equation.

2.5. *Use the GWD-Ansatz to solve the TDSE.*

(a) *Use the differential equations for $q_t, p_t, \alpha_t, \delta_t$ in order to show that the Gaussian wavepacket fulfills the equation of continuity.*
(b) *Solve the differential equations for $q_t, p_t, \alpha_t, \delta_t$ for the free particle case $V(x) = 0$.*
(c) *Solve the differential equations for $q_t, p_t, \alpha_t, \delta_t$ for the harmonic oscillator case $V(x) = \frac{1}{2}m\omega_{\mathrm{e}}^2 x^2$.*
(d) *Calculate $\langle\hat{x}\rangle$, $\langle\hat{p}\rangle$, and $\Delta x = \sqrt{\langle\hat{x}^2\rangle - \langle\hat{x}\rangle^2}$, $\Delta p = \sqrt{\langle\hat{p}^2\rangle - \langle\hat{p}\rangle^2}$ and from this the uncertainty product using the general Gaussian wavepacket. Discuss the special cases from* (b) *and* (c). *What do the results for the harmonic oscillator simplify to in the case $\alpha_{t=0} = m\omega_{\mathrm{e}}/(2\hbar)$?*

The GWD method can also be applied to nonlinear classical problems, however, where it is typically valid only for short times. In this context one often uses the notion of the Ehrenfest time, after which non-Gaussian distortions of a wavepacket become manifest. In general, the solutions of the equations of motion have to be determined numerically in the nonlinear case (see Sect. 2.3.4), and the Taylor expansion and thus the final result is only an approximation.

2.1.4.2 Particle in an Infinite Square Well

As another example for an exactly solvable problem in quantum dynamics, we now consider the evolution of an initial wavepacket in an infinite square well (box potential). This problem has been presented in the December 1995 issue of "Physikalische Blätter" [9] and leads to aesthetically pleasing space-time pictures, which are sometimes referred to as "quantum carpets".

In the work of Kinzel, a Gaussian initial state localized close to the left wall has been investigated. For ease of analytic calculations, let us consider a wavefunction made up of a sum of eigenfunctions with equal weight in the following. The eigenfunctions and eigenvalues of the square well extending from 0 to L are given by

$$\psi_n(x) = \sqrt{\frac{2}{L}} \sin\left\{\frac{n\pi}{L}x\right\}, \tag{2.62}$$

$$E_n = n^2 E_1, \qquad n = 1, 2, 3, \ldots \tag{2.63}$$

with the fundamental energy

$$E_1 = \frac{1}{2m}\left(\frac{\hbar\pi}{L}\right)^2. \tag{2.64}$$

The corresponding frequency and period are

$$\omega = E_1/\hbar; \qquad T = \frac{2\pi}{\omega}. \tag{2.65}$$

2.6. *A particle shall be in the eigenstate ψ_n with energy E_n of an infinite potential well of width L ($0 \le x \le L$). Let us assume that the width of the well is suddenly doubled.*

(a) *Calculate the probability to find a particle in the eigenstate ψ'_m with energy E'_m of the new well.*
(b) *Calculate the probability to find a particle in state ψ'_m whose energy E'_m is equal to E_n.*
(c) *Consider the time evolution for $n = 1$, i.e. at $t = 0$ the wavefunction is the lowest eigenfunction of the small well $\Psi(x, 0) = \psi_1(x)$. Calculate the smallest time $t_{\min}$ for which $\Psi(x, t_{\min}) = \Psi(x, 0)$.*
(d) *Draw a picture of the wavefunction $\Psi(x, t)$ at $t = t_{\min}/2$.*

In the following we will use dimensionless variables for position and time

$$\xi = \frac{x}{L}; \qquad \tau = \frac{t}{T} \tag{2.66}$$

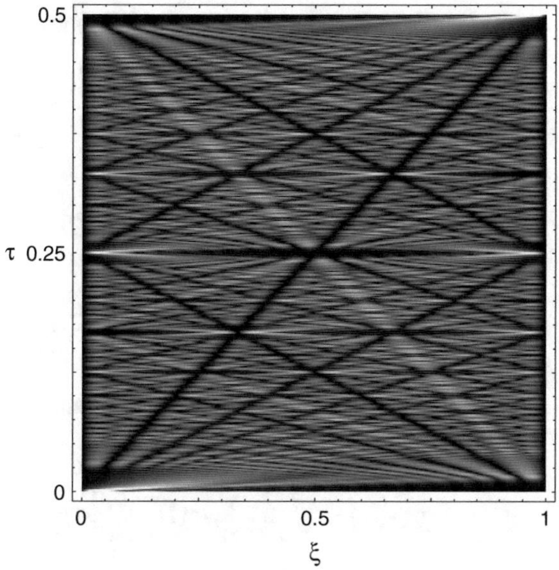

Fig. 2.1 Time-evolution of the absolute value of a superposition of 20 eigenfunctions of the box potential [10]

and will consider an initial wavefunction consisting of N eigenfunctions with equal weights. As can be seen in Fig. 1 of [10], for large N the wavefunctions have the same initial localization properties as the ones used in [9]. Each eigenfunction is evolving with the exponential of the corresponding eigenenergy according to (2.14). The time-evolution of the normalized wavepacket thus is

$$\Psi(\xi, \tau) = \sqrt{\frac{2}{LN}} \sum_{n=1}^{N} \sin(n\pi\xi) \exp(2\pi \mathrm{i} n^2 \tau). \tag{2.67}$$

In Fig. 2.1, we show the absolute value of the time-evolved wavefunction with $N = 20$ in the range $0 \le \xi \le 1; 0 \le \tau \le 0.5$ of the $\xi - \tau$ plane. In this density plot darkness corresponds to a low and brightness to a large value of the plotted function. A detailed explanation of the features of the time-evolved wavefunction has been given in [10]. Let us here concentrate on the suppression of the wavefunction along the lines $\tau = \frac{\xi}{2k}$. In order to make progress, we express the sine function with the help of exponentials and arrive at terms of the form

$$\vartheta(\pm\xi, \tau) = \sum_{n=1}^{N} q^{n^2} \mathrm{e}^{\pm \mathrm{i} n\pi\xi}, \tag{2.68}$$

with $q \equiv \mathrm{e}^{2\pi \mathrm{i}\tau}$. We now shift the argument ξ by $2k\tau$, with positive $k = 1, 2, 3, \ldots$. Using $n = n + k/2 - k/2$ we get

$$
\begin{aligned}
\vartheta(\xi + 2k\tau, \tau) &= \sum_{n=1}^{N} q^{n^2} \mathrm{e}^{\mathrm{i}n\pi(\xi+2k\tau)} \\
&= q^{-\left(\frac{k}{2}\right)^2} \mathrm{e}^{-\mathrm{i}\frac{k}{2}\pi\xi} \sum_{n=1}^{N} q^{\left(n+\frac{k}{2}\right)^2} \mathrm{e}^{\mathrm{i}\left(n+\frac{k}{2}\right)\pi\xi},
\end{aligned} \tag{2.69}
$$

$$
\begin{aligned}
\vartheta(-(\xi + 2k\tau), \tau) &= \sum_{n=1}^{N} q^{n^2} \mathrm{e}^{-\mathrm{i}n\pi(\xi+2k\tau)} \\
&= q^{-\left(\frac{k}{2}\right)^2} \mathrm{e}^{-\mathrm{i}\frac{k}{2}\pi\xi} \sum_{n=1}^{N} q^{\left(n-\frac{k}{2}\right)^2} \mathrm{e}^{-\mathrm{i}\left(n-\frac{k}{2}\right)\pi\xi}.
\end{aligned} \tag{2.70}
$$

The wavefunction along the straight lines originating from $\xi = 0$ can thus be written as

$$
\begin{aligned}
\Psi(2k\tau, \tau) &= \sqrt{\frac{2}{LN}} \frac{1}{2\mathrm{i}} q^{-\left(\frac{k}{2}\right)^2} \sum_{n=1}^{N} \{ q^{\left(n+\frac{k}{2}\right)^2} - q^{\left(n-\frac{k}{2}\right)^2} \} \\
&= \sqrt{\frac{2}{LN}} \frac{1}{2\mathrm{i}} q^{-\left(\frac{k}{2}\right)^2} \\
&\quad \{ q^{\left(N+\frac{k}{2}\right)^2} \ldots + q^{\left(N+1-\frac{k}{2}\right)^2} - q^{\left(\frac{k}{2}\right)^2} \ldots - q^{\left(1-\frac{k}{2}\right)^2} \}.
\end{aligned} \tag{2.71}
$$

Using the analogy with two combs shifted against each other, as shown in Fig. 2.2, it is obvious that from the expression in curly brackets above, only $2k$ terms of order 1 do survive, due to the fact that the major part of the sum cancels term-wise. In the case $k \ll \sqrt{N}$, we thus can conclude

$$
\Psi(2k\tau, \tau) \sim k/\sqrt{N} \approx 0. \tag{2.72}
$$

These considerations explain the "channels" of near extinction of the wavefunction along lines of slope 1/2k, emanating from $(\xi = 0, \tau = 0)$ to the right. For a given value of N, the higher the value of k, fewer terms cancel each other and the less visible the "channel effect" becomes.

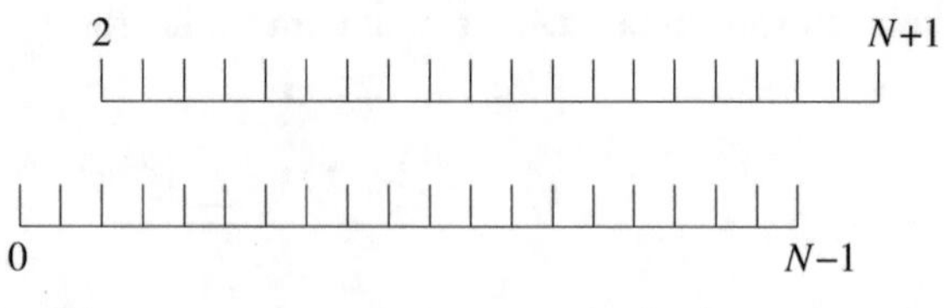

Fig. 2.2 Comb analogy for $N = 20$ and $k = 2$. The overlapping parts of the two combs are representing the terms that cancel each other. Only the terms without partner (here 4) do not cancel

2.2 Analytical Approaches to Solve the TDSE

A review of some exactly analytically solvable cases in nonrelativistic quantum dynamics is given in [11], and we have seen two examples for wavepacket solutions at the end of the previous section. For most problems of interest, however, an exact analytical solution of the time-dependent Schrödinger equation cannot be found. It is therefore of quite some interest to devise alternative approaches to quantum dynamics and/or some approximate or exact Ansätze that are generally applicable and lead to viable approximate and/or numerical schemes.

A notable reformulation of the time-dependent Schrödinger equation is the Feynman path integral expression for the propagator [12]. This is of utmost importance in the following, because from the path integral a much used approximation can be derived: the time-dependent semiclassical formulation of quantum theory. Furthermore, in the case of small external perturbations, time-dependent perturbation theory may be the method of choice for the solution of the time-dependent Schrödinger equation. Moreover, for systems with many degrees of freedom, as a first approximation, the wavefunction can be factorized. We thus discuss the so-called Hartree Ansatz and for the first time also the Born-Oppenheimer method in this chapter. Finally, the exact analytical Floquet Ansatz for the treatment of periodically driven quantum systems is reviewed.

The discussion of the numerical implementation of some of these concepts will be postponed to Sect. 2.3.

2.2.1 Feynman's Path Integral

For time-dependent quantum problems, which occur naturally if we want to describe the interaction of a system with a laser field, as we will see in Chap. 3, an approach that deals with the propagator is very well suited. With the propagator at hand, we can calculate the time evolution of every wavefunction according to (2.36).

2.2.1.1 The Propagator as a Path Integral

A very elegant approach that gives an explicit formula for the propagator goes back to an idea that can be found in later editions of the famous text book by Dirac [13], and has finally been formulated by Feynman [12]. The derivation of the path integral is a prime example for the new quantum mechanical reasoning in terms of probability amplitudes in contrast to the classical way of thinking in probabilities. The famous double slit experiment serves as the chief parable to understand the new twist of quantum mechanical thinking. The postulates that form the basis for the derivation of the path integral are:

Postulate 1: If, in the particle picture, an event (e.g., an electron hitting the screen after passing a double slit) can have occurred in two mutually exclusive ways, the corresponding amplitudes have to be added to find the total amplitude

Postulate 2: If an event consists of two successive events, the corresponding amplitudes do multiply.

In the book by Feynman and Hibbs [6] it is shown how the usage of the two postulates above leads to the composition (or semigroup) property of the propagator, which we have already stated in (2.38). At this point, however, only physical intuition helps. How can the probability amplitude $K(\boldsymbol{r}_f, t; \boldsymbol{r}_i, 0)$ itself be determined?

Inspired by an idea of Dirac, and using the first postulate above, Feynman in 1948 [12] expressed the propagator as a sum over all paths from $\boldsymbol{r}_i$ to $\boldsymbol{r}_f$ in time t. Each of these paths contributes with a phase factor to the sum. The phase is the ratio of the classical action of the respective path and $\hbar$. Mathematically this can be written as

$$K(\boldsymbol{r}_f, t; \boldsymbol{r}_i, 0) = \int_{\boldsymbol{r}(0)=\boldsymbol{r}_i}^{\boldsymbol{r}(t)=\boldsymbol{r}_f} \mathrm{d}[\boldsymbol{r}] \exp\left\{\frac{\mathrm{i}}{\hbar} S[\boldsymbol{r}]\right\}, \tag{2.73}$$

with Hamilton's principal function (being a functional of the path, which is expressed by the square brackets)

$$S[\boldsymbol{r}] = \int_0^t \mathrm{d}t' L. \tag{2.74}$$

The classical Lagrangian L has already appeared in (2.58). The symbol $\int \mathrm{d}[\boldsymbol{r}]$ is denoting the integral over all paths (functional integral). In contrast to standard integration, where one sums a function over a certain range of a *variable*, in a path integral one sums a function of a function (a so-called functional) over a certain class of *functions* that are parametrized by t and that obey the boundary conditions $\boldsymbol{r}(0) = \boldsymbol{r}_i, \boldsymbol{r}(t) = \boldsymbol{r}_f$. Feynman's path integral is therefore sometimes referred to as a "sum over histories". In Fig. 2.3, for an arbitrary one-dimensional potential, we depict the classical path and some other equally important nonclassical paths.

The exact analytic calculation of the path integral, apart from a few exceptions involving quadratic Lagrangians, is not possible. One therefore frequently resorts to approximate solutions. A principal way to calculate the path integral shall, however, be hinted at. In order to make progress, the time interval $[0, t]$ is divided into N equal parts of length Δt, analogously to the procedure in Sect. 2.1.2. This "time-slicing" is depicted in Fig. 2.3. In this way the path integral is discretized and in the limit $N \to \infty$ can be written as an $(N-1)$-dimensional Riemann integral times a normalization constant B_N. In one spatial dimension this reads

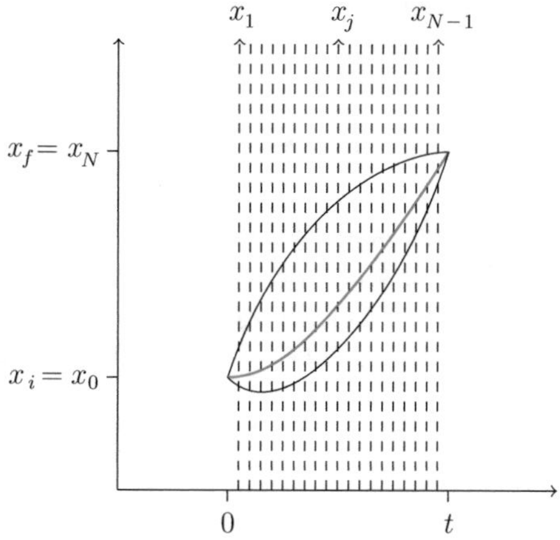

Fig. 2.3 Paths in space-time. Time-slicing and the classical path (*thick red line*) are also depicted

$$K(x_f, t; x_i, 0) = \lim_{N\to\infty} B_N \int \mathrm{d}x_1 \int \cdots \int \mathrm{d}x_{N-1}$$
$$\exp\left\{\frac{\mathrm{i}}{\hbar}\sum_{j=1}^{N}\left[\frac{m(x_j - x_{j-1})^2}{2\Delta t} - V\left(\frac{x_j + x_{j-1}}{2}\right)\Delta t\right]\right\}. \tag{2.75}$$

Proving this expression and deriving B_N is most elegantly done by using a Weyl transformation and will be performed explicitly in Appendix 2.A.

Obviously, the expression above can be interpreted as the successive application of the closure relation (2.38), concatenating short-time propagators

$$K(x_j, \Delta t; x_{j-1}, 0) \sim \exp\left\{\frac{\mathrm{i}}{\hbar}\left[\frac{m(x_j - x_{j-1})^2}{2\Delta t} - V\left(\frac{x_j + x_{j-1}}{2}\right)\Delta t\right]\right\}. \tag{2.76}$$

It considerably deepens one's understanding to derive the short-time propagator directly from the infinitesimal time-evolution operator of (2.32). The interesting question how the Hamilton operator "mutates" into the classical Lagrangian is answered in Exercise 2.7. In the second part of this exercise, the time-dependent Schrödinger equation can be derived. To this end a simplified version of the short-time propagator with a simple end point rule for the discretization of the potential part of the action by replacing $V\left(\frac{x_j+x_{j-1}}{2}\right)$ by $V(x_{j-1})$ in (2.76) is sufficient.

2.7. *Study the short-time propagator and use it to derive the TDSE.*

(a) *Derive the short-time propagator starting from*

$$\hat{U}(\Delta t) = \exp\{-\mathrm{i}\hat{H}\Delta t/\hbar\}$$

for the infinitesimal time-evolution operator.
Hint: Use first order Taylor expansion of the exponential function.

(b) *Use the short-time propagator in order to propagate an arbitrary wavefunction $\Psi(x,t)$ over an infinitesimal time interval Δt via*

$$\Psi(x, t+\Delta t) = \int \mathrm{d}y K(x, \Delta t; y, 0)\Psi(y, t)$$

and derive the TDSE!
Hint: To this integral only a small intervall of y centered around x is contributing. Expansion of the expression above to first order in Δt and up to second order in $\eta = y - x$ leads to a linear partial differential equation for $\Psi(x,t)$.

2.2.2 Stationary Phase Approximation

By inspection of (2.75) it is obvious that the calculation of the propagator for arbitrary potentials becomes arbitrarily complicated. In the case of maximally quadratic potentials all integrals are Gaussian integrals, however, and thus can be done exactly analytically. There are some additional examples, for which exact analytic results for the path integral are known. These are collected in the supplement section of the Dover edition of the textbook by Schulman [14].

In general, however, approximate solutions for the path integral are sought for. The idea is to approximate the exponent in such a fashion that only quadratic terms survive. The corresponding approximation is the stationary phase approximation (SPA). It shall be introduced by first looking at a simple 1D integral of the form

$$\int_{-\infty}^{+\infty} \mathrm{d}x \exp\{\mathrm{i}f(x)/\delta\}g(x). \tag{2.77}$$

To proceed, we perform a Taylor expansion of the function in the exponent up to second order according to $f(x) \approx f(x_0) + f'(x_0)(x - x_0) + 1/2 f''(x_0)(x - x_0)^2$ under the condition of stationarity of the phase, i.e.,

$$f'(x_0) = 0. \tag{2.78}$$

Then with the help of the formula[7]

$$\int_{-\infty}^{+\infty} \mathrm{d}x \exp(\mathrm{i}\alpha x^2) = \sqrt{\frac{\mathrm{i}\pi}{\alpha}}, \tag{2.79}$$

[7]This is a Fresnel integral, i.e., the Gaussian integral of (1.29) with purely imaginary a.

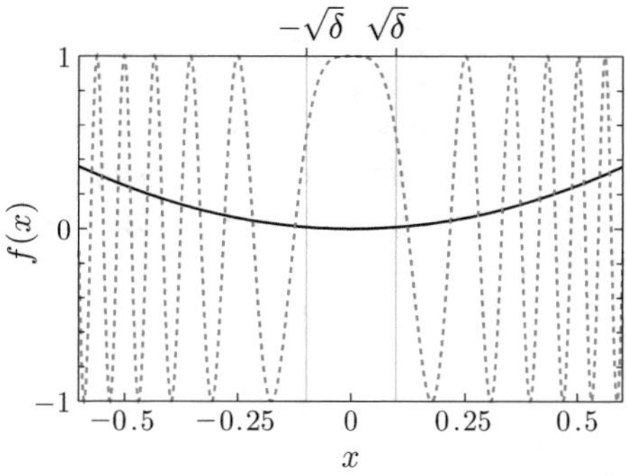

Fig. 2.4 The function $f(x) = x^2$ (*solid black line*) and the real part of $\exp\{ix^2/\delta\}$ with $\delta = 0.01$ (*dashed red line*)

we get

$$\int_{-\infty}^{+\infty} dx \exp\{i f(x)/\delta\} g(x) \overset{\delta \to 0}{=} \sqrt{\frac{2\pi i \delta}{f''(x_0)}} \exp\{i f(x_0)/\delta\} g(x_0) \tag{2.80}$$

for the integral above. This approximation becomes the better, the faster the exponent oscillates, which is determined by the smallness of the parameter δ. In order to demonstrate this fact, in Fig. 2.4 the function $\exp\{ix^2/\delta\}$ is displayed using $\delta = 0.01$. The fast oscillations of the function at x values further away than $\sqrt{\delta}$ from the point of stationary phase lead to the mutual cancellation of the positive and negative contributions to the integral. Around the stationary phase point (which here is $x = 0$) this argument does not apply and therefore the major contribution to the integral is determined by the properties of the function $f(x)$ around that point.

In the next subsection, the notion of stationary phase integration will be extended to the path integral, being an infinite dimensional "normal" integral. Before doing so, a remark on the direct numerical approach to the path integral is in order. As can be seen already by looking at the integrand of our 1D toy problem, a numerical attack to calculate the integral of a highly oscillatory function will be problematic due to the near cancellation of terms. This is even more true for the full fledged path integral and the associated problem is sometimes referred to as the *sign problem*, which is a topic at the forefront of present day research. Much more well-behaved with respect to numerical treatment are imaginary time path integrals, which will not be dealt with herein, however.

2.2.3 Semiclassical Approximation

The semiclassical approximation of the propagator goes back to van Vleck [15]. Its direct derivation from the path integral followed many years later, however, and finally led to the generalization of the van Vleck formula by Gutzwiller [16]. We will later-on use semiclassical arguments quite frequently, because they allow for a qualitative and often also for a quantitative explanation of many interesting quantum phenomena. For this reason we will go through the derivation of the so-called van Vleck-Gutzwiller (VVG) propagator in some detail.

In order to derive a time-dependent semiclassical expression, the SPA will be applied to the path integral (2.73). In this case the exponent $S[x]$ is a functional. Therefore, we need the definition of the variation of a functional (see, e.g., [17]). The analog of (2.78) is

$$\delta S[x_{\rm cl}] = 0. \tag{2.81}$$

This, however, is Hamilton's principle of classical mechanics. The SPA is thus based on the expansion of the exponent of the path integral around the classical path with boundary conditions $x(0) = x_i$ and $x(t) = x_f$. This is also the reason why we have highlighted the classical path in Fig. 2.3. By defining the deviation from the classical path as

$$\eta(t') = x(t') - x_{\rm cl}(t') \qquad \eta(0) = \eta(t) = 0, \tag{2.82}$$

the second order expansion needed for the SPA is given by

$$S[x] = S[x_{\rm cl}] + \frac{1}{2}\int_0^t {\rm d}t'\eta(t')\hat{O}\eta(t'), \tag{2.83}$$

with the stability operator

$$\hat{O} = -m\frac{{\rm d}^2}{{\rm d}t^2} - V''\big|_{x=x_{\rm cl}(t)}. \tag{2.84}$$

More details about the underlying variational calculus can be found in Chap. 12 of [14] and in Appendix 2.B.

For the time being, we assume that only one single point of stationary phase exists. In SPA the propagator can then be written as

$$K(x_f, t; x_i, 0) \approx \int_{\eta(0)=0}^{\eta(t)=0} {\rm d}[\eta]\exp\left\{\frac{\rm i}{2\hbar}\int_0^t {\rm d}t'\eta(t')\hat{O}\eta(t')\right\} \exp\left\{\frac{\rm i}{\hbar}S[x_{\rm cl}]\right\}. \tag{2.85}$$

Due to the additive nature of the action in (2.83), the propagator factorizes into a trivial factor coming from the zeroth order term in the expansion and a so-called prefactor coming from the fluctuations around the classical path. Due to its boundary conditions, this prefactor is also referred to as 0–0-propagator.

The condition for the applicability of the SPA is that the exponent oscillates rapidly. For the functional integral this means that $\hbar$ must be small compared to the action of the classical trajectory. The main contribution to the propagator in the SPA thus stems from the classical path that solves the boundary value problem defined by the propagator labels and from a small region surrounding the classical path. In this context the SPA is therefore also called the semiclassical approximation. There are several ways to derive an explicit expression for the prefactor in (2.85), see, e.g., [14]. As shown in detail in this textbook the final expression for the propagator is given by

$$K(x_f, t; x_i, 0) \approx \sqrt{\frac{\mathrm{i}}{2\pi\hbar}\frac{\partial^2 S[x_{\mathrm{cl}}]}{\partial x_f \partial x_i}} \exp\left\{\frac{\mathrm{i}}{\hbar}S[x_{\mathrm{cl}}]\right\}. \tag{2.86}$$

The classical information that enters the expression above can be gained by solving the root search problem defined by the propagator labels and calculating the corresponding action and its second derivative with respect to the initial and final position.

As mentioned at the beginning of this section, van Vleck succeeded already in 1928 in finding the expression above in a more "heuristic" manner [15]. He started in his derivation from the observation that the insertion of the Ansatz[8]

$$K \sim \exp\{\mathrm{i}S(x, \alpha, t)/\hbar\}, \tag{2.87}$$

with an integration constant α, into the time-dependent Schrödinger equation, after cancellation of all $\hbar$-dependencies, leads to the classical mechanical Hamilton-Jacobi equation [18]

$$H\left(\frac{\partial S}{\partial x}, x\right) + \frac{\partial S}{\partial t} = 0. \tag{2.88}$$

Invoking the correspondence principle, S must thus be a generator of a canonical transformation, the classical action functional, and we have found the exponential part of the propagator.

The yet undetermined prefactor of the propagator follows from a more involved reasoning: the classical probability density for reaching point x_f after starting from point x_i can be determined by integrating over all possible initial momenta (vertical line in the phase space plot in Fig. 2.5) and is given by

$$\rho_{\mathrm{cl}}(x_f, t; x_i, 0) = \frac{1}{h}\int \mathrm{d}p_i \delta[x_f - x_t(x_i, p_i)] = \frac{1}{h}\left|\left.\frac{\partial x_f}{\partial p_i}\right|_{x_i}\right|^{-1}, \tag{2.89}$$

[8] This Ansatz is also the starting point of so-called Bohmian mechanics approaches to quantum dynamics.

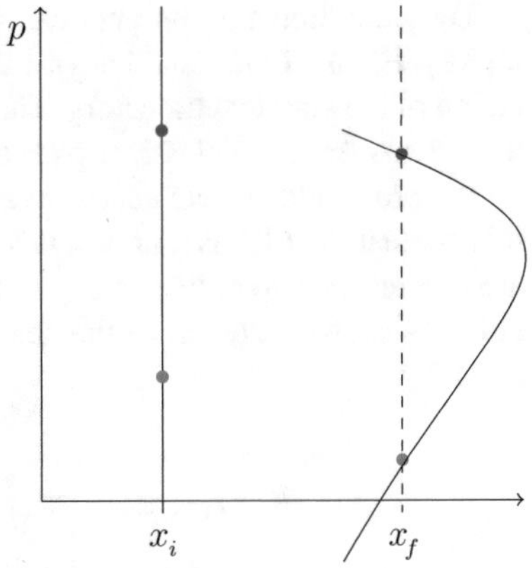

Fig. 2.5 The case of two classical solutions of the boundary value problem $x(0) = x_i$; $x(t) = x_f$. The phase space manifold of initial conditions is indicated by the *solid vertical line*. This manifold evolves into the *bent curve*, whereby two of the initial conditions, indicated by the *colored dots*, fulfill the boundary value problem

where $|_{x_i}$ denotes that x_i is kept fixed and for the time being, we have assumed that there is just one single solution of the double-ended boundary value problem. Furthermore, Planck's constant enters for dimensionality reasons in a similar fashion as in classical statistical mechanics. Converting to quantum mechanical amplitudes, the square root of the classical probability density has to be taken to arrive at the correct semiclassical prefactor. From basic classical mechanics we have the identity

$$\frac{\partial^2 S[x_{\rm cl}]}{\partial x_f \partial x_i} = -\frac{\partial p_i}{\partial x_f}, \tag{2.90}$$

for the so-called van Vleck determinant,[9] however. Thus up to the absolute value in (2.89), reflecting the fact that the probability density is a positive definite quantity, the semiclassical propagator of van Vleck and the one from the SPA to the path integral are proportional.

Almost 40 years later, Gutzwiller has extended the validity of the van Vleck expression to longer times [16]. First of all for finite times there may be several solutions to the classical root search problem. Such a situation is depicted graphically in Fig. 2.5. In the case of multiple solutions an additional summation therefore has to be introduced in (2.86). Using the path integral derivation together with the summation over several points of stationary phase one finally arrives at the van Vleck-Gutzwiller expression

$$K^{\rm VVG}(x_f, t; x_i, 0) \equiv \sum_{\rm cl} \sqrt{\frac{\rm i}{2\pi\hbar} \left| \frac{\partial^2 S[x_{\rm cl}]}{\partial x_f \partial x_i} \right|} \exp\left\{ \frac{\rm i}{\hbar} S[x_{\rm cl}] - {\rm i}\nu\pi/2 \right\}, \tag{2.91}$$

with the so-called Maslov (or Morse) index ν, introduced into the semiclassical propagator by Gutzwiller and counting the caustics a path has gone through. The Maslov phase allows one to use the absolute value inside the square root.

[9] In general, x_f and x_i have to be replaced by vectors!

This final expression has interference effects built in, because of the summation over classical trajectories and is very elegant and intuitive, because it relies solely on classical trajectories. However, it has also a major drawback, especially for systems with several degrees of freedom. Then the underlying root search problem becomes extremely hard to solve and a semiclassical propagator using the solution of classical *initial value problems* would be much needed. Such a reformulation of the semiclassical expression is possible and will be discussed in Sect. 2.3.4 on numerical methods.

2.2.4 *Pictures of Quantum Mechanics and Time-Dependent Perturbation Theory*

Another approximate way to solve the time-dependent Schrödinger equation starts directly from the infinite sum of time-ordered operator products in (2.31) for the time-evolution operator. Considering this series as a perturbation expansion and taking only a few terms into account will lead to reasonable results only for short times. In the case of additive Hamiltonians,

$$\hat{H}(t) = \hat{H}_0(t) + \hat{W}(t), \tag{2.92}$$

we can, however, split the problem into parts and can possibly treat the first one analytically and the rest perturbatively. Please note that in (2.92) both the first and second term may depend on time. This will come in handy when we discuss the different pictures of quantum mechanics in a unified manner.

The time-evolution operator for the unperturbed problem is formally given by

$$\hat{U}_0(t,0) = \hat{T}\exp\left\{-\frac{\mathrm{i}}{\hbar}\int_0^t \mathrm{d}t'\hat{H}_0(t')\right\}. \tag{2.93}$$

With its help we define a wavefunction in the interaction picture, indicated by the index I,

$$|\Psi_\mathrm{S}(t)\rangle =: \hat{U}_0(t,0)|\Psi_\mathrm{I}(t)\rangle, \tag{2.94}$$

where the wavefunction $|\Psi_\mathrm{S}(t)\rangle$ is the one in the Schrödinger picture, which we have considered up to now. Inserting (2.94) into the time-dependent Schrödinger equation and using

$$\mathrm{i}\hbar\dot{\hat{U}}_0(t,0) = \hat{H}_0(t)\hat{U}_0(t,0) \tag{2.95}$$

yields

$$i\hbar|\dot{\Psi}_{\mathrm{I}}(t)\rangle = \hat{W}_{\mathrm{I}}(t)|\Psi_{\mathrm{I}}(t)\rangle, \tag{2.96}$$

i.e., the Schrödinger equation in the interaction picture, where the perturbation Hamiltonian in the interaction picture is given by

$$\hat{W}_{\mathrm{I}}(t) := \hat{U}_0^\dagger(t,0)\hat{W}(t)\hat{U}_0(t,0). \tag{2.97}$$

The time-evolution operator in the interaction picture is

$$\hat{U}_{\mathrm{I}}(t,0) := \hat{T}\exp\left\{-\frac{i}{\hbar}\int_0^t dt'\hat{W}_{\mathrm{I}}(t')\right\}. \tag{2.98}$$

Using (2.94), we note that at $t = 0$ the wavefunctions are identical, i.e. $|\Psi_{\mathrm{S}}(0)\rangle = |\Psi_{\mathrm{I}}(0)\rangle = |\Psi(0)\rangle$. Furthermore, calculating an expectation value in the Schrödinger picture and using (2.94), we get

$$\begin{aligned}\langle\hat{A}\rangle(t) &= \langle\Psi_{\mathrm{S}}(t)|\hat{A}_{\mathrm{S}}|\Psi_{\mathrm{S}}(t)\rangle \\ &= \langle\Psi_{\mathrm{I}}(t)|\hat{U}_0^\dagger(t,0)\hat{A}_{\mathrm{S}}\hat{U}_0(t,0)|\Psi_{\mathrm{I}}(t)\rangle.\end{aligned} \tag{2.99}$$

With the definition of an operator in the interaction picture

$$\hat{A}_{\mathrm{I}}(t) := \hat{U}_0^\dagger(t,0)\hat{A}_{\mathrm{S}}\hat{U}_0(t,0), \tag{2.100}$$

the expectation value becomes

$$\langle\hat{A}\rangle(t) = \langle\Psi_{\mathrm{I}}(t)|\hat{A}_{\mathrm{I}}(t)|\Psi_{\mathrm{I}}(t)\rangle, \tag{2.101}$$

which is identical to the Schrödinger picture expectation value.

2.8. *Verify that the time evolution operator in the interaction picture* $\hat{U}_{\mathrm{I}}(t,0) = \hat{U}_0^+(t,0)\hat{U}(t,0)$ *fulfills the appropriate differential equation.*

The third picture that is frequently applied is the one named after Heisenberg. By an appropriate choice of the unperturbed Hamiltonian in (2.93) and the perturbation all three pictures can be dealt with in the same framework:

- $\hat{H} = \hat{H}_0 + \hat{W}$ leads to the interaction picture
- $\hat{H}_0 = 0$ und $\hat{W} = \hat{H}$ leads to the Schrödinger picture
- $\hat{H}_0 = \hat{H}$ and $\hat{W} = 0$ leads to the Heisenberg picture

The relations between the different cases are given in Table 2.1 for the wavefunctions and in Table 2.2 for the operators. In Schrödinger's original representation, the wavefunction is time-dependent, whereas the operators are time-independent. In the Heisenberg picture it is the other way around. The interaction picture has both wavefunction and operators being time-dependent.

Table 2.1 Relations between the wavefunctions in the different pictures of quantum mechanics

	$\|\Psi_S(t)\rangle$	$\|\Psi_H\rangle$	$\|\Psi_I(t)\rangle$
$\|\Psi_S(t)\rangle$		$\hat{U}(t,0)\|\Psi_H\rangle$	$\hat{U}_0(t,0)\|\Psi_I(t)\rangle$
$\|\Psi_H\rangle$	$\hat{U}^\dagger(t,0)\|\Psi_S(t)\rangle$		$\hat{U}_I^\dagger(t,0)\|\Psi_I(t)\rangle$
$\|\Psi_I(t)\rangle$	$\hat{U}_0^\dagger(t,0)\|\Psi_S(t)\rangle$	$\hat{U}_I(t,0)\|\Psi_H\rangle$	

Table 2.2 Relations between the operators in the different pictures of quantum mechanics

	$\hat{A}_S$	$\hat{A}_H(t)$	$\hat{A}_I(t)$
$\hat{A}_S$		$\hat{U}(t,0)\hat{A}_H(t)\hat{U}^\dagger(t,0)$	$\hat{U}_0(t,0)\hat{A}_I(t)\hat{U}_0^\dagger(t,0)$
$\hat{A}_H(t)$	$\hat{U}^\dagger(t,0)\hat{A}_S\hat{U}(t,0)$		$\hat{U}_I^\dagger(t,0)\hat{A}_I(t)\hat{U}_I(t,0)$
$\hat{A}_I(t)$	$\hat{U}_0^\dagger(t,0)\hat{A}_S\hat{U}_0(t,0)$	$\hat{U}_I(t,0)\hat{A}_H(t)\hat{U}_I^\dagger(t,0)$	

With the interaction picture defined, we can now derive time-dependent perturbation theory for small perturbations $\hat{W}(t)$. Iterative solution of the corresponding time-dependent Schrödinger equation leads to an expression for the propagator in the interaction picture

$$\hat{U}_I(t,0) = \hat{1} + \sum_{n=1}^{\infty}\left(\frac{-i}{\hbar}\right)^n \int_0^t dt_n \int_0^{t_n} dt_{n-1}\cdots\int_0^{t_2} dt_1 \\ \hat{W}_I(t_n)\hat{W}_I(t_{n-1})\cdots\hat{W}_I(t_1), \tag{2.102}$$

analogous to (2.31). In perturbation theory, the series is truncated at finite n and in case of $n = 1$, we get

$$|\Psi_I^1(t)\rangle = |\Psi(0)\rangle - \frac{i}{\hbar}\int_0^t dt'\hat{U}_0^\dagger(t',0)\hat{W}(t')\hat{U}_0(t',0)|\Psi(0)\rangle. \tag{2.103}$$

Going back to the Schrödinger picture, by using (2.94), we get

$$|\Psi_S^1(t)\rangle = \hat{U}_0(t,0)|\Psi(0)\rangle - \frac{i}{\hbar}\int_0^t dt'\hat{U}_0(t,t')\hat{W}(t')\hat{U}_0(t',0)|\Psi(0)\rangle. \tag{2.104}$$

in first order. We will interpret and use expressions of this kind in the discussion of pump-probe spectroscopy in Chap. 5. Terms of higher order will contain multiple, nested integrals but will not be needed there.

2.2.5 Magnus Expansion

Another approach to solve the time-dependent Schrödinger equation starting from the time-evolution operator is the so-called Magnus expansion. The basic idea of this method is resummation and can be understood by considering a function depending on a parameter λ. Expanding this function in powers of the parameter leads to

$$A = A_0(1 + \lambda A_1 + \lambda^2 A_2 + \cdots) . \tag{2.105}$$

Alternatively, the function can be written as a prefactor times an exponential

$$A = A_0 \exp(F) . \tag{2.106}$$

Also the exponent F can be expanded in powers of the parameter

$$F = \lambda F_1 + \lambda^2 F_2 + \cdots . \tag{2.107}$$

The exponential function can now be Taylor expanded and after comparing the coefficients of equal powers of λ, the F_n can be expressed in terms of the A_n. Now If we now truncate the series in (2.105) after $n = 2$, by using the coefficients in the exponent via

$$A \approx A_0 \exp(\lambda A_1 + \lambda^2(A_2 - A_1^2/2)) , \tag{2.108}$$

an expression of infinite order in λ has been gained.

In quantum dynamics, the technique presented above is used for the infinite sum in (2.102), representing the time-evolution operator. The parameter λ is equal to $-\mathrm{i}/\hbar$ in this case. The final result for the time-evolution operator is

$$\begin{aligned} \hat{U}_\mathrm{I}(t, 0) &= \hat{T} \exp\left\{-\frac{\mathrm{i}}{\hbar}\int_0^t \mathrm{d}t'\, \hat{W}_\mathrm{I}(t')\right\} \\ &= \exp\left\{\sum_{n=1}^{\infty} \frac{1}{n!}\left(-\frac{\mathrm{i}}{\hbar}\right)^n \hat{\mathrm{H}}_n(t, 0)\right\} , \end{aligned} \tag{2.109}$$

where

$$\hat{\mathrm{H}}_1 = \int_0^t \mathrm{d}t'\, \hat{W}_\mathrm{I}(t'), \tag{2.110}$$

$$\hat{\mathrm{H}}_2 = \int_0^t \mathrm{d}t_2 \int_0^{t_2} \mathrm{d}t_1 [\hat{W}_\mathrm{I}(t_2), \hat{W}_\mathrm{I}(t_1)], \tag{2.111}$$

$$\hat{\mathrm{H}}_3 = \int_0^t \mathrm{d}t_3 \int_0^{t_3} \mathrm{d}t_2 \int_0^{t_2} \mathrm{d}t_1 \big([\hat{W}_\mathrm{I}(t_3), [\hat{W}_\mathrm{I}(t_2), \hat{W}_\mathrm{I}(t_1)]] + [[\hat{W}_\mathrm{I}(t_3), \hat{W}_\mathrm{I}(t_2)], \hat{W}_\mathrm{I}(t_1)]\big) \tag{2.112}$$

are the first three terms in the expansion of the exponent.

2.9. *Verify the second order expression* $\hat{\mathrm{H}}_2$ *of the Magnus expansion in the exponent of the time-evolution operator in the interaction picture.*

The main advantage of the expression in (2.109) is that, in principle, it is an exact result and that it does not contain the time-ordering operator any more. In numerical applications the summation in the exponent will be terminated at finite n, however, and leads to a unitary propagation scheme at any order. If one would truncate the expansion after $n = 1$, then the time-ordering operator in (2.109) would have been ignored altogether. Although this seems to be rather a crude approximation, in Chap. 4 we will see that it leads to reasonable results in the case of atoms subject to extremely short pulses. Furthermore, it has turned out that in the interaction picture with a suitable choice of $\hat{H}_0$, truncating the Magnus expansion is a successful numerical approach [19].

2.2.6 *Time-Dependent Hartree Method*

Especially in Chap. 5, we will investigate systems with several coupled degrees of freedom. The factorization of the total wavefunction is a first very crude approximative way to solve the time-dependent Schrödinger equation for such systems. It shall therefore be discussed here for the simplest case of two degrees of freedom corresponding to distinguishable particles.

The total Hamiltonian shall be of the form

$$\hat{H} = \sum_{n=1}^{2} \hat{H}_n(x_n) + V_{12}(x_1, x_2), \tag{2.113}$$

with single particle operators

$$\hat{H}_n(x_n) = -\frac{\hbar^2}{2m}\frac{\partial^2}{\partial x_n^2} + V_n(x_n) \tag{2.114}$$

and the coupling potential V_{12} depending on the two coordinates in a non-additive manner. The so-called Hartree Ansatz for the wavefunction is of the form

$$\Psi(x_1, x_2, t) = \Psi_1(x_1, t)\Psi_2(x_2, t) \tag{2.115}$$

of a product of single particle wavefunctions.

This Ansatz is exact in the case that the coupling V_{12} vanishes. The single particle functions then fulfill

$$\mathrm{i}\hbar\dot{\Psi}_n(x_n,t) = \hat{H}_n\Psi_n(x_n,t). \tag{2.116}$$

We now plug the Hartree Ansatz into the full time-dependent Schrödinger equation and find

$$\mathrm{i}\hbar\left(\Psi_2\dot{\Psi}_1 + \Psi_1\dot{\Psi}_2\right) = \Psi_2\hat{H}_1\Psi_1 + \Psi_1\hat{H}_2\Psi_2 + V_{12}\Psi_1\Psi_2\,. \tag{2.117}$$

Multiplying this equation with Ψ_2^* and integrating over the coordinate of particle 2 yields

$$\mathrm{i}\hbar\left(\dot{\Psi}_1 + \Psi_1\langle\Psi_2|\dot{\Psi}_2\rangle_2\right) = \hat{H}_1\Psi_1 + \Psi_1\langle\Psi_2|\hat{H}_2|\Psi_2\rangle_2 + \langle\Psi_2|V_{12}|\Psi_2\rangle_2\Psi_1. \tag{2.118}$$

By using the single particle equations of zeroth order with the index 2, the second terms on the LHS and the RHS cancel each other and one finds

$$\mathrm{i}\hbar\dot{\Psi}_1(x_1,t) = \left(-\frac{\hbar^2}{2m}\Delta_1 + V_{1,\mathrm{eff}}(x_1,t)\right)\Psi_1(x_1,t), \tag{2.119}$$

with an effective, time-dependent potential

$$V_{1,\mathrm{eff}}(x_1,t) = V_1(x_1) + \langle\Psi_2(t)|V_{12}|\Psi_2(t)\rangle_2. \tag{2.120}$$

An analogous equation can be derived for particle 2

$$\mathrm{i}\hbar\dot{\Psi}_2(x_2,t) = \left(-\frac{\hbar^2}{2m}\Delta_2 + V_{2,\mathrm{eff}}(x_2,t)\right)\Psi_2(x_2,t), \tag{2.121}$$

by multiplying the time-dependent Schrödinger equation with Ψ_1^* and integrating over x_1.

The particles move in effective "mean" fields that are determined by the dynamics of the other particle. The coupled equations have to be solved self-consistently. This is the reason that the Hartree method sometimes is called a TDSCF (time-dependent self consistent field) method. The multi-configuration time-dependent Hartree (MCTDH) method [20] goes far beyond what has been presented here and in principle allows for an exact numerical solution of the time-dependent Schrödinger equation.

2.2.7 *Quantum-Classical Methods*

In quantum-classical methods, the degrees of freedom are separated into a subset that shall be dealt with on the basis of classical mechanics and a subset to be described fully quantum mechanically. Analogously to the Hartree method, the classical degrees of freedom will move in an effective potential that is determined by the solution of the quantum problem.

For reasons of convenience we start the discussion with the time-independent Schrödinger equation and restrict it to the case of two degrees of freedom: a light particle with coordinate x and mass m and a heavy one with X and M, respectively. Having the full solution of the time-independent Schrödinger equation

$$\hat{H}\psi_n(x, X) = E_n\psi_n(x, X), \tag{2.122}$$

with the Hamiltonian

$$\hat{H} = \frac{\hat{p}^2}{2m} + \frac{\hat{P}^2}{2M} + v(x, X) + V(X) \tag{2.123}$$

at hand would allow us to construct a time-dependent solution according to

$$\Psi(x, X, t) = \sum_n c_n \exp\left[-\frac{\mathrm{i}}{\hbar}E_n t\right]\psi_n(x, X). \tag{2.124}$$

A way to treat the coupled system approximately is intimately related to the separation Ansatz of the previous section and will be discussed in much more detail later-on in Sect. 5.3.1 on the Born-Oppenheimer approximation. The idea is simple: one replaces the coupled problem by a pair of uncoupled single particle problems. In order to do so, first the light particle is dealt with under certain (fixed) conditions for the heavy particle (denoted by $|X$)

$$\hat{H}^0(x|X)\phi_j(x|X) = \epsilon_j^0(X)\phi_j(x|X), \tag{2.125}$$

where

$$\hat{H}^0(x|X) = \frac{\hat{p}^2}{2m} + v(x, X) \tag{2.126}$$

depends parametrically on X and j is the quantum number of the light particle.

Using the product Ansatz

$$\psi_n(x, X) \approx \phi_j(x|X)\chi_{l,j}(X) \tag{2.127}$$

in the full time-independent Schrödinger equation and integrating out the coordinate of the light particle in the same way as in Sect. 2.2.6, one arrives at equations of the form

$$\hat{H}_j^1(X)\chi_{l,j}(X) = \epsilon_{l,j}^1\chi_{l,j}(X), \tag{2.128}$$

with the Hamiltonian

$$\hat{H}_j^1(X) = \frac{\hat{P}^2}{2M} + V(X) + \epsilon_j^0(X) \tag{2.129}$$

and eigenvalues, which are approximately given by

$$E_n \approx \epsilon_{l,j}^1, \tag{2.130}$$

depending on j as well as on l, which is the quantum number of the heavy particle. The heavy particle is thus governed by an effective potential, $V(X) + \epsilon_j^0(X)$, depending on the quantum state of the light particle.

Let us now turn to dynamics. In the Ehrenfest method one postulates the classical treatment of the heavy particle. Analogous to the effective potential an effective force can be derived which reads

$$F_{\text{eff}} = -\frac{\partial}{\partial X}\left\{V + \int \mathrm{d}x\, \Phi^* \hat{H}^0 \Phi\right\}, \tag{2.131}$$

and where the wavefunction of the light particle fulfills the time-dependent Schrödinger equation

$$\mathrm{i}\hbar\dot{\Phi}(x|X(t),t) = \hat{H}^0(x|X(t))\Phi(x|X(t)). \tag{2.132}$$

Expanding this wavefunction in eigenfunctions of the light particle

$$\Phi(x|X(t),t) = \sum_j c_j(t)\phi_j(x|X(t)) \tag{2.133}$$

yields coupled differential equations for the coefficients

$$\mathrm{i}\hbar\dot{c}_j(t) = \epsilon_j^0 c_j - \mathrm{i}\hbar\dot{X}\sum_k d_{jk}c_k, \tag{2.134}$$

where $d_{jk} = \int \mathrm{d}x \phi_j \frac{\partial \phi_k}{\partial X}$. Together with the explicit expression

$$F_{\text{eff}} = -\frac{\partial V}{\partial X} - \sum_j |c_j|^2 \frac{\partial \epsilon_j^0}{\partial X} + \sum_{j,k<j}[c_j^* c_k + c_k^* c_j][\epsilon_j^0 - \epsilon_k^0]d_{jk} \tag{2.135}$$

for the effective force, this can be proven by solving Exercise 2.10.

2.10. *Verify the fundamental equations of the Ehrenfest method.*

(a) *First prove the validity of the coupled differential equations for the coefficients* c_j *(Use the product and the chain rule of differentiation).*
(b) *Calculate the effective force by using* $d_{kj} = -d_{jk}$ *(Proof?)*

The first term in the expression of the force is the so-called external force, whereas the second one describes adiabatic and the third one nonadiabatic dynamics.[10] The last two terms have to be determined by solving the quantum problem of the light particle. An alternative quantum-classical approach is the so-called surface hopping technique. Its relation to the Ehrenfest approach, and which method is suited under which circumstances is discussed in [21].

2.2.8 Floquet Theory

For the description of cw-laser driven systems, the problem of time-periodic Hamiltonians is of central importance. In this case we have

$$\hat{H}(t+T) = \hat{H}(t), \tag{2.136}$$

with the period $T = 2\pi/\omega$ of the external perturbation.

As in the general time-dependent case, the time evolution operator can be used to propagate a wavefunction. In addition to the properties discussed in Sect. 2.1.2, we can now make use of the property

$$\hat{U}(t+T, s+T) = \hat{U}(t, s). \tag{2.137}$$

In order to solve the time-dependent Schrödinger equation, we prove that the Hamiltonian extended by the time derivative

$$\hat{\mathcal{H}}(t) \equiv \hat{H}(t) - \mathrm{i}\hbar\partial_t \tag{2.138}$$

commutes with the time-evolution operator over one period. The time-dependent Schrödinger equation can be rewritten by using the above definition and we apply the time-evolution operator from the left

$$\hat{U}(t+T,t)\hat{\mathcal{H}}(t)|\Psi(t)\rangle = 0, \tag{2.139}$$

$$\hat{U}(t+T,t)\hat{\mathcal{H}}(t)\hat{U}^{-1}(t+T,t)\hat{U}(t+T,t)|\Psi(t)\rangle = 0, \tag{2.140}$$

$$\hat{\mathcal{H}}(t)\hat{U}(t+T,t)|\Psi(t)\rangle = 0. \tag{2.141}$$

The last, decisive step follows form the periodicity of the Hamiltonian and with the help of the chain rule.[11] We can thus conclude that

$$[\hat{U}(t+T,t), \hat{\mathcal{H}}(t)] = 0 \tag{2.142}$$

---

[10]The explanation of these terms follows in Chap. 5.

[11]We have used $\hat{U}(t+T,t)\hat{\mathcal{H}}(t)\hat{U}^{-1}(t+T,t) = \hat{\mathcal{H}}(t+T) = \hat{H}(t+T) - \mathrm{i}\hbar\partial_{t+T} = \hat{H}(t) - \mathrm{i}\hbar\partial_t$.

holds. The two operators thus have a common system of eigenfunctions, which shall be denoted by $|\Psi_\epsilon(t)\rangle$.

From the composition property (2.27) and with (2.137) it follows that

$$\hat{U}(2T,0) = \hat{U}(2T,T)\hat{U}(T,0) = \hat{U}^2(T,0). \tag{2.143}$$

The group of time-evolution operators over one period therefore is an Abelian group. Its eigenfunctions have to transform according to a one-dimensional irreducible representation [22]

$$\hat{U}(T,0)|\Psi_\epsilon(0)\rangle = \exp\left\{-\frac{\mathrm{i}}{\hbar}\epsilon T\right\}|\Psi_\epsilon(0)\rangle. \tag{2.144}$$

Comparing this equation with the time-evolution over one period

$$\hat{U}(T,0)|\Psi_\epsilon(0)\rangle = |\Psi_\epsilon(T)\rangle \tag{2.145}$$

leads to the Floquet theorem for the solution of the time-dependent Schrödinger equation

$$\Psi_\epsilon(x,t) = \exp\left\{-\frac{\mathrm{i}}{\hbar}\epsilon t\right\}\psi_\epsilon(x,t), \tag{2.146}$$

$$\psi_\epsilon(x,t) = \psi_\epsilon(x,t+T). \tag{2.147}$$

The wavefunction is a product of an exponential factor times a periodic function.[12] The factor ϵ in the exponent of (2.146) is sometimes referred to as *Floquet exponent* and the corresponding periodic function ψ_ϵ is called *Floquet function* in order to honor the french mathematician Gaston Floquet, who worked on differential equations with periodic coefficients in the 19th century.

The product in (2.146) is formally analogous to the separation (2.9) in the stationary case. In order to use this analogy, we rewrite the time-dependent Schrödinger equation as in (2.139), to read

$$\hat{\mathcal{H}}(x,t)\Psi(x,t) = 0. \tag{2.148}$$

Inserting the Floquet solution (2.146) and performing the time-derivative of the exponential part yields

$$\hat{\mathcal{H}}(x,t)\psi_\alpha(x,t) = \epsilon_\alpha\psi_\alpha(x,t), \tag{2.149}$$

where the quantum number index α has been introduced. This "Floquet type Schrödinger equation" has the same formal structure as the time-independent Schrödinger equation. Therefore the Floquet exponents are also called *quasi-energies*

[12] Formally this Ansatz is equivalent to the Bloch theorem of solid state physics.

and the Floquet functions are referred to as *quasi-eigenfunctions*. In the case of a Hermitian Hamiltonian the quasi-energies are real.

2.11. *Use the extended scalar product*

$$\langle\langle u_\alpha | v_\beta \rangle\rangle := \frac{1}{T}\int_0^T \mathrm{d}t \int_{-\infty}^{\infty} \mathrm{d}x u_\alpha^*(x,t) v_\beta(x,t)$$

and the extended Hamiltonian $\hat{\mathcal{H}} = \hat{H}(t) - \mathrm{i}\hbar\partial_t$, *with* $\hat{H}(t+T) = \hat{H}(t)$, *in order to show that the Floquet energies* ϵ_α *are real in case of Hermitian Hamiltonias* $\hat{H}(t)$.

In the case of vanishing external field the Hamiltonian becomes time-independent. This implies that also ψ_α is time-independent. The index α therefore is related to the quantum numbers of the unperturbed problem. It is a special feature of the Floquet solution (2.146) and (2.147) that the modified quasi-eigenfunctions and corresponding energies

$$\psi_{\alpha'}(x,t) := \psi_\alpha(x,t)\exp(\mathrm{i}k\omega t), \tag{2.150}$$

$$\epsilon_{\alpha'} := \epsilon_\alpha + k\hbar\omega, \tag{2.151}$$

with $k = 0, \pm 1, \pm 2, \ldots$ [13] are equivalent to $\psi_\alpha(x,t)$, ϵ_α, due to the fact that they correspond to the same total solution $\Psi_\alpha(x,t)$. The index

$$\alpha' := (\alpha, k) \quad k = 0, \pm 1, \pm 2, \ldots \tag{2.152}$$

denotes a class of infinitely many solutions. Out of each class only one lays in a so-called Brillouin zone of width $\hbar\omega$, however. The discussion above and more details on the underlying Hilbert space theory can be found in [23]. Without proving the completeness, we will use

$$\hat{1} = \sum_{\alpha'} |\psi_{\alpha'}(0)\rangle\langle\psi_{\alpha'}(0)| \tag{2.153}$$

as the representation of unity in terms of (discrete) Floquet states. In order for this representation to be true also in the nondriven case, it is clear that only one member of the class of equivalent eigensolutions may contribute to the sum above. A solution of the time-dependent Schrödinger equation can therefore be written as a superposition of quasi-eigenfunctions with appropriate coefficients

$$|\Psi(t)\rangle = \sum_{\alpha'} c_{\alpha'}|\psi_{\alpha'}(t)\rangle \exp\left\{-\frac{\mathrm{i}}{\hbar}\epsilon_{\alpha'}t\right\}, \tag{2.154}$$

$$c_{\alpha'} = \langle\psi_{\alpha'}(0)|\Psi(0)\rangle. \tag{2.155}$$

[13] Note that k has to be an integer in order for the modified quasi-eigenfunction to be periodic.

This equation exhibits clearly that the quasi-energies determine the long-term time-evolution of a periodically driven quantum system. The behavior of the quasi-energies as a function of an external parameter (e.g., the field strength or the frequency) will therefore be very important. In order to study this behavior, the symmetry of the Hamiltonian will be decisive. We will come back to this point in Appendix 3.A to the Chap. 3.

2.3 Numerical Methods

Apart from special two-level problems that will be dealt with in the next chapter and systems with maximally quadratic potentials (and problems that can be mapped onto such cases) there are only a few exactly analytically solvable problems in quantum dynamics, as can be seen by studying the review by Kleber [11].

Almost all interesting problems of atomic and molecular physics with and without the presence of laser fields classically display nonlinear dynamics, however, and the Gaussian wavepacket dynamics of Sect. 2.1.4 will be valid only for short times. Exact numerical solutions of the quantum dynamics are therefore sought for. Apart from time-dependent information that is, e.g., needed for the description of pump-probe experiments, to be discussed in Chap. 5, also spectral information for systems with autonomous Hamiltonians can be gained from time series, as was shown in Sect. 2.1.2.

In the following, different ways to solve the time-dependent Schrödinger equation numerically will be described. First, we will review some numerically exact methods, and in the end the implementation of the semiclassical theory lined out in Sect. 2.2.3 by so-called initial value methods will be discussed, thereby also touching the numerical solution of the underlying classical equations of motion.

Most methods to solve the TDSE that we discuss can be characterized according to two formal criteria, which will be called problem (a) and problem (b) in the following:

(a) Which (finite) basis is used to represent the wavefunction?
(b) In which (approximate) way is the time-evolution performed?

We will distinguish the methods according to their different approach to the solution of the problems above.

2.3.1 Orthogonal Basis Expansion

All methods to solve the time-dependent Schrödinger equation numerically have to deal with a way to express the wavefunction in a certain representation. Here we shall consider the expansion of the wavefunction in a set of orthogonal basis functions, which are eigenfunctions of a certain (simple) Hamiltonian like, e.g., the 1D harmonic oscillator one

$$\hat{H}_{\mathrm{HO}} = -\frac{\hbar^2}{2m}\frac{\partial^2}{\partial x^2} + \frac{1}{2}m\omega_{\mathrm{e}}^2 x^2. \tag{2.156}$$

Its eigenfunctions are given by

$$\psi_n(x) = \sqrt{\frac{\sigma}{n!2^n\sqrt{\pi}}} H_n(\sigma x) \exp\left\{-\frac{1}{2}\sigma^2 x^2\right\}, \tag{2.157}$$

where H_n with $n = 0, 1, 2, \ldots$ are Hermite polynomials [56], the first three of which are

$$H_0(x) = 1, \tag{2.158}$$
$$H_1(x) = 2x, \tag{2.159}$$
$$H_2(x) = 4x^2 - 2, \tag{2.160}$$

and $\sigma = \sqrt{m\omega_{\mathrm{e}}/\hbar}$.

The alternative representation of the harmonic oscillator Hamiltonian

$$\hat{H}_{\mathrm{HO}} = \hbar\omega_{\mathrm{e}}\left(\hat{a}^\dagger\hat{a} + \frac{1}{2}\right) \tag{2.161}$$

in terms of so-called creation and annihilation operators

$$\hat{a}^\dagger = \frac{1}{\sqrt{2}}\left(\sigma\hat{x} - \frac{1}{\sigma}\frac{\partial}{\partial x}\right), \tag{2.162}$$
$$\hat{a} = \frac{1}{\sqrt{2}}\left(\sigma\hat{x} + \frac{1}{\sigma}\frac{\partial}{\partial x}\right) \tag{2.163}$$

is very helpful. These operators fulfill the commutation relation

$$[\hat{a}, \hat{a}^\dagger] = \hat{1} \tag{2.164}$$

and have the properties

$$\hat{a}^\dagger|n\rangle = \sqrt{n+1}|n+1\rangle \qquad n = 0, 1, 2, \ldots, \tag{2.165}$$
$$\hat{a}|n\rangle = \sqrt{n}|n-1\rangle \qquad n = 1, 2, \ldots, \tag{2.166}$$

i.e., they either allow to "climb up" or "climb down" the ladder of harmonic oscillator states (in short hand notation just denoted by the index n) and are therefore also referred to as ladder operators.

By resolving the definition of the ladder operators in terms of the position and the derivative operator

$$\hat{x} = \frac{1}{\sqrt{2}\sigma}(\hat{a} + \hat{a}^\dagger), \tag{2.167}$$

$$\frac{\partial}{\partial x} = \frac{\sigma}{\sqrt{2}}(\hat{a} - \hat{a}^\dagger), \tag{2.168}$$

one can express arbitrary powers of these operators in terms of products of $\hat{a}^\dagger$ and $\hat{a}$. Matrix elements of any operator between harmonic oscillator states can then be calculated by employing (2.165) and (2.166).

An arbitrary time-dependent wavefunction can now be expanded into eigenfunctions of the harmonic oscillator according to

$$|\Psi(t)\rangle = \sum_{l=0}^{\infty} d_l(t)|l\rangle. \tag{2.169}$$

After insertion of this expression into the time-dependent Schrödinger equation governed by the Hamiltonian $\hat{H}$, and multiplication from the left with $\langle n|$, an infinite linear system of coupled ordinary differential equations for the expansion coefficients

$$\mathrm{i}\hbar\dot{d}_n(t) = \sum_{l=0}^{\infty} d_l(t)\langle n|\hat{H}|l\rangle \tag{2.170}$$

is gained. This system can (in principle) be solved if the initial conditions were known. At this point, however, we should address our "problems" as stated in the introduction to this section:

(a) The basis problem is solved by truncating the expansion at a large $l = L - 1$, which is determined by the initial state that shall be described. One thus uses a "Finite Basis Representation". Convergence of the results can be checked by increasing the size L of the finite basis.
(b) The numerical integration of the linear system of differential equations could be performed with the help of a suitable integration routine like the Runge-Kutta method [25].

Solving the system of coupled differential equations can be circumvented, however, by finding the (first N) eigenvalues E_n and eigenfunctions $|n_H\rangle$ of the Hamiltonian $\hat{H}$ in case this is autonomous. By determining the (now time-independent) expansion coefficients of the wavefunction in terms of these eigenfunctions, the wavefunction is in principle exactly[14] time-evolved by using the corresponding eigenenergies according to

---

[14]This would be true, if the energies were exact, which is prohibited by problem (a).

$$|\Psi(t)\rangle = \sum_{n=0}^{N-1} c_n |n_H\rangle \mathrm{e}^{-\mathrm{i}E_n t/\hbar}. \tag{2.171}$$

We should keep in mind, however, that the solution of the eigenvalue problem requires a numerical effort of order L^3 if $L \times L$ is the size of the matrix to be diagonalized [25] and is therefore only suitable if L can be kept small.

2.3.1.1 The Floquet Matrix

The alternative method to tackle problem (b) by solving the eigenvalue problem is fortunately not restricted to autonomous Hamiltonians. It also works for periodically driven systems that have been discussed in Sect. 2.2.8 and leads to the calculation of the quasi-energies and quasi-eigenfunctions. We start from the Floquet type Schrödinger equation in Dirac notation

$$\hat{\mathcal{H}}(t)|\psi_\alpha(t)\rangle = \epsilon_\alpha |\psi_\alpha(t)\rangle, \tag{2.172}$$

with the extended Hamiltonian defined in (2.138). Due to the periodic time-dependence of the Floquet functions, they can be Fourier expanded according to

$$|\psi_\alpha(t)\rangle = \sum_{n=-\infty}^{\infty} |\psi_\alpha^n\rangle \mathrm{e}^{-in\omega t}. \tag{2.173}$$

The Fourier coefficients on the RHS of this expression can in turn be expanded in a complete orthogonal system of basis functions $\{|k\rangle\}$ via

$$|\psi_\alpha^n\rangle = \sum_{k=0}^{\infty} \psi_{k,\alpha}^n |k\rangle \tag{2.174}$$

and the Schrödinger equation thus is given by

$$\sum_{n=-\infty}^{\infty}\sum_{k=0}^{\infty} \hat{\mathcal{H}}\psi_{k,\alpha}^n |k\rangle \mathrm{e}^{-in\omega t} = \sum_{n=-\infty}^{\infty}\sum_{k=0}^{\infty} \epsilon_\alpha \psi_{k,\alpha}^n |k\rangle \mathrm{e}^{-in\omega t}. \tag{2.175}$$

Multiplying this equation with $\langle lm| := \langle \psi_l \mathrm{e}^{-im\omega t}|$ from the left and integrating over one period of the external force yields

$$\sum_{n=-\infty}^{\infty}\sum_{k=0}^{\infty} \langle\langle lm|\hat{\mathcal{H}}|kn\rangle\rangle \psi_{k,\alpha}^n = \epsilon_\alpha \psi_{l,\alpha}^m, \tag{2.176}$$

where we have used the definition $\langle\langle\cdots\rangle\rangle := \frac{1}{T}\int_0^T \mathrm{d}t\langle\cdots\rangle$ of the extended scalar product that has already been used in Exercise 2.11.

Equation (2.176) has first been given by Shirley [26] in the case of a two-level system. Shirley then transformed the equations in order to recover an eigenvalue problem, whose solutions are the quasi-energies. This procedure can also be used for an infinite dimensional Hilbert space, however. In order to do so, one rewrites the equation above according to

$$\sum_{n=-\infty}^{\infty}\sum_{k=0}^{\infty}\left\{\langle l|\hat{H}^{[m-n]}|k\rangle - n\hbar\omega\delta_{mn}\delta_{lk}\right\}\psi_{k,\alpha}^{n} = \epsilon_\alpha\psi_{l,\alpha}^{m}, \tag{2.177}$$

where the definition

$$\hat{H}^{[m-n]} = \frac{1}{T}\int_0^T \mathrm{d}t\,\hat{H}(t)\exp(\mathrm{i}[m-n]\omega t) \tag{2.178}$$

has been introduced and we have used

$$\frac{1}{T}\int_0^T \mathrm{d}t\,\exp((\mathrm{i}[m-n]\omega t) = \delta_{mn}. \tag{2.179}$$

In the case of a monochromatic perturbation,

$$\hat{H}(t) \equiv \hat{H}_0 + \hat{H}_1\sin(\omega t), \tag{2.180}$$

time integration yields

$$\hat{H}^{[m-n]} = \hat{H}_0\delta_{mn} + \frac{\hat{H}_1}{2\mathrm{i}}\left\{\delta_{m,n-1} - \delta_{m,n+1}\right\}. \tag{2.181}$$

Equation (2.176) is the eigenvalue problem of the extended Hamiltonian $\hat{\mathcal{H}}$, whose matrix elements are given by

$$\langle l|\hat{H}^{[m-n]}|k\rangle - n\hbar\omega\delta_{mn}\delta_{lk}. \tag{2.182}$$

The Fourier expansion (2.173) has rendered the problem time-independent. One has to cope with an additional "dimension" ($n = 0, \pm 1, \pm 2\ldots$), however.

After choosing a basis (e.g., the harmonic oscillator basis) and calculating the matrix elements, the quasi-energies are the eigenvalues and the quasi-eigenfunctions are the eigenvectors of (2.177). The Floquet matrix to be diagonalized is

$$\begin{pmatrix} \ddots & & & & & & \\ & \mathbf{H}_0 - 2\hbar\omega\mathbf{1} & \frac{1}{2i}\mathbf{H}_1 & \mathbf{0} & \mathbf{0} & \mathbf{0} & \\ & -\frac{1}{2i}\mathbf{H}_1 & \mathbf{H}_0 - 1\hbar\omega\mathbf{1} & \frac{1}{2i}\mathbf{H}_1 & \mathbf{0} & & \\ & \mathbf{0} & -\frac{1}{2i}\mathbf{H}_1 & \mathbf{H}_0 & \frac{1}{2i}\mathbf{H}_1 & \mathbf{0} & \\ & \mathbf{0} & \mathbf{0} & -\frac{1}{2i}\mathbf{H}_1 & \mathbf{H}_0 + 1\hbar\omega\mathbf{1} & \frac{1}{2i}\mathbf{H}_1 & \\ & \mathbf{0} & \mathbf{0} & \mathbf{0} & -\frac{1}{2i}\mathbf{H}_1 & \mathbf{H}_0 + 2\hbar\omega\mathbf{1} & \\ & & & & & & \ddots \end{pmatrix}. \quad (2.183)$$

Here $\mathbf{1}, \mathbf{0}$ are unit and zero matrices, and in principle the block matrices have to be added ad infinitum, i.e., $n \to \infty$. In the numerics, however, one uses matrices $\mathbf{H}_0$ and $\mathbf{H}_1$ of finite size $L \times L$ as well as a finite number $2M + 1$ of Fourier terms. Convergence can be checked by increasing L as well as M.

In general the basis function expansion method is only a viable approach if the matrix elements of the Hamiltonian can be calculated easily. If the basis is the harmonic oscillator one, this is the case if the potential is given by a polynomial of low order. In other cases or if the potential is multidimensional, so-called "discrete variable representations" (DVR) [27] are frequently used. Finally, it should be noted that the diagonalization of the Floquet matrix becomes much more difficult, if the system under consideration contains a continuum of states. Then the method of complex rotation can be employed [28].

2.3.2 *Split-Operator Method*

The split-operator method for the solution of the time-dependent Schrödinger equation is based on the approximate representation of the time-evolution operator, i.e., the treatment of problem (b) by using the Zassenhaus formula [29][15]

$$e^{\hat{x}+\hat{y}} = e^{\hat{x}} e^{\hat{y}} e^{-1/2[\hat{x},\hat{y}]} e^{1/3[\hat{y},[\hat{x},\hat{y}]]+1/6[\hat{x},[\hat{x},\hat{y}]]} \cdots . \quad (2.184)$$

In the following, we restrict the discussion to a particle moving in one spatial dimension under a Hamiltonian of the usual form

$$\hat{H} = \hat{T}_{\mathrm{k}}(\hat{p}) + \hat{V}(\hat{x}). \quad (2.185)$$

For very short but finite time steps Δt, one then finds from the Zassenhaus formula that

$$e^{-i\hat{H}\Delta t/\hbar} \approx e^{-i\hat{T}_{\mathrm{k}}\Delta t/\hbar} e^{-i\hat{V}\Delta t/\hbar} \quad (2.186)$$

[15] The Baker-Campbell-Haussdorff (BCH) formula is the dual relation and reads $\exp\{\hat{x}\}\exp\{\hat{y}\} = \exp\{\hat{x} + \hat{y} + 1/2[\hat{x}, \hat{y}] + 1/12([\hat{x}, [\hat{x}, \hat{y}]] + [\hat{y}, [\hat{y}, \hat{x}]]) + \cdots\}$.

is accurate to first order in Δt. This approximation is also the basis for the (exact) Trotter product formula [14]

$$\mathrm{e}^{-\mathrm{i}\hat{H}t/\hbar} = \lim_{N\to\infty}\left[\mathrm{e}^{-\mathrm{i}\hat{T}_{\mathrm{k}}t/(N\hbar)}\mathrm{e}^{-\mathrm{i}\hat{V}t/(N\hbar)}\right]^N. \tag{2.187}$$

By working through Exercise 2.12 one can prove that a Strang splitting according to

$$\mathrm{e}^{-\mathrm{i}\hat{H}\Delta t/\hbar} = \mathrm{e}^{-\mathrm{i}\hat{V}\Delta t/(2\hbar)}\mathrm{e}^{-\mathrm{i}\hat{T}_{\mathrm{k}}\Delta t/\hbar}\mathrm{e}^{-\mathrm{i}\hat{V}\Delta t/(2\hbar)} + O(\Delta t^3) \tag{2.188}$$

leads to an approximation of higher accuracy.

2.12. *Show that the Strang splitting of the time-evolution operator leads to a second order method.*
Hint: use the Zassenhaus as well as the BCH formula

Problem (a) is now dealt with by representing the wavefunction at $t = 0$ on an equidistant position space grid $x_n \in [x_{\mathrm{min}}, x_{\mathrm{max}}]$, $n = 1, \ldots, N$. The wavefunction propagated for a time Δt at the grid point x_n is then given by

$$\begin{aligned}\Psi(x_n, \Delta t) &= \langle x_n|\mathrm{e}^{-\mathrm{i}\hat{H}\Delta t/\hbar}|\Psi(0)\rangle \\ &\approx \langle x_n|\mathrm{e}^{-\mathrm{i}\hat{V}\Delta t/(2\hbar)}\mathrm{e}^{-\mathrm{i}\hat{T}_{\mathrm{k}}\Delta t/\hbar}\mathrm{e}^{-\mathrm{i}\hat{V}\Delta t/(2\hbar)}|\Psi(0)\rangle.\end{aligned} \tag{2.189}$$

By inserting unity twice in terms of position states and once in terms of momentum states, the threefold integral (for the numerics, the integrations are discretized due to the grid based representation of the wavefunction)

$$\begin{aligned}\Psi(x_n, \Delta t) \approx \int \mathrm{d}x' \int \mathrm{d}p' \int \mathrm{d}x'' \langle x_n|\mathrm{e}^{-\mathrm{i}\hat{V}\Delta t/2\hbar}|x''\rangle \\ \langle x''|\mathrm{e}^{-\mathrm{i}\hat{T}_{\mathrm{k}}\Delta t/\hbar}|p'\rangle\langle p'|\mathrm{e}^{-\mathrm{i}\hat{V}\Delta t/2\hbar}|x'\rangle\langle x'|\Psi(0)\rangle\end{aligned} \tag{2.190}$$

emerges. The integral over x'' can be performed immediately due to the diagonal nature of the potential operator in position space and the δ-function appearing in

$$\langle x_n|\mathrm{e}^{-\mathrm{i}\hat{V}\Delta t/(2\hbar)}|x''\rangle = \mathrm{e}^{-\mathrm{i}V(x_n)\Delta t/(2\hbar)}\delta(x'' - x_n). \tag{2.191}$$

Also the second exponentiated potential term simplifies, yielding

$$\begin{aligned}\langle p'|\mathrm{e}^{-\mathrm{i}\hat{V}\Delta t/(2\hbar)}|x'\rangle &= \langle p'|x'\rangle\mathrm{e}^{-\mathrm{i}V(x')\Delta t/(2\hbar)} \\ &= \frac{1}{\sqrt{2\pi\hbar}}\mathrm{e}^{-\mathrm{i}p'x'/\hbar}\mathrm{e}^{-\mathrm{i}V(x')\Delta t/(2\hbar)}.\end{aligned} \tag{2.192}$$

The x' integration is a Fourier transformation of the intermediate wavefunction

$$\tilde{\Psi}(x', 0) = \mathrm{e}^{-\mathrm{i}V(x')\Delta t/(2\hbar)}\Psi(x', 0) \tag{2.193}$$

into momentum space. This leads to the fact that the exponentiated operator of the kinetic energy becomes a multiplicative factor and can be applied easily via

$$\begin{aligned}\langle x''|\mathrm{e}^{-\mathrm{i}\hat{T}_\mathrm{k}\Delta t/\hbar}|p'\rangle &= \langle x''|p'\rangle \mathrm{e}^{-\mathrm{i}T_\mathrm{k}(p')\Delta t/\hbar} \\ &= \frac{1}{\sqrt{2\pi\hbar}}\mathrm{e}^{\mathrm{i}p'x''/\hbar}\mathrm{e}^{-\mathrm{i}T_\mathrm{k}(p')\Delta t/\hbar}. \end{aligned} \tag{2.194}$$

The p' integration transforms the wavefunction back into position space.

The main numerical effort is the need to perform two Fourier transforms of the wavefunction during the propagation over one time step. These can be performed by using the fast Fourier transformation (FFT) algorithm [25], however. The implementation of the split-operator based FFT method[16] can therefore be summarized as follows:

1. Represent the initial wavefunction on a position space grid
2. Apply the operator $\mathrm{e}^{-\mathrm{i}\hat{V}\Delta t/(2\hbar)}$
3. Perform a FFT into momentum space
4. Apply the operator $\mathrm{e}^{-\mathrm{i}\hat{T}_\mathrm{k}\Delta t/\hbar}$
5. Perform an inverse FFT back into position space
6. Apply the operator $\mathrm{e}^{-\mathrm{i}\hat{V}\Delta t/(2\hbar)}$.

This procedure is applied for the propagation over a small time step. For the propagation over long times it will be repeated frequently and if the intermediate values of the wavefunction are not needed, the two half time steps of potential propagation can be combined (apart from the first and the last one). Furthermore, we stress that to propagate the wavefunction over the next time step, we will need its value not only at x_n but at all values of x. This is reflecting the *nonlocal* nature of quantum theory. For the calculation of the new wavefunction the old one is needed everywhere. This is in contrast to classical mechanics. A trajectory only depends on its own initial conditions; classical mechanics is a local theory.

A nice review of the details of FFT and a corresponding subroutine can be found in [25]. Some facts will be briefly repeated here. A function $\Phi(x_n)$ can be written as a discrete Fourier transform according to

$$\Phi(x_n) = \sum_{k=-N/2+1}^{N/2} a_k \mathrm{e}^{2\pi\mathrm{i}kx_n/X}, \tag{2.195}$$

with the Fourier coefficients

$$a_k = \frac{1}{N}\sum_{n=1}^{N} \Phi(x_n)\mathrm{e}^{-2\pi\mathrm{i}kx_n/X}. \tag{2.196}$$

[16]Originally this approach was proposed by Fleck, Morris and Feit for the solution of the Maxwell wave-equation [30].

For the implementation it is important to note that:

- $N = 2^j$ has to be an integer power of 2
- The grid length is $X = x_{max} - x_{min}$ and x_n are equidistant with $\Delta x = X/N$
- The numerical effort scales with $N \ln N$ [25]
- The maximal momentum that can be described is

$$p_{max} = h/(2\Delta x) = Nh/(2X),$$

  and $p_{min} = -p_{max}$
- The covered phase space volume is $V_P = 2Xp_{max} = Nh$
- The time step should fulfill $\Delta t < \hbar\pi/(3V_{\rm max})$, with $V_{\rm max}$ the maximal excursion of the potential [31]. For very long propagation times, see also [32].
- If calculated using (2.45), energy resolution is given by $\Delta E_{\rm min} = \hbar\pi/T_{\rm t}$, with $T_{\rm t}$ the total propagation time.

There are more recent implementations of FFT which do not have the restriction to integer powers of 2 and which, through adaption to the platform that is used for the calculations can have considerable advantages in speed (FFTW: fastest Fourier transformation in the West [33]).

The usage of the time-evolution operator for constant Hamiltonians at the beginning of our discussion is no restriction of the presented methodology to time-independent Hamiltonians. As in the case of the infinitesimal time-evolution operator (2.32), one can use a constant Hamiltonian for the propagation over a small time interval Δt. At the beginning of the next time step a slightly changed Hamiltonian is employed.

Finally, one drawback of the method that will not come into play in the present book, however, shall be mentioned. The split-operator idea only succeeds in producing simple multiplicative exponentials if there are no products of $\hat{p}$ and $\hat{x}$ in the Hamiltonian. These would appear in the treatment of dissipative quantum problems, which are outside the scope of this presentation.

2.3.2.1 Negative Imaginary Absorbing Potentials

Another possible drawback of a grid based method like the split-operator FFT method shall be dealt with in a bit more detail. It can be cast in the form of a question: What happens to a wavepacket, when it hits the grid boundaries? It would reenter on the other side of the grid, leading to nonphysical results! This can be avoided by adding a negative imaginary potential of the form

$$V(x) = -\mathrm{i}f(x)\Theta(x - x_a), \tag{2.197}$$

which is nonzero for values $x > x_a$, close to the right grid boundary $x_{\rm max}$ and a similar term at the left side of the grid. In Sect. 2.1 we have made use of the fact that the potential is real valued in order to show that the norm of any wavefunction is conserved. The fact that the total potential is now complex leads to a loss of norm. This may, however, be not as disturbing as the re-entrance phenomenon, especially in situations where the wavepacket would just move on in “free-space” like in a scattering situation after the scattering event is over.

The choice of the functional form of $f(x)$ in (2.197) is crucial. It turns out that the potential has to rise smoothly and rather slowly in order that there do not occur unphysical reflections of the wavefunction induced by the negative imaginary potential. A detailed study of several different functional forms of the imaginary potential can be found in [34].

2.3.3 *Alternative Methods of Time-Evolution*

In the material presented so far we have dealt both, with the solution of problem (a) as well as problem (b). In the following some alternative ways of treating the time-evolution, i.e., problem (b) shall be reviewed.

2.3.3.1 Method of Finite Differences

The discretization of the time-derivative in (2.22) with the help of the first order formula

$$|\dot{\Psi}(t)\rangle \approx \frac{|\Psi(t+\Delta t)\rangle - |\Psi(t)\rangle}{\Delta t} \tag{2.198}$$

leads to an *explicit* numerical method if the RHS is evaluated with the “old” wavefunction $|\Psi(t)\rangle$. This means that the wavefunction at a later time is explicitly given by the wavefunction at the earlier time. Unfortunately, however, this so-called Euler method is numerically instable.

An at least conditionally stable method can be constructed by application of the second-order formula

$$|\dot{\Psi}(t)\rangle \approx \frac{|\Psi(t+\Delta t)\rangle - |\Psi(t-\Delta t)\rangle}{2\Delta t} \tag{2.199}$$

for the time-derivative. The corresponding method is referred to as second order differencing (SOD) and has been advocated for the solution of the time-dependent Schrödinger equation by Askar and Cakmak [35]. The method can be shown to be energy and norm conserving. The condition under which it is stable can be derived by considering the eigenvalues of the propagation matrix that appears by using the discrete form of the time-derivative

$$\begin{pmatrix} |\Psi_{n+1}\rangle \\ |\Psi_n\rangle \end{pmatrix} = \begin{pmatrix} \hat{1} - 4\hat{H}^2\Delta t^2/\hbar^2 & -2\mathrm{i}\hat{H}\Delta t/\hbar \\ -2\mathrm{i}\hat{H}\Delta t/\hbar & \hat{1} \end{pmatrix} \begin{pmatrix} |\Psi_{n-1}\rangle \\ |\Psi_{n-2}\rangle \end{pmatrix}. \tag{2.200}$$

The eigenvalues of the matrix are (replacing $\hat{H}$ by E)

$$\lambda_{1,2} = 1 - 2E^2\Delta t^2/\hbar^2 \pm \frac{2E\Delta t}{\hbar}\sqrt{\frac{E^2\Delta t^2}{\hbar^2} - 1}. \tag{2.201}$$

The discrete mapping is norm conserving due to $\lambda_1\lambda_2 = 1$. For stability, the radicand in the expression above has to be negative, such that the eigenvalues become complex. Otherwise after the n-th iteration an exponential increase of numerical instabilities would occur. Thus for stability $\Delta t < \hbar/E_{\max}$ has to hold, where $E_{\max}$ is the largest eigenvalue of $\hat{H}$ taking part in the dynamics [36]. A slightly different look at the second order differencing method method is taken in Exercise 2.13.

2.13. *Show that in the second order differencing method the following holds if $\hat{H}$ is Hermitian (and time-independent)*

(a) $\mathrm{Re}\langle\Psi(t-\Delta t)|\Psi(t)\rangle = \mathrm{Re}\langle\Psi(t)|\Psi(t+\Delta t)\rangle = \mathrm{const}$
(b) $\mathrm{Re}\langle\Psi(t-\Delta t)|\hat{H}|\Psi(t)\rangle = \mathrm{Re}\langle\Psi(t)|\hat{H}|\Psi(t+\Delta t)\rangle = \mathrm{const}$
(c) *Interprete the results gained above.*
(d) *Consider the time-evolution of an eigenstate ψ of the Hamiltonian with eigenvalue E and derive a criterion for the maximally allowed time step Δt.*
Hint: Insert the exact time-evolution into the SOD scheme and distinguish the exact eigenvalue from the approximate E_{app} due to SOD time evolution.

2.3.3.2 Crank-Nicolson Method

An alternative possibility to circumvent the problem of instability of the Euler method is given by the Crank-Nicolson procedure. Here the first order formula

$$\hat{U}(\Delta t) \approx \hat{1} - \mathrm{i}\hat{H}\Delta t/\hbar, \tag{2.202}$$

representing the short-time evolution operator is used forward as well as backward in time

$$|\Psi_{n+1}\rangle = \hat{U}(\Delta t)|\Psi_n\rangle, \tag{2.203}$$
$$|\Psi_{n-1}\rangle = \hat{U}(-\Delta t)|\Psi_n\rangle. \tag{2.204}$$

In order to make progress one resolves both equations for $|\Psi_n\rangle$ by multiplying with the corresponding inverse operators. Equating the gained expressions yields

$$(\hat{1} + \mathrm{i}\hat{H}\Delta t/\hbar)|\Psi_{n+1}\rangle = (\hat{1} - \mathrm{i}\hat{H}\Delta t/\hbar)|\Psi_{n-1}\rangle. \tag{2.205}$$

The procedure now is an implicit one, that is stable as well as norm conserving.

Due to its implicit nature, the method requires a matrix inversion and formally leads to the Cayley approximation (see also Chap. 19.2 of [25])

$$|\Psi_{n+1}\rangle = \frac{\hat{1} - \mathrm{i}\hat{H}\Delta t/(2\hbar)}{\hat{1} + \mathrm{i}\hat{H}\Delta t/(2\hbar)}|\Psi_n\rangle \tag{2.206}$$

for the propagated wavefunction. The CN scheme is second order accurate in Δt, as can be seen by a Taylor expansion of the final expression and comparing to the Taylor expansion of the exact time-evolution operator.

2.3.3.3 Polynomial Methods

The idea behind polynomial methods is the expansion of the time-evolution operator in terms of polynomials, according to

$$\mathrm{e}^{-\mathrm{i}\hat{H}t/\hbar} = \sum_n a_n P_n(\hat{H}) \, . \tag{2.207}$$

Two different approaches are commonly used:

- In the Chebyshev method, the polynomials are fixed to be the complex valued Chebyshev ones. A first application to the problem of wavefunction propagation has been presented by Tal-Ezer and Kosloff [37]. These authors have shown that the approach is up to six times more efficient than the SOD method, presented above. It allows for evolution over relatively long time steps. Drawbacks are that intermediate time information is not readily available and, even worse in the present context, that time-dependent Hamiltonians cannot be treated.
- In contrast to the first approach, in the Lanczos method, the polynomials are not fixed but are generated in the course of the propagation. A very profound introduction to the commonly applied short iterative Lanczos method can be found in [38].

2.3.4 *Semiclassical Initial Value Representations*

As the final prerequisite before we deal with the physics of laser-matter interaction, a reformulation of the semiclassical van Vleck-Gutzwiller propagator presented in Sect. 2.2.3 shall be discussed. We have already mentioned that the VVG method is based on the solution of classical *boundary value* (or root search) problems, which makes it hard to implement. A much more user friendly approach would be based on classical *initial value* solutions and is therefore termed initial value representation.

We start the discussion of a specific initial value representation of the semiclassical propagator with a short introduction to commonly used symplectic integration procedures for the solution of the underlying classical dynamics.

2.3.4.1 Symplectic Integration

Positions and momenta of a classical Hamiltonian system with N degrees of freedom obey the equations of motion

$$\dot{q}_n = \frac{\partial H(\boldsymbol{p}, \boldsymbol{q}, t)}{\partial p_n}, \tag{2.208}$$

$$\dot{p}_n = -\frac{\partial H(\boldsymbol{p}, \boldsymbol{q}, t)}{\partial q_n}. \tag{2.209}$$

Using the Poisson bracket

$$\{a, b\} = \sum_{n=1}^{N} \left(\frac{\partial a}{\partial q_n} \frac{\partial b}{\partial p_n} - \frac{\partial a}{\partial p_n} \frac{\partial b}{\partial q_n} \right), \tag{2.210}$$

the equations above can also be cast into an equation for the $2N$- dimensional phase-space vector $\boldsymbol{\eta}^{\mathrm{T}} = (\boldsymbol{q}^{\mathrm{T}}, \boldsymbol{p}^{\mathrm{T}})$, reading

$$\dot{\boldsymbol{\eta}} = -\{H, \boldsymbol{\eta}\} =: -\hat{H}\boldsymbol{\eta}\,. \tag{2.211}$$

Although we are dealing with classical mechanics an operator, $\hat{H}$, appears here. In the present subsection this operator stands for the application of the Poisson bracket with the Hamilton function.

Formally, the equation above can be integrated over a small time step, yielding

$$\boldsymbol{\eta}(t + \Delta t) = \exp\{-\Delta t \hat{H}\}\boldsymbol{\eta}(t). \tag{2.212}$$

Now we can again use the split-operator method, i.e. an "effective Hamiltonian" can be introduced according to

$$\exp\{-\Delta t \hat{H}_{\mathrm{eff}}\} := \exp\{-\Delta t \hat{T}_{\mathrm{k}}\}\exp\{-\Delta t \hat{V}\}. \tag{2.213}$$

This effective operator is only an approximation to the true one. The total phase space volume is conserved, however, i.e., Liouville's theorem holds also for the approximate dynamics [39]. More generally, it can be shown that symplectic integration methods preserve the N Poincaré invariants of the Hamiltonian system [40].

2.14. *By expanding up to second order in* Δt *show that there is a difference between* $\exp\{-\Delta t \hat{H}\}$ *and* $\exp\{-\Delta t \hat{T}_{\mathrm{k}}\}\exp\{-\Delta t \hat{V}\}$.

Hint: The Jacobi identity $\{A, \{B, C\}\} + \{B, \{C, A\}\} + \{C, \{A, B\}\} = 0$ *might be helpful.*

In the numerics, the splitting of the exponentiated Hamiltonian into the kinetic and the potential part means that first one solves a problem in which only $\hat{V}$ operates, i.e., the momentum is altered due to the effect of the potential. This is the so-called *kick* step. Then the position is shifted using constant updated momentum. This is the so-called *drift* step. For very short times (expand the exponential to first order) one gets

$$\boldsymbol{p}^1 = \boldsymbol{p}^0 + \Delta t \boldsymbol{F}_{q=q^0}, \tag{2.214}$$

$$\boldsymbol{q}^1 = \boldsymbol{q}^0 + \Delta t \boldsymbol{G}_{p=p^1}, \tag{2.215}$$

where the superscript denotes the iteration step and the abbreviations

$$\boldsymbol{G} = \frac{\partial T_{\mathrm{k}}}{\partial \boldsymbol{p}} \tag{2.216}$$

$$\boldsymbol{F} = -\frac{\partial V}{\partial \boldsymbol{q}} \tag{2.217}$$

have been used. This procedure is a variant of the symplectic Euler method. It performs much better than the highly unstable "standard" Euler method, for which in the second line the old momentum $\boldsymbol{p}^0$ is used.

Analogously to the discussion of the split-operator procedure in quantum mechanics in Sect. 2.3.2, a split-operator procedure of higher order can be used. By employing the Strang splitting

$$\exp\{-\Delta t \hat{H}_{\mathrm{eff}}\} := \exp\{-\Delta t \hat{T}_{\mathrm{k}}/2\} \exp\{-\Delta t \hat{V}\} \exp\{-\Delta t \hat{T}_{\mathrm{k}}/2\}, \tag{2.218}$$

the so-called leap frog method

$$\boldsymbol{q}^1 = \boldsymbol{q}^0 + \frac{\Delta t}{2} \boldsymbol{G}(\boldsymbol{p} = \boldsymbol{p}^0), \tag{2.219}$$

$$\boldsymbol{p}^2 = \boldsymbol{p}^0 + \Delta t \boldsymbol{F}(\boldsymbol{q} = \boldsymbol{q}^1), \tag{2.220}$$

$$\boldsymbol{q}^2 = \boldsymbol{q}^1 + \frac{\Delta t}{2} \boldsymbol{G}(\boldsymbol{p} = \boldsymbol{p}^2) \tag{2.221}$$

arises. In general, any symplectic integration scheme (where the kick step comes first) can be cast into the following form:

$$
\begin{aligned}
&\text{do} \quad k = 1, M \\
&\qquad \boldsymbol{p}^k = \boldsymbol{p}^{k-1} + b_k \Delta t \boldsymbol{F}(\boldsymbol{q}^{k-1}) \\
&\qquad \boldsymbol{q}^k = \boldsymbol{q}^{k-1} + a_k \Delta t \boldsymbol{G}(\boldsymbol{p}^k) \\
&\text{enddo}
\end{aligned}
$$

Coefficients a_k, b_k of different symplectic methods are gathered in Table 2.3. Additional coefficients can be found in [39]. In order to keep the numerical effort of force calculation rather low, it is desirable to have as many b-coefficients equal to zero as possible.

In a thorough study of the numerical accuracy of symplectic integrators it has been found that they yield very stable trajectories and that, for autonomous Hamiltonians, the standard deviation of the energy and the energy drift are comparatively small [39].

2.3.4.2 Coherent States

Having set the stage with the discussion of the solution of the classical equations of motion, we now come to the central ingredients for the reformulation of the semiclassical propagator expression. These are the so-called coherent states, which are discussed in detail in the textbook of Louisell [43] and in Heller's Les Houches lecture notes [44]. In Dirac notation they are given by

$$|z\rangle = \mathrm{e}^{-1/2|z|^2} \mathrm{e}^{z\cdot\hat{\boldsymbol{a}}^\dagger} |\boldsymbol{0}\rangle, \tag{2.222}$$

where $|\boldsymbol{0}\rangle$ is the ground state of N uncoupled 1D harmonic oscillators of mass m and frequency ω_e. Furthermore, we have used the multi dimensional analog of (2.162)

$$\hat{\boldsymbol{a}}^\dagger = \frac{1}{\sqrt{2}}\left(\frac{\hat{\boldsymbol{q}}}{b} - \mathrm{i}\frac{\hat{\boldsymbol{p}}}{c}\right), \tag{2.223}$$

for the vector of creation operators with $b = \sqrt{\hbar/m\omega_\mathrm{e}}$, $c = \sqrt{\hbar m \omega_\mathrm{e}}$ and

Table 2.3 Coefficients for some symplectic integration methods of increasing order

Ruth's leap frog (position Verlet)	$a_1 = 1/2$ $a_2 = 1/2$	$b_1 = 0$ $b_2 = 1$
Fourth-order Gray [41]	$a_1 = (1-\sqrt{1/3})/2$	$b_1 = 0$
	$a_2 = \sqrt{1/3}$	$b_2 = (1/2+\sqrt{1/3})/2$
	$a_3 = -a_2$	$b_3 = 1/2$
	$a_4 = (1+\sqrt{1/3})/2$	$b_4 = (1/2-\sqrt{1/3})/2$
Sixth-order Yoshida [42]	$a_1 = 0.78451361047756$	$b_1 = 0.39225680523878$
	$a_2 = 0.23557321335936$	$b_2 = 0.51004341191846$
	$a_3 = -1.1776799841789$	$b_3 = -0.47105338540976$
	$a_4 = 1.3151863206839$	$b_4 = 0.068753168252520$
	$a_5 = a_3$	$b_5 = b_4$
	$a_6 = a_2$	$b_6 = b_3$
	$a_7 = a_1$	$b_7 = b_2$
	$a_8 = 0$	$b_8 = b_1$

$$z = \frac{1}{\sqrt{2}}\left(\frac{\boldsymbol{q}}{b} + \mathrm{i}\frac{\boldsymbol{p}}{c}\right), \tag{2.224}$$

with the expectation values $\boldsymbol{q}$ and $\boldsymbol{p}$ of the operators $\hat{\boldsymbol{q}}$ and $\hat{\boldsymbol{p}}$.

In position representation, the coherent states are N-dimensional Gaussian wavepackets of the form

$$\langle \boldsymbol{x}|z\rangle = \left(\frac{1}{\pi b^2}\right)^{N/4} \exp\left\{-\frac{1}{2b^2}(\boldsymbol{x}-\boldsymbol{q})^2 + \frac{\mathrm{i}}{bc}\boldsymbol{p}\cdot\left(\boldsymbol{x}-\frac{\boldsymbol{q}}{2}\right)\right\}. \tag{2.225}$$

Coherent states form an (over-)complete set of basis states [43] and can be used to represent unity according to

$$\hat{1} = \int \frac{\mathrm{d}^{2N}z}{\pi^N}|z\rangle\langle z| = \int \frac{\mathrm{d}^N p\, \mathrm{d}^N q}{(2\pi\hbar)^N}|z\rangle\langle z|. \tag{2.226}$$

2.15. *Restricting the discussion to $N = 1$, prove that the coherent states form a complete set by expressing them as a sum over harmonic oscillator eigenstates.*

The basis for the reformulation of the semiclassical propagator is the matrix element of the time-evolution operator[17] between coherent states

$$K(z_f, t; z_i, 0) \equiv \langle z_f|\mathrm{e}^{-\mathrm{i}\hat{H}t/\hbar}|z_i\rangle. \tag{2.227}$$

The semiclassical approximation for this object can be performed quite analogously to the derivation of the van Vleck-Gutzwiller propagator by starting from the appropriate path integral [45]. However, it turns out that the final expression contains a classical over-determination problem due to the fact that not only the position is fixed at the initial and the final time but also the momentum! This problem is solved by the complexification of phase space. We will not dwell on that rather involved topic any longer. Fortunately the over-determination problem will be resolved rather elegantly in the following.

2.3.4.3 Herman-Kluk Propagator

The next step in order to make progress is to consider the time-evolution operator in position representation. It can be expressed via the coherent state propagator, by inserting unity in terms of coherent states twice, according to

[17] For notational convenience, we assume the Hamiltonian to be time-independent; the following results are also valid in the general case of a time-dependent Hamiltonian, however.

$$K(\boldsymbol{x}_f, t; \boldsymbol{x}_i, 0) = \langle \boldsymbol{x}_f | \mathrm{e}^{-\mathrm{i}\hat{H}t/\hbar} | \boldsymbol{x}_i \rangle$$
$$= \int \frac{\mathrm{d}^{2N} z_f}{\pi^N} \int \frac{\mathrm{d}^{2N} z_i}{\pi^N} \langle \boldsymbol{x}_f | z_f \rangle \langle z_f | \mathrm{e}^{-\mathrm{i}\hat{H}t/\hbar} | z_i \rangle \langle z_i | \boldsymbol{x}_i \rangle. \quad (2.228)$$

If we now replace the coherent state matrix element of the propagator by its semiclassical approximation and perform the final phase space integration in the stationary phase approximation, the over-determination problem is resolved and the semiclassical propagator is reformulated in terms of real classical initial value solutions [46]. This procedure yields

$$K^{\mathrm{HK}}(\boldsymbol{x}_f, t; \boldsymbol{x}_i, 0) \equiv \int \frac{\mathrm{d}^N p_i \mathrm{d}^N q_i}{(2\pi\hbar)^N} \langle \boldsymbol{x}_f | \tilde{z}_t \rangle R(\boldsymbol{p}_i, \boldsymbol{q}_i, t)$$
$$\exp\left\{ \frac{\mathrm{i}}{\hbar} S(\boldsymbol{p}_i, \boldsymbol{q}_i, t) \right\} \langle \tilde{z}_i | \boldsymbol{x}_i \rangle, \quad (2.229)$$

which is the so-called Herman-Kluk propagator. By a different reasoning it has first been derived by Herman and Kluk [47], based on previous work by Heller [48]. Definitions that are used in the expression above are the classical action *functional*, that depends on the initial phase space variables, and time and for this reason is written (and denoted) as a *function* here, according to

$$S(\boldsymbol{p}_i, \boldsymbol{q}_i, t) \equiv \int_0^t \mathrm{d}t' \left[\boldsymbol{p}_{t'} \cdot \dot{\boldsymbol{q}}_{t'} - H \right]. \quad (2.230)$$

Furthermore,

$$R(\boldsymbol{p}_i, \boldsymbol{q}_i, t) \equiv \left| \frac{1}{2} \left(\mathbf{m}_{11} + \mathbf{m}_{22} - \mathrm{i}\hbar\gamma \mathbf{m}_{21} - \frac{1}{\mathrm{i}\hbar\gamma} \mathbf{m}_{12} \right) \right|^{1/2}, \quad (2.231)$$

with $\gamma = m\omega_{\mathrm{e}}/\hbar$, denotes the Herman-Kluk determinantal prefactor, which contains classical stability (monodromy) block-matrices $\mathbf{m}_{ij}$. They are solutions to the linearized Hamilton equations and for reference they are defined in Appendix 2.C.

The Gaussian wavepackets in (2.229) have a slightly different phase convention than the ones of (2.225). For this reason a new symbol with a tilde,

$$\langle \boldsymbol{x} | \tilde{z} \rangle = \left(\frac{\gamma}{\pi} \right)^{N/4} \exp\left\{ -\frac{\gamma}{2} (\boldsymbol{x} - \boldsymbol{q})^2 + \frac{\mathrm{i}}{\hbar} \boldsymbol{p} \cdot (\boldsymbol{x} - \boldsymbol{q}) \right\}, \quad (2.232)$$

has been introduced. To complete the explanation of all abbreviations, the centers of the final Gaussians in phase space are $\{\boldsymbol{p}_t(\boldsymbol{p}_i, \boldsymbol{q}_i), \boldsymbol{q}_t(\boldsymbol{p}_i, \boldsymbol{q}_i)\}$, which are initial value solutions of the classical Hamilton equations.

In contrast to the VVG prefactor, the expression (2.231) does not exhibit singularities at caustics. Recently, it has been proven that the Herman-Kluk method is a uniform semiclassical method [49]. Furthermore, for the numerics it is important that the square root in the prefactor has to be taken in such a fashion that the result is continuous as a function of time [50]. This is reminiscent of the Maslov phase in the van Vleck-Gutzwiller expression (2.91), which does not have to be calculated explicitly, however.

A final remark on the connection between the two semiclassical expressions for the propagator, which we have discussed so far, shall be made. After performing the integration over the initial phase space variables in (2.229) in the stationary phase approximation, the van Vleck-Gutzwiller expression will emerge. One can also turn around that reasoning and derive the Herman-Kluk prefactor, by demanding that the SPA applied to the phase space integral yields the van Vleck-Gutzwiller expression [50]. For the derivation of a more general form of the prefactor in this way, see [51]. An even simpler way to derive the VVG propagator from the Herman-Kluk expression by taking the limit $\gamma \to \infty$ is explicitly given in Appendix 2.D.

2.3.4.4 Semiclassical Propagation of Gaussian Wavepackets

The pure HK propagator is a clumsy object, due to the need to integrate over all of phase space. Fortunately, however, in the focus of our interest will not be the bare propagator but its application to an initial Gaussian wavepacket. Let us therefore consider the mixed matrix element

$$\begin{aligned} K(\boldsymbol{x}_f, t; \tilde{z}_\alpha, 0) &\equiv \langle \boldsymbol{x}_f | \mathrm{e}^{-\mathrm{i}\hat{H}t/\hbar} | \tilde{z}_\alpha \rangle \\ &= \int \mathrm{d}^N x_i \, K(\boldsymbol{x}_f, t; \boldsymbol{x}_i, 0) \langle \boldsymbol{x}_i | \tilde{z}_\alpha \rangle \end{aligned} \tag{2.233}$$

of the time-evolution operator, where $K(\boldsymbol{x}_f, t; \boldsymbol{x}_i, 0)$ shall be replaced by the HK approximation of (2.229).

The Gaussian to be propagated $\langle \boldsymbol{x}_i | \tilde{z}_\alpha \rangle$ is determined by its phase space center $(\boldsymbol{q}_\alpha, \boldsymbol{p}_\alpha)$ and shall have the same inverse width parameter γ as the coherent state basis functions. The calculation of the overlap in (2.233) can be done analytically and yields the simple result

$$\begin{aligned} \langle \tilde{z}_i | \tilde{z}_\alpha \rangle = \exp\Bigg\{ &-\frac{\gamma}{4}(\boldsymbol{q}_i - \boldsymbol{q}_\alpha)^2 + \frac{\mathrm{i}}{2\hbar}(\boldsymbol{q}_i - \boldsymbol{q}_\alpha)\cdot(\boldsymbol{p}_i + \boldsymbol{p}_\alpha) \\ &- \frac{1}{4\gamma\hbar^2}(\boldsymbol{p}_i - \boldsymbol{p}_\alpha)^2 \Bigg\}. \end{aligned} \tag{2.234}$$

In (2.233), the integration over initial phase space still has to be done. It is, however, much more user friendly than in the case of the bare propagator, due to the fact that the overlap just calculated is effectively cutting off the integrand too far away from the initial center in phase space. In numerical applications the phase space integration is often performed by using Monte Carlo methods [52]. Pictorially, the application of the Herman-Kluk propagator to a Gaussian can be represented as shown in Fig. 2.6.

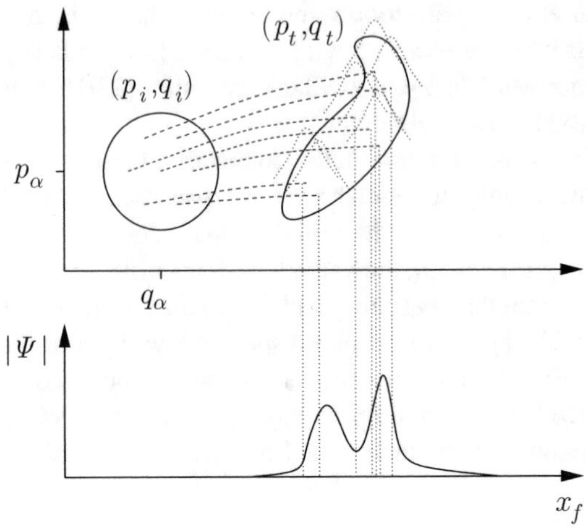

Fig. 2.6 Pictorial representation of the semiclassical initial value procedure to propagate a Gaussian wavepacket à la Herman and Kluk in one dimension: The propagated wavepacket $\langle x_f|\Psi\rangle$ is a weighted sum over many Gaussians. The weights are the product of a prefactor times a complex exponential function $R\exp\{iS/\hbar\}\langle\tilde{z}_i|\tilde{z}_\alpha\rangle$; adapted from [53]

How is all that related to the thawed GWD of Heller that we have used in Sect. 2.1.4? There the Gaussian wavepacket has been propagated using its center trajectory alone. The GWD therefore is much more crude than the HK method! There should be a way to derive GWD from the more general expression, however. This is indeed the case. To this end one has to expand the exponent in the integral over initial phase space around the center $(\boldsymbol{p}_\alpha, \boldsymbol{q}_\alpha)$ of the initial Gaussian up to second order. The integration is then a Gaussian integration and can be performed analytically.[18] One finally gets [51]

$$K^{\text{GWD}}(\boldsymbol{x}_f, t; \tilde{\boldsymbol{z}}_\alpha, 0) \equiv \left(\frac{\gamma}{\pi}\right)^{N/4} |(\mathbf{m}_{22} + \mathrm{i}\hbar\gamma\mathbf{m}_{21})|^{-1/2}$$
$$\exp\left\{-\frac{1}{2}(\boldsymbol{x}_f - \boldsymbol{q}_{\alpha t})\cdot\gamma_t(\boldsymbol{x}_f - \boldsymbol{q}_{\alpha t}) + \frac{\mathrm{i}}{\hbar}\boldsymbol{p}_{\alpha t}\cdot(\boldsymbol{x}_f - \boldsymbol{q}_{\alpha t}) + \frac{\mathrm{i}}{\hbar}S\right\}, \tag{2.235}$$

with the time-dependent $N \times N$ inverse width parameter matrix

$$\gamma_t = \gamma\big(\mathbf{m}_{11} + \frac{1}{\mathrm{i}\gamma\hbar}\mathbf{m}_{12}\big)\big(\mathbf{m}_{22} + \mathrm{i}\gamma\hbar\mathbf{m}_{21}\big)^{-1}. \tag{2.236}$$

The width of the single Gaussian can thus change in the course of time, in contrast to the widths of the many Gaussians in the case of the HK propagator. This is the reason, why the more simple single Gaussian method is called "thawed" Gaussian wavepacket dynamics, whereas the more complex, multiple Gaussian HK method applied to a Gaussian initial state is closely related to the "frozen" Gaussian

[18]Note that this procedure is more approximative than a stationary phase approximation.

wavepacket dynamics of Heller [48]. When speaking of frozen Gaussians propagators, strictly, this refers to a HK-like approach but with a unit prefactor [54].

Finally, it is worthwhile to check that the time-dependent parameter γ_t fulfills a nonlinear Riccati differential equation similar to (2.61). To this end, in Exercise 2.16, the equations of motion of the stability matrix elements given in Appendix 2.C should be used.

2.16. *The TGWD inverse width parameter allows for a reformulation of the HK prefactor. For reasons of simplicity, consider the* 1D *case.*

(a) *Show that the nonlinear Riccati differential equation*

$$\dot{\gamma}_t = -\frac{\mathrm{i}\hbar}{m}\gamma_t^2 - \frac{1}{\mathrm{i}\hbar}V''$$

is fulfilled by the inverse width parameter γ_t.

(b) *Writing the inverse width parameter in the log-derivative form*

$$\gamma_t = \frac{m}{\mathrm{i}\hbar}\frac{\dot{Q}}{Q},$$

with $Q = m_{22} + \mathrm{i}\hbar\gamma m_{21}$, *show that the complex conjugate of the HK prefactor can be entirely formulated in terms of* γ_t *via*

$$R^* = \sqrt{\frac{1}{2}(1+\gamma_t/\gamma)\exp\left\{\frac{1}{2}\int_0^t \mathrm{d}t'\frac{\mathrm{i}\hbar}{m}\gamma_{t'}\right\}}.$$

The different level of accuracy of the two approximations is illustrated in Fig. 2.7, where a comparison of the multi-trajectory HK method and the single trajectory GWD are contrasted with exact numerical results, gained by using the split-operator FFT method of Sect. 2.3.2. The displayed quantity is the real part of the auto-correlation function

$$c_{\alpha\alpha}(t) \equiv \langle \Psi_\alpha(0)|\Psi_\alpha(t)\rangle \tag{2.237}$$

of an initial Gaussian wavepacket in a Morse potential with dimensionless Hamiltonian

$$\hat{H} = \frac{\hat{p}^2}{2} + D(1-\exp\{-\lambda x\})^2, \tag{2.238}$$

with parameters $D = 30$, $\lambda = 0.08$.[19] GWD describes the envelope of the quantum curve rather well. The fine oscillations that are captured almost perfectly by the multiple trajectory method are not described at all, however.

---

[19] In Sect. 5.1.2 the physical background of the Morse oscillator will be elucidated.

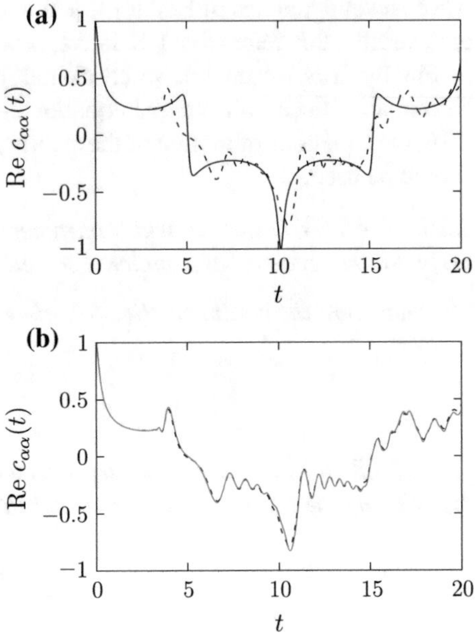

Fig. 2.7 Different trajectory-based and fully quantum mechanical auto-correlation functions of a Gaussian wavepacket with dimensionless parameters $q_\alpha, p_\alpha = 0$ und $\gamma = 12$ in a Morse potential. **a** thawed GWD (*solid line*) versus exact quantum mechanics (*dashed line*) **b** HK (*solid red line*) versus exact quantum mechanics (*dashed line*); adapted from [55]

2.4 Notes and Further Reading

TDSE and Time-Evolution Operator

The restriction to the use of the nonrelativistic TDSE for the description of laser-matter interaction is due to the focus of this book on intermediate field strengths, maximally of the order of an atomic unit (see also Appendix 4.A), and wavelengths around the visible range. If stronger fields are considered (and for longer wavelengths), in the case of electrons, the Dirac equation has to be solved. A concise discussion of the Dirac equation and relativistic corrections to the Schrödinger equation can be found in Appendix 7 of [56]. The spinor character of the wavefunction leads to a considerable complexification of the problem, which may not be necessary if the dipole approximation (see Chap. 3) is still applicable [57].

The very detailed book by Schleich [58] contains a lot of additional information on Schrödinger type time-evolution operators and also the time-evolution of the density operator is discussed therein. More on time-dependent and energy-dependent Green's function can be found in Economou's book [59]. The extraction of spectral information from time-dependent quantum information goes back to work of Heller, which is reviewed, e.g., in [44] and is covered also in the book by Tannor [38]. In both previous citations many additional references concerning Gaussian wavepacket dynamics can be found. A recent book by Schuch is focusing on the nonlinear Riccati equation, appearing in the GWD, with applications to quantum theory and irreversible processes [60].

Analytical Methods

A lot of additional material on the path integral formulation of quantum mechanics is contained in the tutorial article by Ingold [61] and the books by Feynman and Hibbs [6] and by Schulman [14]. The second book has more details on variational calculus and on exactly solvable path integrals (especially in the new Dover edition). It has been pointed out by Makri and Miller that a simple short-time propagator, based on the trapezoidal rule for the discretization of the potential part of the Lagrangian (or taking a midpoint rule), is not correct through first order in Δt, even in the case of the harmonic oscillator [62], although it leads to the correct propagator (2.75). Improved short-time propagators have been proposed there.

The significance of generating functions of canonical transformations in the (semi-)classical limit of quantum mechanics is discussed in depth by Miller [63]. The book by Reichl [64] contains chapters on semiclassical methods and on time-periodic systems, dealing with Floquet theory. These methods are then used in the context of quantum chaology. The book by Billing contains more material on semiclassics and on mixed quantum classical methods [65]. A book devoted to the semiclassical approach to the solution of the TDSE, as well as to the understanding of quantum mechanics, using this approach, is the one by Heller [66].

The Magnus expansion of the time-evolution operator is formally closely analogous to the so-called cumulant expansion, known from statistical physics [67].

Numerical Methods

The book by Tannor [38] contains more information on methodological and numerical approaches to solve the time-dependent Schrödinger equation. Polynomial and DVR methods are dealt with in detail there. A book that is fun to read, although it covers a seemingly dry topic is the classic "Numerical Recipes" [25]. Among many other things, more details on FFT and on finite difference methods to solve the TDSE can be found therein. Methods for the solution of the TDSE in the case of strong field driving are discussed in [68]. More recently, a phase space approach has been devised that is taylored for the solution of the TDSE for laser-driven electronic wavepacket propagation [69]. In [70], molecular quantum dynamics is discussed from the viewpoint of the MCTDH method.

A review of different semiclassical approximations based on Gaussian wavepackets is given in [55], whereas a combination of the Herman-Kluk method with thawed GWD for correlated many-particle systems, termed semiclassical hybrid dynamcis (SCHD), is laid out in [51]. The coupled coherent states (CCS) method of Shalashilin and Child [71] allows in principle for an exact numerical solution of the TDSE and has the Herman-Kluk method as a limiting case, as shown very elegantly in [72]. Finally, we have not discussed Bohmian mechanics that is recently being used not only as an interpretational tool, but in a synthetic way, in order to generate solutions of the TDSE using (nonclassical) trajectories [73]. For another very good discussion of this topic, see Chap. 4 of [38]. The nonlocality of quantum mechanics is especially apparent in Bohmian mechanics, due to the presence of the so-called quantum potential that couples the motion of individual trajectories.

2.A The Royal Road to the Path Integral

An elegant proof of the representation of the path integral as an infinite dimensional Riemann integral can be performed by using the Weyl transformation [74]

$$A(p,q) = \int \mathrm{d}u\, \mathrm{e}^{\mathrm{i}qu/\hbar} \langle p + u/2|\hat{A}|p - u/2\rangle\,, \tag{2.239}$$

that transforms an operator into a phase space function. The inverse transformation is

$$\hat{A} = \frac{1}{h}\int \mathrm{d}p\,\mathrm{d}q\, A(p,q)\hat{\Delta}(p,q), \tag{2.240}$$

$$\hat{\Delta} = \int \mathrm{d}v\, \mathrm{e}^{\mathrm{i}pv/\hbar}|q + v/2\rangle\langle q - v/2|\,. \tag{2.241}$$

Using the time-slicing procedure introduced in Sect. 2.2.1, the propagator can be represented by a product of short-time propagators according to

$$K(x_f, t; x_i, 0) = \int \mathrm{d}q_1 \cdots \int \mathrm{d}q_{N-1} \prod_{k=0}^{N-1} \langle q_{k+1}|\mathrm{e}^{-\frac{\mathrm{i}}{\hbar}\hat{H}(\hat{p},\hat{q})\Delta t}|q_k\rangle\,, \tag{2.242}$$

where $q_N = x_f$ and $q_0 = x_i$ and where we have switched from x to q notation for reasons of convenience. Using the Weyl transform of the Hamiltonian

$$H(p,q) = \frac{p^2}{2m} + V(q) \tag{2.243}$$

and (2.240), (2.241), the short-time propagators can be written as an integral

$$\begin{aligned}
&\langle q_{k+1}| \exp\left\{-\frac{\mathrm{i}}{\hbar}\hat{H}(\hat{p},\hat{q})\Delta t\right\}|q_k\rangle \\
&= \frac{1}{h}\langle q_{k+1}| \int \mathrm{d}p\,\mathrm{d}q\, \exp\left\{-\frac{\mathrm{i}}{\hbar}H(p,q)\Delta t\right\} \int \mathrm{d}v\, \mathrm{e}^{\mathrm{i}pv/\hbar}|q + v/2\rangle\langle q - v/2|q_k\rangle \\
&= \frac{1}{h}\int \mathrm{d}p\,\mathrm{d}q\, \exp\left\{-\frac{\mathrm{i}}{\hbar}H(p,q)\Delta t\right\} \mathrm{e}^{\mathrm{i}p(q_{k+1}-q_k)/\hbar}\delta\left\{\frac{q_{k+1}+q_k}{2} - q\right\} \\
&= \frac{1}{h}\int \mathrm{d}p\, \exp\left\{\frac{\mathrm{i}}{\hbar}\left[p\frac{q_{k+1}-q_k}{\Delta t} - H\left(p, \frac{q_{k+1}+q_k}{2}\right)\right]\Delta t\right\}\,.
\end{aligned} \tag{2.244}$$

Inserting this expression into (2.242), the phase-space integral form of the propagator [74]

$$K(x_f, t; x_i, 0) = \int \frac{\mathrm{d}p_0 \cdots \mathrm{d}p_{N-1}\mathrm{d}q_1 \cdots \mathrm{d}q_{N-1}}{(2\pi\hbar)^N}$$
$$\exp\left\{\frac{\mathrm{i}}{\hbar}\sum_{k=0}^{N-1}\left[p_k\left(\frac{q_{k+1}-q_k}{\Delta t}\right) - H\left(p_k, \frac{q_{k+1}+q_k}{2}\right)\right]\Delta t\right\} \quad (2.245)$$

can be derived. Please note that the number of p-integrations is higher by one than the number of q-integrations. Furthermore, the expression above does not contain operators any more!

Due to the quadratic form of the Weyl transform (2.243) in p, the p integrals are Gaussian, however, and can be done exactly analytically. For $N \to \infty$ and $N\Delta t = t$, this leads to the expression in the main text,

$$K(x_f, t; x_i, 0) = \lim_{N\to\infty}\left[\frac{m}{2\pi\mathrm{i}\hbar\Delta t}\right]^{N/2}\int\prod_{k=1}^{N-1}\mathrm{d}q_k$$
$$\exp\left\{\frac{\mathrm{i}}{\hbar}\sum_{k=0}^{N-1}\left[\frac{m(q_{k+1}-q_k)^2}{2\Delta t} - V\left(\frac{q_{k+1}+q_k}{2}\right)\Delta t\right]\right\}, \quad (2.246)$$

where the classical action appears in the exponent. Furthermore, the normalization constant (or measure factor) that was still undetermined in (2.75) is given by

$$B_N = \left[\frac{m}{2\pi\mathrm{i}\hbar\Delta t}\right]^{N/2}. \quad (2.247)$$

2.B Variational Calculus

In general the variation of a *functional* Φ depending on a *function* $h(x)$ is defined via

$$\delta\Phi \equiv \Phi[h+\delta h] - \Phi[h] = \int \mathrm{d}x \frac{\delta\Phi}{\delta h(x)}\delta h(x)\,. \quad (2.248)$$

For the specific cases

$$\Phi_1[h] = \int_a^b \mathrm{d}x\, h(x) f(x), \quad (2.249)$$

$$\Phi_2[h] = \int_a^b \mathrm{d}x\, F(x, h(x)), \quad (2.250)$$

$$\Phi_3[h] = \int_a^b \mathrm{d}x\, F\left(x, h(x), \frac{\mathrm{d}h(x)}{\mathrm{d}x}\right), \quad (2.251)$$

from the definition above

$$\frac{\delta \Phi_1}{\delta h(x)} = f(x), \tag{2.252}$$

$$\frac{\delta \Phi_2}{\delta h(x)} = \frac{\partial F}{\partial h}, \tag{2.253}$$

$$\frac{\delta \Phi_3}{\delta h(x)} = \frac{\partial F}{\partial h} - \frac{\mathrm{d}}{\mathrm{d}x}\frac{\partial F}{\partial h'} \tag{2.254}$$

can be deduced for the functional derivatives and the variation is reduced to the calculation of well-known partial derivatives.

To perform the stationary phase approximation to the path integral, the second variation of the classical action functional S at the classical path is needed. To calculate this quantity, we are using the first equation in (2.248) and are considering the classical path $x_{\mathrm{cl}}(t)$ and deviations η that vanish at the initial and final time $t' = 0, t$.

The first variation of the action is then given by

$$\begin{aligned}
\delta S[x_{\mathrm{cl}}] &= S[x_{\mathrm{cl}} + \eta] - S[x_{\mathrm{cl}}] \\
&= \int_0^t \mathrm{d}t' \left\{ \frac{m}{2}\left[\frac{\mathrm{d}}{\mathrm{d}t'}(x_{\mathrm{cl}} + \eta)\right]^2 - V(x_{\mathrm{cl}} + \eta) \right\} - S[x_{\mathrm{cl}}] \\
&= \int_0^t \mathrm{d}t' \left\{ \frac{m}{2}\dot{x}_{\mathrm{cl}}^2 + m\dot{x}_{\mathrm{cl}}\dot{\eta} - V(x_{\mathrm{cl}}) - V'(x_{\mathrm{cl}})\eta \right\} - S[x_{\mathrm{cl}}] \\
&= \int_0^t \mathrm{d}t' \left\{ m\dot{x}_{\mathrm{cl}}\dot{\eta} - V'(x_{\mathrm{cl}})\eta \right\} \\
&= \int_0^t \mathrm{d}t' \left\{ -m\ddot{x}_{\mathrm{cl}}\eta - V'(x_{\mathrm{cl}})\eta \right\} + m\dot{x}_{\mathrm{cl}}\eta|_0^t \\
&= \int_0^t \mathrm{d}t' \left\{ -m\ddot{x}_{\mathrm{cl}} - V'(x_{\mathrm{cl}}) \right\} \eta \,.
\end{aligned} \tag{2.255}$$

The vanishing of the first variation of the action, $\delta S[x_{\mathrm{cl}}] = 0$, is Hamilton's principle of classical mechanics and in the simple 1D case considered here, it leads to Newton's equation

$$m\ddot{x}_{\mathrm{cl}} + V'(x_{\mathrm{cl}}) = 0 \tag{2.256}$$

for the classical path, which we could have also concluded directly from (2.254) by identifying the independent variable with t and F with $L(\dot{x}, x, t) = T(\dot{x}) - V(x)$.

Up to second order we get

$$S = S[x_{\mathrm{cl}}] + \frac{1}{2}\delta^2 S[x_{\mathrm{cl}}]. \tag{2.257}$$

Here the second variation is given by

$$\begin{aligned}\delta^2 S[x_{\rm cl}] &= \delta S[x_{\rm cl}+\eta] - \delta S[x_{\rm cl}] \\ &= \int_0^t {\rm d}t' \left\{ -m\frac{{\rm d}^2}{{\rm d}t'^2}(x_{\rm cl}+\eta) - V'(x_{\rm cl}+\eta)\right\}\eta \\ &\quad - \int_0^t {\rm d}t' \left\{-m\ddot{x}_{\rm cl} - V'(x_{\rm cl})\right\}\eta \\ &= \int_0^t {\rm d}t' \left\{-m\ddot{\eta} - V''(x_{\rm cl})\eta\right\}\eta \\ &= \int_0^t {\rm d}t' \eta \hat{O}\eta, \end{aligned} \tag{2.258}$$

with the stability operator $\hat{O}$ from (2.84) of Sect. 2.2.1.

2.C Stability Matrix

We have seen that the central ingredient of the prefactor of the semiclassical Herman-Kluk propagator is the stability (or monodromy) matrix defined by

$$\mathbf{M} = \begin{pmatrix} \mathbf{m}_{11} & \mathbf{m}_{12} \\ \mathbf{m}_{21} & \mathbf{m}_{22} \end{pmatrix} = \begin{pmatrix} \frac{\partial \boldsymbol{p}_t}{\partial \boldsymbol{p}_i^{\rm T}} & \frac{\partial \boldsymbol{p}_t}{\partial \boldsymbol{q}_i^{\rm T}} \\ \frac{\partial \boldsymbol{q}_t}{\partial \boldsymbol{p}_i^{\rm T}} & \frac{\partial \boldsymbol{q}_t}{\partial \boldsymbol{q}_i^{\rm T}} \end{pmatrix}. \tag{2.259}$$

This matrix determines the time-evolution of small deviations in the initial conditions of a specific trajectory according to

$$\begin{pmatrix} \delta \boldsymbol{p}_t \\ \delta \boldsymbol{q}_t \end{pmatrix} = \mathbf{M} \begin{pmatrix} \delta \boldsymbol{p}_i \\ \delta \boldsymbol{q}_i \end{pmatrix}, \tag{2.260}$$

where $\delta \boldsymbol{p}_t = \tilde{\boldsymbol{p}}_t - \boldsymbol{p}_t$ and $\delta \boldsymbol{q}_t = \tilde{\boldsymbol{q}}_t - \boldsymbol{q}_t$. Pictorially, this is represented for one spatial dimension in Fig. 2.8. The area that is spanned by the deviation vectors is not changing in the course of time.

The equations of motion for the stability matrix can be gained by linearizing Hamilton's equations for the deviations. After Taylor expansion, we get

$$\delta\dot{\boldsymbol{p}}_t = -\frac{\partial H}{\partial \boldsymbol{q}}(\tilde{\boldsymbol{p}}_t, \tilde{\boldsymbol{q}}_t) + \frac{\partial H}{\partial \boldsymbol{q}}(\boldsymbol{p}_t, \boldsymbol{q}_t) = -\frac{\partial^2 H}{\partial \boldsymbol{q}\partial \boldsymbol{p}^{\rm T}}\delta \boldsymbol{p}_t - \frac{\partial^2 H}{\partial \boldsymbol{q}\partial \boldsymbol{q}^{\rm T}}\delta \boldsymbol{q}_t, \tag{2.261}$$

$$\delta\dot{\boldsymbol{q}}_t = \frac{\partial H}{\partial \boldsymbol{p}}(\tilde{\boldsymbol{p}}_t, \tilde{\boldsymbol{q}}_t) - \frac{\partial H}{\partial \boldsymbol{p}}(\boldsymbol{p}_t, \boldsymbol{q}_t) = \frac{\partial^2 H}{\partial \boldsymbol{p}\partial \boldsymbol{p}^{\rm T}}\delta \boldsymbol{p}_t + \frac{\partial^2 H}{\partial \boldsymbol{p}\partial \boldsymbol{q}^{\rm T}}\delta \boldsymbol{q}_t, \tag{2.262}$$

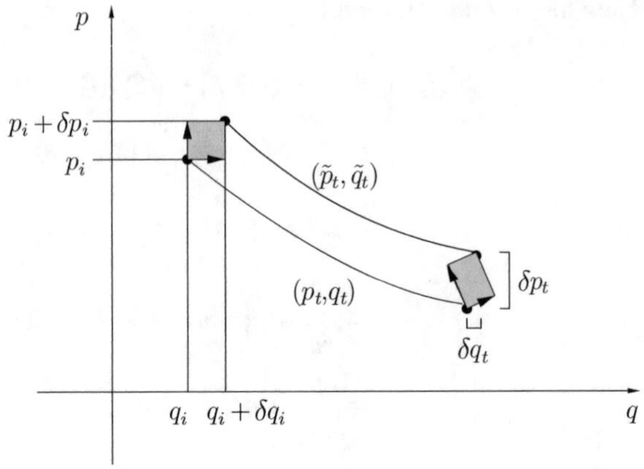

Fig. 2.8 Variation of initial conditions in phase space for one degree of freedom. The *shaded area* is constant in time (Liouville's theorem), adapted from [44]

where the vectors are column vectors and their transposes are row vectors. The linear differential equation of first order for the stability matrix thus reads

$$\frac{d}{dt}\mathbf{M} = -\mathbf{JHM}\,, \tag{2.263}$$

with the skew symmetric matrix

$$\mathbf{J} = \begin{pmatrix} \mathbf{0} & \mathbf{1} \\ -\mathbf{1} & \mathbf{0} \end{pmatrix}, \tag{2.264}$$

and the Hessian of the Hamiltonian

$$\mathbf{H} = \begin{pmatrix} \frac{\partial^2 H}{\partial \boldsymbol{p} \partial \boldsymbol{p}^{\mathrm{T}}} & \frac{\partial^2 H}{\partial \boldsymbol{p} \partial \boldsymbol{q}^{\mathrm{T}}} \\ \frac{\partial^2 H}{\partial \boldsymbol{q} \partial \boldsymbol{p}^{\mathrm{T}}} & \frac{\partial^2 H}{\partial \boldsymbol{q} \partial \boldsymbol{q}^{\mathrm{T}}} \end{pmatrix}. \tag{2.265}$$

The initial conditions follow directly from the definition (2.259) to be

$$\mathbf{M}(0) = \begin{pmatrix} \mathbf{m}_{11}(0) & \mathbf{m}_{12}(0) \\ \mathbf{m}_{21}(0) & \mathbf{m}_{22}(0) \end{pmatrix} = \begin{pmatrix} \mathbf{1} & \mathbf{0} \\ \mathbf{0} & \mathbf{1} \end{pmatrix}. \tag{2.266}$$

Using the equations of motion (2.263) and the initial conditions, it can be shown that $\frac{d}{dt}\mathbf{M}^{\mathrm{T}}\mathbf{JM} = 0$ and therefore $\mathbf{M}^{\mathrm{T}}\mathbf{JM} = \mathbf{M}^{\mathrm{T}}(t=0)\mathbf{JM}(t=0) = \mathbf{J}$ for all times, i.e., $\mathbf{M}$ is a symplectic matrix.

Written out more explicitly, the previous statement reads

$$\mathbf{m}_{22}^{\mathrm{T}}\mathbf{m}_{11} - \mathbf{m}_{12}^{\mathrm{T}}\mathbf{m}_{21} = \mathbf{1} \quad \forall t, \tag{2.267}$$

$$\mathbf{m}_{11}^{\mathrm{T}}\mathbf{m}_{21} - \mathbf{m}_{21}^{\mathrm{T}}\mathbf{m}_{11} = \mathbf{0} \quad \forall t, \tag{2.268}$$

$$\mathbf{m}_{22}^{\mathrm{T}}\mathbf{m}_{12} - \mathbf{m}_{12}^{\mathrm{T}}\mathbf{m}_{22} = \mathbf{0} \quad \forall t. \tag{2.269}$$

The determinant of the stability matrix is equal to unity as can be shown by using the relations above.[20] This is equivalent to the conservation of phase space volume mentioned previously. In 1D one does not need to take care of block matrix calculus, and in addition Liouville's theorem can be checked by calculating the vector product of the vectors drawn in Fig. 2.8.

Finally it should be mentioned that the matrix element $\mathbf{m}_{21}$, up to a sign, is the inverse of the second derivative of the action appearing in the prefactor of the VVG propagator (2.91)

$$\mathbf{m}_{21} = -\left(\frac{\partial^2 S}{\partial \boldsymbol{q}_t \partial \boldsymbol{q}_i^{\mathrm{T}}}\right)^{-1} . \tag{2.270}$$

2.D From the HK- to the VVG-Propagator

The most straightforward way to derive the van Vleck-Gutzwiller from the Herman-Kluk propagator is by taking the limit $\gamma \to \infty$ in

$$K^{\mathrm{HK}}(\boldsymbol{x}_f, t; \boldsymbol{x}_i, 0) = \int \frac{\mathrm{d}^N p_i \mathrm{d}^N q_i}{(2\pi\hbar)^N} \langle \boldsymbol{x}_f | \tilde{z}_t \rangle R(\boldsymbol{p}_i, \boldsymbol{q}_i, t) \exp\left\{\frac{\mathrm{i}}{\hbar} S(\boldsymbol{p}_i, \boldsymbol{q}_i, t)\right\} \langle \tilde{z}_i | \boldsymbol{x}_i \rangle , \tag{2.271}$$

which is reproduced here for convenience. The Gaussian wavepackets in that limit are "almost" δ-functions. δ-functions are normalized differently, however, and therefore in order to make use of their properties we note that in the limit $\gamma \to \infty$

$$\lim_{\gamma\to\infty} \left(\frac{4\pi}{\gamma}\right)^{-N/4} \langle \boldsymbol{x}_f | \tilde{z}_t \rangle = \delta(\boldsymbol{x}_f - \boldsymbol{q}_t), \tag{2.272}$$

$$\lim_{\gamma\to\infty} \left(\frac{4\pi}{\gamma}\right)^{-N/4} \langle \tilde{z}_i | \boldsymbol{x}_i \rangle = \delta(\boldsymbol{x}_i - \boldsymbol{q}_i). \tag{2.273}$$

The inverse of the γ-dependent factors in front of the Gaussians together with the prefactor (2.231) give

$$\lim_{\gamma\to\infty} \left(\frac{4\pi}{\gamma}\right)^{N/2} R(\boldsymbol{p}_i, \boldsymbol{q}_i, t) = (2\pi\hbar\mathrm{i})^{N/2} \sqrt{-|\mathbf{m}_{21}|} , \tag{2.274}$$

[20] Be careful to use the formula $\det \mathbf{M} = \det \mathbf{m}_{22} \det(\mathbf{m}_{11} - \mathbf{m}_{12}\mathbf{m}_{22}^{-1}\mathbf{m}_{21})$ valid for block matrices!

where the bars stand for taking the determinant of the stability submatrix. The $\boldsymbol{q}_i$ integration can be performed immediately allowing us to replace $\boldsymbol{q}_i$ by $\boldsymbol{x}_i$. For the $\boldsymbol{p}_i$ integration we use the relation

$$\delta[\boldsymbol{x}_f - \boldsymbol{q}_t(\boldsymbol{p}_i, \boldsymbol{x}_i)] = \sum_j \frac{1}{||\partial \boldsymbol{q}_t/\partial \boldsymbol{p}_i||}\delta(\boldsymbol{p}_i - \boldsymbol{p}_j) \tag{2.275}$$

for δ-functions of functions of the integration variable. Here we have to sum over all momenta $\boldsymbol{p}_j$ leading to zeros of $\boldsymbol{x}_f - \boldsymbol{q}_t(\boldsymbol{p}_i, \boldsymbol{x}_i)$ and the double bars denote the absolute value of the determinant.

Due to $\partial \boldsymbol{q}_t/\partial \boldsymbol{p}_i = \mathbf{m}_{21}$ and with (2.270) from Appendix 2.C, we finally arrive at the N degree of freedom form of the VVG expression

$$K^{\rm VVG}(\boldsymbol{x}_f, t; \boldsymbol{x}_i, 0) = \left(\frac{\rm i}{2\pi\hbar}\right)^{\frac{N}{2}} \sum_j \sqrt{\det\left(\frac{\partial^2 S_j}{\partial \boldsymbol{x}_f \partial \boldsymbol{x}_i^{\rm T}}\right)} \exp\left\{\frac{\rm i}{\hbar} S_j(\boldsymbol{x}_f, \boldsymbol{x}_i)\right\} . \tag{2.276}$$

References

1. E. Schrödinger, Ann. Phys. (Leipzig) **81**, 109 (1926)
2. E. Schrödinger, Ann. Phys. (Leipzig) **79**, 489 (1926)
3. J.S. Briggs, J.M. Rost, Eur. Phys. J. D **10**, 311 (2000)
4. F. Dyson, Phys. Rev. **75**, 486 (1949)
5. V.A. Mandelshtam, J. Chem. Phys. **108**, 9999 (1998)
6. R.P. Feynman, A.R. Hibbs, *Quantum Mechanics and Path Integrals*, emended edn. (Dover, Mineola, 2010)
7. E. Schrödinger, Die Naturwissenschaften **14**, 664 (1926)
8. E.J. Heller, J. Chem. Phys. **62**, 1544 (1975)
9. W. Kinzel, Physikalische Blätter **51**, 1190 (1995)
10. F. Grossmann, J.M. Rost, W.P. Schleich, J. Phys. A Math. Gen. **30**, L277 (1997)
11. M. Kleber, Phys. Rep. **236**, 331 (1994)
12. R.P. Feynman, Rev. Mod. Phys. **20**, 367 (1948)
13. P.A.M. Dirac, *The Principles of Quantum Mechanics*, 4th edn. (Oxford, London, 1958)
14. L.S. Schulman, *Techniques and Applications of Path Integration* (Dover, Mineola, 2005)
15. J.H. van Vleck, Proc. Acad. Nat. Sci. USA **14**, 178 (1928)
16. M.C. Gutzwiller, J. Math. Phys. **8**, 1979 (1967)
17. S. Grossmann, *Funktionalanalysis II* (Akademie Verlag, Wiesbaden, 1977)
18. H. Goldstein, C. Poole, J. Safko, *Classical Mechanics*, 3rd edn. (Addison Wesley, San Francisco, 2002)
19. W.R. Salzman, J. Chem. Phys. **85**, 4605 (1986)
20. M.H. Beck, A. Jäckle, G.A. Worth, H.D. Meyer, Phys. Rep. **324**, 1 (2000)
21. D. Kohen, F. Stillinger, J.C. Tully, J. Chem. Phys. **109**, 4713 (1998)
22. M. Tinkham, *Group Theory and Quantum Mechanics* (McGraw-Hill, New York, 1964)
23. H. Sambe, Phys. Rev. A **7**, 2203 (1973)
24. M. Abramowitz, I.A. Stegun, *Handbook of Mathematical Functions* (Dover Publications, New York, 1965)

25. W.H. Press, S.A. Teukolsky, W.T. Vetterling, B.P. Flannery, *Numerical Recipes in Fortran*, 2nd edn. (Cambridge University Press, Cambridge, 1992)
26. J.H. Shirley, Phys. Rev. **138**, B979 (1965)
27. J.C. Light, *Time-Dependent Quantum Molecular Dynamics*, ed. by J. Broeckhove, L. Lathouwers (Plenum Press, New York, 1992), p. 185
28. U. Peskin, N. Moiseyev, Phys. Rev. A **49**, 3712 (1994)
29. W. Witschel, J. Phys. A **8**, 143 (1975)
30. J.A. Fleck, J.R. Morris, M.D. Feit, Appl. Phys. **10**, 129 (1976)
31. M.D. Feit, J.A. Fleck, A. Steiger, J. Comp. Phys. **47**, 412 (1982)
32. M. Braun, C. Meier, V. Engel, Comp. Phys. Comm. **93**, 152 (1996)
33. M. Frigo, S.G. Johnson, Proceedings of the IEEE **93**(2), 216 (2005). Special issue on *Program Generation, Optimization, and Platform Adaptation*
34. A. Vibok, G.G. Balint-Kurti, J. Phys. Chem **96**, 8712 (1992)
35. A. Askar, A.S. Cakmak, J. Chem. Phys. **68**, 2794 (1978)
36. C. Leforestier, R.H. Bisseling, C. Cerjan, M.D. Feit, R. Friesner, A. Guldberg, A. Hammerich, G. Jolicard, W. Karrlein, H.D. Meyer, N. Lipkin, O. Roncero, R. Kosloff, J. Comp. Phys. **94**, 59 (1991)
37. H. Tal-Ezer, R. Kosloff, J. Chem. Phys. **81**, 3967 (1984)
38. D.J. Tannor, *Introduction to Quantum Mechanics: A Time-Dependent Perspective* (University Science Books, Sausalito, 2007)
39. S.K. Gray, D.W. Noid, B.G. Sumpter, J. Chem. Phys. **101**, 4062 (1994)
40. J.D. Meiss, Rev. Mod. Phys. **64**, 795 (1992)
41. M.L. Brewer, J.S. Hulme, D.E. Manolopoulos, J. Chem. Phys. **106**, 4832 (1997)
42. H. Yoshida, Phys. Lett. A **150**, 262 (1990)
43. W.H. Louisell, *Quantum Statistical Properties of Radiation* (Wiley, New York, 1990)
44. E.J. Heller, *Chaos and Quantum Physics*, ed. by M.J. Giannoni, A. Voros, J. Zinn-Justin. Les Houches Session LII (Elsevier, Amsterdam, 1991), pp. 549–661
45. J.R. Klauder, *Random Media*, ed. by G. Papanicolauou (Springer, New York, 1987), p. 163
46. F. Grossmann, J.A.L. Xavier, Phys. Lett. A **243**, 243 (1998)
47. M.F. Herman, E. Kluk, Chem. Phys. **91**, 27 (1984)
48. E.J. Heller, J. Chem. Phys. **75**, 2923 (1981)
49. K.G. Kay, Chem. Phys. **322**, 3 (2006)
50. K.G. Kay, J. Chem. Phys. **100**, 4377 (1994)
51. F. Grossmann, J. Chem. Phys. **125**, 014111 (2006)
52. E. Kluk, M.F. Herman, H.L. Davis, J. Chem. Phys. **84**, 326 (1986)
53. F. Grossmann, M.F. Herman, J. Phys. A Math. Gen. **35**, 9489 (2002)
54. S. Zhang, E. Pollak, J. Chem. Phys. **121**(8), 3384 (2004)
55. F. Grossmann, Comm. At. Mol. Phys. **34**, 141 (1999)
56. B.H. Bransden, C.J. Joachain, *Physics of Atoms and Molecules*, 2nd edn. (Pearson Education, Harlow, 2003)
57. S. Selstø, E. Lindroth, J. Bengtsson, Phys. Rev. A **79**, 043418 (2009)
58. W.P. Schleich, *Quantum Optics in Phase Space* (Wiley-VCH, Berlin, 2001)
59. E.N. Economou, *Green's Functions in Quantum Physics*, 3rd edn. (Springer, Berlin, 2006)
60. D. Schuch, *Quantum Theory from a Nonlinear Perspective: Riccati Equations in Fundamental Physics* (Springer International Publishing, 2018)
61. G.L. Ingold, Coherent Evolution, in *Noisy Environments*, ed. by A. Buchleitner, K. Hornberger, Lecture Notes, in Physics, (Springer, Berlin, 2002), pp. 1–53
62. N. Makri, W.H. Miller, Chem. Phys. Lett. **151**, 1 (1988)
63. W.H. Miller, Adv. Chem. Phys. **25**, 69 (1974)
64. L.E. Reichl, *The Transition to Chaos*, 2nd edn. (Springer, New York, 2004)
65. G.D. Billing, *The Quantum Classical Theory* (Oxford University Press, New York, 2003)
66. E.J. Heller, *The Semiclassical Way to Dynamics and Spectroscopy* (Princeton University Press, Princeton, 2018)
67. A. Nitzan, *Chemical Dynamics in Condensed Phases* (Oxford University Press, Oxford, 2006)

68. K.J. Schafer, *Strong Field Laser Physics*, ed. by T. Brabec. Springer Series in Optical Sciences, vol. 134 (Springer, Berlin, 2009), chap. 6, pp. 111–145
69. N. Takemoto, A. Shimshovitz, D.J. Tannor, J. Chem. Phys. **137**, 011102 (2012)
70. F. Gatti, B. Lasorne, H.D. Meyer, A. Nauts, *Applications of Quantum Dynamics in Chemistry*. Lecture Notes in Chemistry, vol. 98 (Springer International Publishing, 2017)
71. D.V. Shalashilin, M.S. Child, J. Chem. Phys. **113**, 10028 (2000)
72. W.H. Miller, J. Phys. Chem. B **106**, 8132 (2002)
73. R.E. Wyatt, *Quantum Dynamics with Trajectories: Introduction to Quantum Hydrodynamics, Interdisciplinary Applied Mathematics*, vol. 28 (Springer, New York, 2005)
74. M. Mizrahi, J. Math. Phys. **16**, 2201 (1975)

Part II
Applications

Chapter 3
Field-Matter Coupling and Two-Level Systems

With this chapter, we start the applications part of this book by considering the interaction between lasers and matter. Lasers have already been discussed in Chap. 1. Therefore, we begin immediately with the theoretical description of the coupling of a given classical light field realized, e.g., by a laser, to a quantum mechanical system.[1] After the discussion of different gauges or frames, related by unitary transformations, the Volkov solution for the laser-driven free particle is reviewed.

Due to their simplicity and the fact that they serve as paradigms for many phenomena observed in more complex systems, some analytically solvable two-level systems will be discussed in the remainder of this chapter. We will first look at Rabi oscillations mediated by a static electric field and after the introduction of the rotating wave approximation, the laser-driven case will be reviewed.

3.1 Light-Matter Interaction

The interaction of a single quantum particle with an electromagnetic field shall be considered in the following. Starting from the principle of minimal coupling and using several unitary transformations, some commonly used ways of setting up a field driven Hamiltonian will be presented.

[1] In the literature this is frequently called semiclassical laser matter interaction [1]. We have, however, used the expression "semiclassics" already differently in Chap. 2.

F. Grossmann, *Theoretical Femtosecond Physics*, Graduate Texts in Physics,
https://doi.org/10.1007/978-3-319-74542-8_3

3.1.1 Minimal Coupling

The most straightforward approach to the coupling of a charged particle with charge q to an electromagnetic field is given by the principle of "minimal coupling". In classical mechanics this principle aims at producing Newton's equation with the Lorentz force by constructing a corresponding Lagrangian.

3.1. *Study classical minimal coupling by answering the following questions:*

(a) *Under which conditions for the potentials* $\mathbf{A}$ *and* Φ *does the classical Lagrangian*

$$L(\dot{\boldsymbol{r}}, \boldsymbol{r}, t) = \frac{m}{2}\dot{\boldsymbol{r}}^2 - q\Phi(\boldsymbol{r}, t) + q\dot{\boldsymbol{r}} \cdot \boldsymbol{A}(\boldsymbol{r}, t)$$

lead to Newton's equation of motion with the Lorentz force?

(b) *Give explicit expressions for the canonical momentum* $\boldsymbol{p} = \partial L/\partial \dot{\boldsymbol{r}}$ *and for the mechanical momentum* $\boldsymbol{p}_{\mathrm{m}} = m\dot{\boldsymbol{r}}$.

(c) *What is the explicit form of the Hamiltonian* $H(\boldsymbol{p}, \boldsymbol{r}, t) = \dot{\boldsymbol{r}} \cdot \boldsymbol{p} - L(\dot{\boldsymbol{r}}, \boldsymbol{r}, t)$?

To arrive at the quantum version of minimal coupling, we could just use the classical result and invoke the correspondence principle. More instructive is a direct approach to quantum minimal coupling, however, which shall be discussed in some detail now.

Let us first consider the effect of a local unitary transformation with the scalar field $\chi(\boldsymbol{r}, t)$

$$\Psi'(\boldsymbol{r}, t) = \mathrm{e}^{\mathrm{i}\frac{q}{\hbar}\chi(\boldsymbol{r},t)}\Psi(\boldsymbol{r}, t) \tag{3.1}$$

on the time-dependent Schrödinger equation [2]. For the transformed wavefunction the transformed equation

$$\mathrm{i}\hbar\dot{\Psi}'(\boldsymbol{r}, t) = \hat{H}'\Psi'(\boldsymbol{r}, t) \tag{3.2}$$

holds, where the primed Hamiltonian is given by

$$\hat{H}' = \mathrm{e}^{\mathrm{i}\frac{q}{\hbar}\chi(\boldsymbol{r},t)}\hat{H}\mathrm{e}^{-\mathrm{i}\frac{q}{\hbar}\chi(\boldsymbol{r},t)} - q\dot{\chi}(\boldsymbol{r}, t), \tag{3.3}$$

with

$$\hat{H} = \frac{\hat{\boldsymbol{p}}^2}{2m} + V(\boldsymbol{r}). \tag{3.4}$$

Shifting the momentum operator $\hat{\boldsymbol{p}} = \frac{\hbar}{\mathrm{i}}\nabla$ twice past the exponential factor of the unitary transformation, we get the identity

$$\mathrm{e}^{\mathrm{i}\frac{q}{\hbar}\chi(\boldsymbol{r},t)}\hat{\boldsymbol{p}}^2\mathrm{e}^{-\mathrm{i}\frac{q}{\hbar}\chi(\boldsymbol{r},t)}\Psi'(\boldsymbol{r}, t) = (\hat{\boldsymbol{p}} - q\nabla\chi)^2\Psi'(\boldsymbol{r}, t), \tag{3.5}$$

and therefore

$$\hat{H}' = \frac{1}{2m}\left(\frac{\hbar}{\mathrm{i}}\nabla - q\nabla\chi\right)^2 + V(\boldsymbol{r}) - q\frac{\partial\chi}{\partial t} \tag{3.6}$$

holds for the primed Hamiltonian.

Following Weyl [3], the time-dependent Schrödinger equation has to be invariant under the unitary transformation (3.1) introduced above. To satisfy this requirement, the original time-dependent Schrödinger equation has to be modified slightly, however, according to

$$\mathrm{i}\hbar\dot{\Psi}(\boldsymbol{r},t) = \left[\frac{1}{2m}\left(\frac{\hbar}{\mathrm{i}}\nabla - q\boldsymbol{A}(\boldsymbol{r},t)\right)^2 + V(\boldsymbol{r}) + q\Phi(\boldsymbol{r},t)\right]\Psi(\boldsymbol{r},t). \tag{3.7}$$

This equation is now formally equivalent to the transformed time-dependent Schrödinger equation

$$\mathrm{i}\hbar\dot{\Psi}'(\boldsymbol{r},t) = \left[\frac{1}{2m}\left(\frac{\hbar}{\mathrm{i}}\nabla - q\boldsymbol{A}'(\boldsymbol{r},t)\right)^2 + V(\boldsymbol{r}) + q\Phi'(\boldsymbol{r},t)\right]\Psi'(\boldsymbol{r},t) \tag{3.8}$$

if the relations

$$\boldsymbol{A}' = \boldsymbol{A} + \nabla\chi, \qquad \Phi' = \Phi - \dot{\chi} \tag{3.9}$$

hold. These, however, are the gauge transformations of the potentials $\boldsymbol{A}(\boldsymbol{r},t)$ and $\Phi(\boldsymbol{r},t)$ of classical electrodynamics. The electromagnetic fields

$$\boldsymbol{\mathcal{E}} = -\frac{\partial\boldsymbol{A}}{\partial t} - \nabla\Phi, \tag{3.10}$$

$$\boldsymbol{B} = \nabla\times\boldsymbol{A} \tag{3.11}$$

are unchanged by such transformations.

Summarizing, minimal coupling amounts to replacing the canonical momentum $\hat{\boldsymbol{p}}$ by the kinetic momentum $\hat{\boldsymbol{p}} - q\boldsymbol{A}(\boldsymbol{r},t)$ and shifting the potential by $q\Phi(\boldsymbol{r},t)$ in the Hamiltonian. The probability current density in the equation of continuity (2.4) of Chap. 2 has to be changed accordingly, as can be seen by solving Exercise 3.2.

3.2. *Find the modified expression for the probability current density* $\boldsymbol{j}$ *in the case of coupling of the motion of a charged particle to an external field. Show that the expression you gained is gauge invariant.*

Expanding the square of the kinetic momentum, cross terms of the form $\hat{\boldsymbol{p}}\cdot\boldsymbol{A}$ and $\boldsymbol{A}\cdot\hat{\boldsymbol{p}}$ appear. In the Coulomb gauge, which, for sources at infinity, is defined by

$$\Phi(\boldsymbol{r},t) = 0, \qquad \nabla\cdot\boldsymbol{A}(\boldsymbol{r},t) = 0\,, \tag{3.12}$$

we can conclude that $\hat{\boldsymbol{p}}$ commutes with $\boldsymbol{A}$. Therefore the two cross-terms are identical as in classical mechanics.

Furthermore, the energy obviously is not conserved any more in the presence of a time-dependent external field. The question, which operator could be considered as the energy operator, does not have a straightforward answer, however. Studying Exercise 3.3 and the literature given in the solutions sheds more light on this question.

3.3. *Let $\hat{\Theta}(\boldsymbol{A}, \Phi)$ be an operator that depends on the potentials of the electromagnetic field.*

(a) *Show that for the operator $\hat{\Theta}$ to have a gauge invariant expectation value*

$$e^{i\frac{q}{\hbar}\chi}\hat{\Theta}(\boldsymbol{A}, \Phi)e^{-i\frac{q}{\hbar}\chi} = \hat{\Theta}(\boldsymbol{A}', \Phi')$$

has to hold.

(b) *Show that $\hat{H} = \frac{(\hat{\boldsymbol{p}}-q\boldsymbol{A})^2}{2m} + V(\boldsymbol{r}) + q\Phi$ is not a gauge invariant operator and its expectation value cannot be the energy. Discuss an alternative, that may be considered as the energy operator.*

In general, the case of a system of many charged particles which are coupled to a laser field has to be studied. As we will see in Chap. 5, the motion of the center of mass and the relative motion without a laser can be separated. With the laser they do not necessarily separate any more [3]. We will deal with the coupling of an electromagnetic field to a many particle system in more detail in Chap. 5.

3.1.2 Dipole Approximation and Length Gauge

Another well-known form of light matter interaction rests on the dipole approximation, in which case the vector potential is assumed to be independent of position.[2] For an atom of typical size of the order of Angstroms in a field of optical wavelength of several hundred nanometers this is a well founded approximation, as depicted in Fig. 3.1.

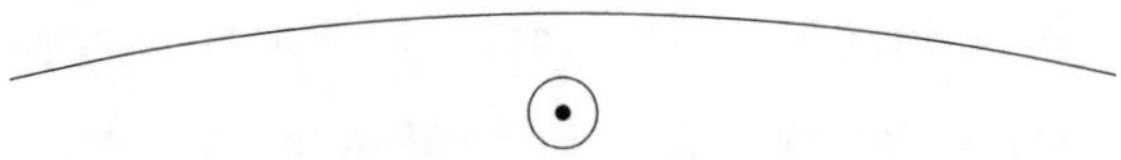

Fig. 3.1 An atom in the field of a light wave with wavelength much longer than the typical extension of the atom

[2]Therefore, due to (3.11), the magnetic induction vanishes in dipole approximation.

The effect of the vector potential in the minimal coupling Hamiltonian of (3.7) in Coulomb gauge and in dipole approximation is a time-dependent shift of the momentum leading to

$$\hat{H}_{\mathrm{v}} = \frac{[\hat{\boldsymbol{p}} - q\boldsymbol{A}(t)]^2}{2m} + V(\boldsymbol{r}). \tag{3.13}$$

Due to the corresponding Lagrangian (vector potential couples to velocity), the term velocity gauge is therefore frequently used.[3] Applying a gauge transformation with the scalar field

$$\chi(\boldsymbol{r}, t) = -\boldsymbol{r} \cdot \boldsymbol{A}(t) \tag{3.14}$$

leads to transformed potentials of the form

$$\boldsymbol{A}' = 0, \qquad \Phi' = -\frac{\partial \chi}{\partial t} = \boldsymbol{r} \cdot \dot{\boldsymbol{A}} = -\boldsymbol{r} \cdot \boldsymbol{\mathcal{E}}(t), \tag{3.15}$$

where the last step follows from (3.10) in Coulomb gauge. The present gauge is thus also called length gauge (electric field couples to the position). The corresponding time-dependent Schrödinger equation then reads

$$\mathrm{i}\hbar\dot{\Psi}_{\mathrm{l}}(\boldsymbol{r}, t) = \left[\frac{\hat{\boldsymbol{p}}^2}{2m} + V(\boldsymbol{r}) - q\,\boldsymbol{r} \cdot \boldsymbol{\mathcal{E}}(t)\right] \Psi_{\mathrm{l}}(\boldsymbol{r}, t) \tag{3.16}$$

and contains the laser-matter interaction in terms of the dipole operator $q\,\hat{\boldsymbol{r}}$. Historically it has been introduced by Göppert-Mayer [5] by using the fact that the Lagrangians in the length as well as in the velocity gauge only differ by a total time-derivative.

3.4. *Switch from the velocity to the length gauge by adding a total time derivative to the Lagrangian.*

(a) *Show first that adding a total time-derivative $\frac{\mathrm{d}}{\mathrm{d}t} f(\boldsymbol{r}, t)$ to the Lagrangian does not alter the equations of motion.*
(b) *In the dipole approximation and the Coulomb gauge ($\Phi = 0, \boldsymbol{A} = \boldsymbol{A}(t)$) add $-q\frac{\mathrm{d}}{\mathrm{d}t}(\boldsymbol{r} \cdot \boldsymbol{A})$ to the velocity gauge Lagrangian and simplify the resulting expression.*

We stress that in dipole approximation and under the Coulomb gauge, the vector potential as well as the electric field are independent of the position vector and the magnetic induction is zero. A coupling to the magnetic field by going beyond the

[3] Synonymously, some authors [3, 4] use the expression $\boldsymbol{A} \cdot \boldsymbol{p}$ gauge.

dipole approximation would become necessary for a large electron quiver velocity (see Sect. 3.1.4) on the order of the speed of light.[4]

Velocity and length gauge are related by a unitary transformation and therefore any measurable quantity may not depend on the gauge used. If one uses approximations during the solution process, however, there may be orders of magnitude and even qualitative [7] differences between the results predicted in the different gauges. A recent investigation of a gauge independent strong field approximation is given in [8]. A more technical question is, in which gauge numerical calculations should be performed. For the investigation of high-order harmonic generation using laser irradiated hydrogen atoms, to be discussed in Sect. 4.5, it was found that for high laser intensities, the velocity gauge seems to be favorable from a numerical perspective [9]. This fact was corroborated in a recent publication on the ionization of hydrogen atoms, see the supplemental material of [10]. The case of very short (down to half cycle) pulses is discussed with respect to gauge invariance in [11].

Finally, it is worthwhile to note that, as shown in Appendix 3.A, the notion of parity, well-known in autonomous Hamiltonian systems, can be generalized to the case of periodically, dipolarly driven systems.

3.1.3 Kramers-Henneberger Transformation

In the case of strong fields, another unitary transformation will turn out to be very useful. We start again from the minimally coupled time-dependent Schrödinger equation (3.7) in the Coulomb gauge and in dipole approximation, leading to

$$
\begin{aligned}
\mathrm{i}\hbar\dot{\Psi}_{\mathrm{v}}(\boldsymbol{r},t) &= \left[\frac{1}{2m}\left(\frac{\hbar}{\mathrm{i}}\nabla - q\boldsymbol{A}(t)\right)^2 + V(\boldsymbol{r})\right]\Psi_{\mathrm{v}}(\boldsymbol{r},t) \\
&= \left[-\frac{\hbar^2}{2m}\nabla^2 + \frac{\mathrm{i}q\hbar}{m}\boldsymbol{A}(t)\cdot\nabla + \frac{q^2}{2m}\boldsymbol{A}^2(t) + V(\boldsymbol{r})\right]\Psi_{\mathrm{v}}(\boldsymbol{r},t). \quad (3.17)
\end{aligned}
$$

Successively performing two unitary transformations

$$
\Psi_{\mathrm{a}}(\boldsymbol{r},t) = \hat{U}_2\hat{U}_1\Psi_{\mathrm{v}}(\boldsymbol{r},t), \quad (3.18)
$$

with

$$
\hat{U}_1 = \exp\left\{\frac{\mathrm{i}q^2}{2m\hbar}\int_0^t \mathrm{d}t'\boldsymbol{A}^2(t')\right\}, \quad (3.19)
$$

$$
\hat{U}_2 = \exp\left\{-\frac{q}{m}\int_0^t \mathrm{d}t'\boldsymbol{A}(t')\cdot\nabla\right\} \quad (3.20)
$$

[4]For a charged particle in a plane electromagnetic wave, the magnetic part of the Lorentz force is smaller by a factor v/c than the electric one [6].

defines a wavefunction in the Kramers-Henneberger frame [12, 13]. The use of the subscript "a" will become apparent below. The first transformation eliminates the squared vector potential, whereas the displacement operator (3.20) moves the coupling into the argument of the potential, as can be seen by solving Exercise 3.5. The time-dependent Schrödinger equation in the Kramers-Henneberger frame is then given by

$$\mathrm{i}\hbar\dot{\Psi}_{\mathrm{a}}(\boldsymbol{r},t) = \left[-\frac{\hbar^2}{2m}\Delta + V[\boldsymbol{r}+\boldsymbol{\alpha}(t)]\right]\Psi_{\mathrm{a}}(\boldsymbol{r},t)\,, \tag{3.21}$$

where

$$\boldsymbol{\alpha}(t) = -\frac{q}{m}\int_0^t \mathrm{d}t'\boldsymbol{A}(t'). \tag{3.22}$$

3.5. *Show that the two unitary transformations into the Kramers-Henneberger frame eleminate the terms proportional to $\mathbf{A}^2$ and $\mathbf{A}$ in the Hamiltonian. Due to the fact that the first transformation is a global phase transformation, it just remains to calculate*

$$\hat{U}_2\hat{V}\hat{U}_2^{-1},$$

to prove the shift in the argument of the potential.
Hint: Use the operator relation known as Baker-Haussdorff (or Hadamard) lemma $\mathrm{e}^{\hat{L}}\hat{M}\mathrm{e}^{-\hat{L}} = \sum_{n=0}^{\infty}\frac{1}{n!}[\hat{L},\hat{M}]_n$, where $[\hat{L},\hat{M}]_n = [\hat{L},[\hat{L},\hat{M}]_{n-1}]$ and $[\hat{L},\hat{M}]_0 = \hat{M}$.

Differentiating (3.22) twice and using (3.10) in the Coulomb gauge, we find that

$$m\ddot{\boldsymbol{\alpha}}(t) = q\boldsymbol{\mathcal{E}} \tag{3.23}$$

holds. The Kramers-Henneberger transformation thus is characterized by a spatial translation into an accelerated frame, corresponding to the oscillatory quiver motion of the charged particle in the electric field. The present case is therefore also frequently referred to as the "acceleration gauge", although the use of the term gauge is misleading, since no gauge transformations of the potentials can be given here [11].

In the high-frequency limit, the TDSE in the Kramers-Henneberger frame can be averaged over a (short) period of the external field. This has, e.g., been done in the calculations in [14] for the case of a periodically driven double-well potential, to be discussed in more detail in Sect. 5.5.1. This way, analytical predictions of the influence of high-frequency driving on the system dynamics can be given.

3.1.4 Volkov Wavepacket and Ponderomotive Energy

To fill the presented formalism with life, we now turn to an important, exactly solvable model. For the case of 1D free motion, i.e., $V(x) = 0$ of an electron with mass $m_{\rm e}$ and charge $q = -e$ in a cw laser field $\mathcal{E} = \mathcal{E}_0 \cos(\omega t)$, the time-dependent Schrödinger equation in length gauge (3.16) can be solved exactly analytically under the assumption of a Gaussian initial state.

Due to the fact that the total time-dependent potential

$$V_{\rm L}(x,t) = ex\mathcal{E}_0 \cos(\omega t) \tag{3.24}$$

is linear, the resulting Volkov wavepacket with initial phase space center $(0, q_0)$ is given by using the GWD of Sect. 2.1.4 for $\alpha_0 = \gamma/2$ according to[5]

$$\Psi(x,t) = \left(\frac{\gamma}{\pi}\right)^{1/4} \sqrt{\frac{1}{1+{\rm i}\gamma\hbar t/m_{\rm e}}} \exp\left\{\frac{\rm i}{\hbar}\left[\frac{U_{\rm p}}{2\omega}\sin(2\omega t) - U_{\rm p}t + xp(t)\right]\right\}$$
$$\exp\left\{-\frac{\gamma}{2(1+{\rm i}\gamma\hbar t/m_{\rm e})}[x-q(t)]^2\right\}, \tag{3.25}$$

where the general solutions

$$p(t) = p_0 - e\mathcal{E}_0 \sin(\omega t)/\omega, \tag{3.26}$$
$$q(t) = q_0 + \frac{p_0 t}{m_{\rm e}} + e\mathcal{E}_0[\cos(\omega t) - 1]/(m_{\rm e}\omega^2) \tag{3.27}$$

of the classical equations of motion for position and momentum have been used with the initial conditions $p(0) = p_0 = 0, q(0) = q_0$.

The amplitude of oscillations of position $e\mathcal{E}_0/(m_e\omega^2)$ is the so-called quiver amplitude. We can convince ourselves of the analytic form of the Volkov solution by solving Exercise 3.6. As a side result it will turn out that the derivative of the kinetic energy averaged over a period of the external field vanishes. A free particle can therefore not accumulate energy from the field.

3.6. *Using Gaussian Wavepacket Dynamics calculate the wavefunction of a free electron in a laser field with the potential*

$$V_{\rm L}(x,t) = ex\mathcal{E}_0 \cos(\omega t)$$

in length gauge.

[5]The gauge index will be mostly suppressed in the remainder of the book, as we will explicitly state which gauge is used.

(a) *Determine the solutions (p_t, q_t) of the classical equations of motion with the initial conditions $(0, q_0)$. Then calculate the classical kinetic energy and its derivative and average the results over one period $T = 2\pi/\omega$ of the external field. Interprete the results.*
(b) *Use the result for α_t from the free particle case (why is this possible?).*
(c) *Employing integration by parts, show that*

$$\int_0^t \mathrm{d}t' L = -\int_0^t \mathrm{d}t' \frac{p_{t'}^2}{2m_\mathrm{e}} + q_t p_t$$

holds. Use this result to determine the phase $\delta_t = \int_0^t \mathrm{d}t'(L - \alpha_{t'})$ and insert everything in the GWD expression. Why is the final result exact?

As could be seen by working through the previous exercise, there is an important quantity hidden in the Volkov solution. This is the average of the kinetic energy over one period, which is given by

$$U_\mathrm{p} := \frac{1}{T}\int_0^T \mathrm{d}t \frac{p^2}{2m_\mathrm{e}} = \frac{e^2\mathcal{E}_0^2}{4m_\mathrm{e}\omega^2}, \tag{3.28}$$

as can be shown by using (3.26). This quantity is called ponderomotive energy (or ponderomotive potential). $2U_\mathrm{p}$ is the maximal kinetic energy that an electron may have at a certain time. It is important to keep in mind that low frequency fields lead to high ponderomotive energies.

A generalization of the results to 3D can be performed in several ways. Alternatively to replacing all occurences of position and momentum in the solution above by the corresponding 3D vectors, we start from the TDSE in velocity gauge with the Hamiltonian of (3.13) in the case of $V(\boldsymbol{r}) = 0$. After a Fourier transformation of the wavefunction to the momentum representation (see also Sect. 2.3.2) via

$$\Psi(\boldsymbol{p}) = \frac{1}{(2\pi\hbar)^{3/2}}\int \mathrm{d}^3 r \exp\left\{-\frac{\mathrm{i}}{\hbar}\boldsymbol{p}\cdot\boldsymbol{r}\right\}\Psi(\boldsymbol{r}), \tag{3.29}$$

we get

$$\mathrm{i}\hbar\dot{\Psi}(\boldsymbol{p}, t) = \frac{1}{2m_\mathrm{e}}\left[\boldsymbol{p} + e\boldsymbol{A}(t)\right]^2 \Psi(\boldsymbol{p}, t) \tag{3.30}$$

for the time-dependent Schrödinger equation. There is no operator hat on the momentum any more because we have used the corresponding eigenvalue equation.

A solution to the ordinary differential equation above is given by

$$\Psi(\boldsymbol{p}, t) = \Psi(\boldsymbol{p}, 0)\exp\left\{-\frac{\mathrm{i}}{\hbar}\int_0^t \mathrm{d}t' \frac{\left[\boldsymbol{p} + e\boldsymbol{A}(t')\right]^2}{2m_\mathrm{e}}\right\}, \tag{3.31}$$

as can be proven by separation of variables and checked by differentiation with respect to time. Inverse Fourier transformation back to position space now gives the Volkov wavepacket

$$\Psi(\boldsymbol{r},t) = \frac{1}{(2\pi\hbar)^{3/2}}\int \mathrm{d}^3p\ \exp\left\{\frac{\mathrm{i}}{\hbar}\boldsymbol{p}\cdot\boldsymbol{r}\right\}$$
$$\exp\left\{-\frac{\mathrm{i}}{\hbar}\int_0^t \mathrm{d}t'\frac{\left[\boldsymbol{p}+e\boldsymbol{A}(t')\right]^2}{2m_\mathrm{e}}\right\}\Psi(\boldsymbol{p},0). \tag{3.32}$$

If we use as a special case a delta function centered around $\boldsymbol{p}_0$,

$$\Psi(\boldsymbol{p},0) = \delta(\boldsymbol{p}-\boldsymbol{p}_0), \tag{3.33}$$

for the initial $\boldsymbol{p}$ state, then the Volkov state

$$\Psi_\mathrm{v}(\boldsymbol{r},t) = \frac{1}{(2\pi\hbar)^{3/2}}\exp\left\{\frac{\mathrm{i}}{\hbar}\boldsymbol{p}_0\cdot\boldsymbol{r} - \frac{\mathrm{i}}{\hbar}\int_0^t \mathrm{d}t'\frac{\left[\boldsymbol{p}_0+e\boldsymbol{A}(t')\right]^2}{2m_\mathrm{e}}\right\} \tag{3.34}$$

in velocity gauge emerges.

Starting from the velocity gauge result and using the gauge transformation from (3.14) leads to

$$\Psi_\mathrm{l}(\boldsymbol{r},t) = \frac{1}{(2\pi\hbar)^{3/2}}\exp\left\{\frac{\mathrm{i}}{\hbar}[\boldsymbol{p}_0+e\boldsymbol{A}(t)]\cdot\boldsymbol{r}\right.$$
$$\left.-\frac{\mathrm{i}}{\hbar}\int_0^t \mathrm{d}t'\frac{\left[\boldsymbol{p}_0+e\boldsymbol{A}(t')\right]^2}{2m_\mathrm{e}}\right\}, \tag{3.35}$$

which is the Volkov state in length gauge.

In the Kramers-Henneberger frame, starting again from (3.34), the dependence on $\boldsymbol{A}(t)$ cancels out and we arrive at the corresponding Volkov state

$$\Psi_\mathrm{a}(\boldsymbol{r},t) = \frac{1}{(2\pi\hbar)^{3/2}}\exp\left\{\frac{\mathrm{i}}{\hbar}\boldsymbol{p}_0\cdot\boldsymbol{r} - \frac{\mathrm{i}}{\hbar}\frac{\boldsymbol{p}_0^2}{2m_\mathrm{e}}t\right\}, \tag{3.36}$$

which is just a plane wave.

A more general initial state, e.g., a Gaussian centered around $\boldsymbol{p}_0$, will lead to a generalization of the GWD solution to 3D. It is a nice exercise in Gaussian integration to rederive the 1D result (3.25) from (3.32).

3.2 Analytically Solvable Two-Level Problems

Driven two-level systems are the easiest realizations of the field-matter coupling formalism just reviewed. Several paradigms in the theory of laser induced dynamics can be found already in the solutions of these simple systems. They shall therefore now be studied in some detail. We will concentrate on analytically solvable cases, which can either be solved exactly or under some approximations.

3.2.1 Dipole Matrix Element

First of all, the Hamilton matrix has to be set up. To this end, we consider two energy levels with the unperturbed orthogonal states $|\psi_1\rangle, |\psi_2\rangle$ and the corresponding energies $E_1 = -\hbar\epsilon, E_2 = \hbar\epsilon$, which are the diagonal elements of the Hamilton matrix.

To write down an expression for the off-diagonal elements of the Hamilton matrix in the case of an external perturbation, we assume that it is due to the coupling to an electric field of the form

$$\boldsymbol{\mathcal{E}}(\boldsymbol{r}, t) = \boldsymbol{\mathcal{E}}_0 \cos(\boldsymbol{k} \cdot \boldsymbol{r} - \omega t). \tag{3.37}$$

We now turn to the dipole approximation of Sect. 3.1.2, i.e., $\lambda = 2\pi/k$ shall be much larger than the size of the quantum system, as depicted in Fig. 3.1. In the argument of the cosine, $\boldsymbol{r}$ can then be replaced by $\boldsymbol{r}_0$ which can be set to zero without loss of generality. The electric field is then purely time-dependent

$$\boldsymbol{\mathcal{E}}(t) = \boldsymbol{\mathcal{E}}_0 \cos(\omega t) \tag{3.38}$$

and the coordinate independent force

$$\boldsymbol{F}(t) = -e\,\boldsymbol{\mathcal{E}}(t) \tag{3.39}$$

acts on the electron. The corresponding potential energy is given by

$$V_{\rm L}(\boldsymbol{r}, t) = e\,\boldsymbol{r} \cdot \boldsymbol{\mathcal{E}}(t). \tag{3.40}$$

Adding this potential energy to the Hamiltonian leads to the length gauge form of the Hamiltonian in (3.16).

If the two levels under consideration have real eigenfunctions with different parity (see Exercise 3.8) then

$$\hbar\nu_{12}(t) \equiv \boldsymbol{\mathcal{E}}(t) \cdot \int \mathrm{d}^3 r\, \psi_1\, e\, \boldsymbol{r} \psi_2 = \boldsymbol{\mu}_{12} \cdot \boldsymbol{\mathcal{E}}(t) = \hbar\nu_{21}(t) \tag{3.41}$$

follows for the non-vanishing off-diagonal elements of the Hamilton matrix, which are proportional to the matrix element

$$\boldsymbol{\mu}_{12} \equiv \int \mathrm{d}^3 r\, \psi_1\, e\, \boldsymbol{r} \psi_2 \tag{3.42}$$

of the (negated) dipole operator. The diagonal matrix elements due to the laser potential are zero.

3.2.2 Rabi Oscillations Induced by a Constant Perturbation

For the following, we assume that the perturbation is time-independent, i.e., we set $\omega \to 0$ and define $\nu := \lim_{\omega \to 0} \nu_{12}(t)$. As an Ansatz for the solution of the time-dependent Schrödinger equation (2.22), a superposition of the unperturbed eigenstates with time-dependent coefficients

$$|\Psi(t)\rangle = c_1(t)|\psi_1\rangle + c_2(t)|\psi_2\rangle \tag{3.43}$$

can be chosen. For the vector $\boldsymbol{c}^{\mathrm{T}} = (c_1, c_2)$ of coefficients, the linear system of coupled ordinary differential equations

$$\mathrm{i}\hbar\dot{\boldsymbol{c}} = \mathbf{H}\boldsymbol{c}\,, \tag{3.44}$$

with the two by two Hamilton matrix

$$\mathbf{H} = \hbar \begin{pmatrix} -\epsilon & \nu \\ \nu & \epsilon \end{pmatrix} \tag{3.45}$$

emerges. This Hamiltonian can be expressed in terms of the Pauli spin matrices, which are discussed in Appendix 3.B.

As mentioned in Sect. 2.3.1, the time-evolution can be determined by solving the eigenvalue problem. The eigenvalues of the matrix in (3.45) are

$$E_\pm = \pm\hbar\sqrt{\epsilon^2 + \nu^2} \tag{3.46}$$

and the corresponding eigenstates are given by

$$|\psi_+\rangle = \sin(\Theta)|\psi_1\rangle + \cos(\Theta)|\psi_2\rangle, \tag{3.47}$$

$$|\psi_-\rangle = \cos(\Theta)|\psi_1\rangle - \sin(\Theta)|\psi_2\rangle, \tag{3.48}$$

where the definition

$$\Theta \equiv \frac{1}{2}\arctan\left(\frac{\nu}{\epsilon}\right) \tag{3.49}$$

has been used.[6] In the case of degeneracy of the unperturbed states ($\epsilon = 0$) $\Theta = \pi/4$, and the eigenstates are the symmetric, respectively antisymmetric combination of the two unperturbed states.

The spectral representation of the time-evolution operator for the solution of the Schrödinger equation is given by

$$\hat{U}(t,0) = \sum_{\pm} |\psi_\pm\rangle \exp\left\{-\frac{\mathrm{i}}{\hbar}E_\pm t\right\}\langle\psi_\pm|, \tag{3.50}$$

as can be seen by comparison with (2.39). In the basis of the eigenvectors (1,0) and (0,1) of the unperturbed Hamilton matrix, the matrix

$$\begin{aligned}\mathbf{U}(t,0) = &\begin{pmatrix} \sin^2(\Theta) & \sin(\Theta)\cos(\Theta) \\ \sin(\Theta)\cos(\Theta) & \cos^2(\Theta)\end{pmatrix}\exp\left\{-\frac{\mathrm{i}}{\hbar}E_+ t\right\} \\ &+ \begin{pmatrix} \cos^2(\Theta) & -\sin(\Theta)\cos(\Theta) \\ -\sin(\Theta)\cos(\Theta) & \sin^2(\Theta)\end{pmatrix}\exp\left\{-\frac{\mathrm{i}}{\hbar}E_- t\right\}\end{aligned} \tag{3.51}$$

for the time-evolution operator can be derived. This matrix allows us to calculate

$$P_{21}(t) = |\langle\psi_2|\hat{U}(t,0)|\psi_1\rangle|^2 = |U_{21}(t,0)|^2, \tag{3.52}$$

which is the probability to find the system in state $|\psi_2\rangle$ at time t, if it was in state $|\psi_1\rangle$ at time zero (in terms of the coefficients this corresponds to the initial conditions $c_1(0) = 1$, $c_2(0) = 0$). From the matrix representation of $\hat{U}$ by using $\sin\Theta\cos\Theta = \frac{1}{2}\sin(2\Theta)$, we get

$$P_{21}(t) = \frac{\nu^2}{\nu^2+\epsilon^2}\sin^2(\Omega_{\mathrm{R}}t/2), \tag{3.53}$$

where

$$\Omega_{\mathrm{R}} \equiv 2\sqrt{\epsilon^2+\nu^2} \tag{3.54}$$

is the definition of the so-called Rabi frequency. $P_{21}(t)$ performs Rabi oscillations with the amplitude

[6]The identities $\arctan(x) = \arccos(1/\sqrt{1+x^2})$ and $\arctan(x) = \arcsin(x/\sqrt{1+x^2})$ can be used to resolve the cosine and sine terms in (3.47) and (3.48).

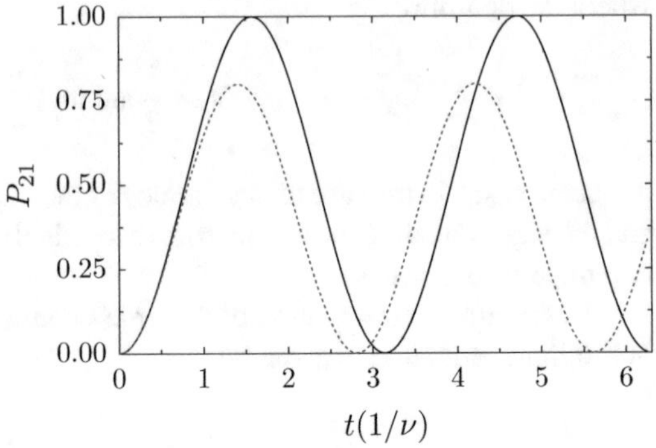

Fig. 3.2 Rabi oscillations of the probability to be in the upper state, starting from the lower state, induced by the perturbation $\nu = 1$. The degenerate $\epsilon = 0$ (*solid black line*), as well as the non-degenerate case $\epsilon = 0.5$ (*dashed blue line*) are depicted as a function of time in units of $1/\nu$; all energies in arbitrary units

$$A = \frac{\nu^2}{\nu^2 + \epsilon^2}, \tag{3.55}$$

which are depicted in Fig. 3.2. Only in case of degeneracy, $\epsilon = 0$, do the oscillations have an amplitude of 1. Furthermore, for non-degenerate systems the oscillations are faster than for degenerate unperturbed levels.

Rabi oscillations are analogous to the tunneling dynamics of the probability amplitude in a symmetric double well, which will be considered in Chap. 5. There the eigenstates of the unperturbed problem are the symmetric, respectively antisymmetric superposition of *localized* states in the left and right well and thus for a localized initial condition $c_1 = \pm c_2 = 1/\sqrt{2}$ has to be chosen.

3.2.3 Time-Dependent Perturbations and Rotating Wave Approximation

In the presence of a time-dependent perturbation $\hat{V}(t) = \hbar\hat{v}(t)$, and with $E_{1,2} = \hbar\epsilon_{1,2}$, the time-dependent Schrödinger equation for the coefficients is

$$\mathrm{i}\dot{c}_1 = c_1\epsilon_1 + c_2 v_{12}(t), \tag{3.56}$$

$$\mathrm{i}\dot{c}_2 = c_2\epsilon_2 + c_1 v_{21}(t). \tag{3.57}$$

In the "strong-coupling" limit, i.e., for $v_{21} \gg \epsilon_2 - \epsilon_1$, these coupled differential equations can be solved perturbatively [15]. There exists, however, another approximate approach to solve the differential equations, starting from the Ansatz

$$c_1(t) = d_1(t)\exp[-\mathrm{i}\epsilon_1 t], \tag{3.58}$$

$$c_2(t) = d_2(t)\exp[-\mathrm{i}\epsilon_2 t], \tag{3.59}$$

which leads to

$$\mathrm{i}\dot{d}_1 = d_2 v_{12}(t) \exp[-\mathrm{i}\omega_{21} t], \tag{3.60}$$

$$\mathrm{i}\dot{d}_2 = d_1 v_{21}(t) \exp[\mathrm{i}\omega_{21} t], \tag{3.61}$$

where the abbreviation $\omega_{21} = \epsilon_2 - \epsilon_1$ has been introduced. Note that the transformation from the vector $\boldsymbol{c}$ to the vector $\boldsymbol{d}$ of coefficients is equivalent to a unitary transformation into the interaction picture.

In the case of a monochromatic (coherent) perturbation $v_{12}(t) \sim \cos(\omega t)$, the system of differential equations can be solved analytically by using the so-called rotating wave approximation, as will be shown in the following. In the case of interaction with incoherent radiation (a random superposition of monochromatic laser fields) we can use perturbation theory and in this way give a microscopic derivation of the B-coefficient of Chap. 1. This last case will be dealt with in Appendix 3.C.

3.2.3.1 Rotating Wave Approximation

For the monochromatic field in (3.38), the Schrödinger equation in the interaction picture (3.60) and (3.61) can be written as

$$\mathrm{i}\dot{d}_1 = d_2 \frac{\boldsymbol{\mu}_{12} \cdot \boldsymbol{\mathcal{E}}_0}{2\hbar} \{\exp[\mathrm{i}(\omega - \omega_{21})t] + \exp[-\mathrm{i}(\omega + \omega_{21})t]\}, \tag{3.62}$$

$$\mathrm{i}\dot{d}_2 = d_1 \frac{\boldsymbol{\mu}_{21} \cdot \boldsymbol{\mathcal{E}}_0}{2\hbar} \{\exp[-\mathrm{i}(\omega - \omega_{21})t] + \exp[\mathrm{i}(\omega + \omega_{21})t]\}. \tag{3.63}$$

In order to introduce the rotating wave approximation (RWA), we define the detuning between the field and the external frequency

$$\Delta_\mathrm{d} \equiv \omega - \omega_{21}. \tag{3.64}$$

For $\Delta_\mathrm{d} \ll \omega_{21}$, the terms that oscillate at about twice the frequency of the external field are the so-called counter-rotating terms. In the differential equations above they can be neglected, if we assume that the coefficients $d_{1,2}$ change on a much longer time scale. To prove this procedure mathematically, one has to average the differential equations over times much larger than $1/(\omega + \omega_{21})$, see Exercise 3.7.

The differential equations in RWA are now dramatically simplified and read

$$\mathrm{i}\dot{d}_1 = d_2 \frac{\mu \mathcal{E}_0}{2\hbar} \exp[\mathrm{i}\Delta_\mathrm{d} t], \tag{3.65}$$

$$\mathrm{i}\dot{d}_2 = d_1 \frac{\mu \mathcal{E}_0}{2\hbar} \exp[-\mathrm{i}\Delta_\mathrm{d} t], \tag{3.66}$$

where we have assumed in addition that the polarization of the field is in the direction of the dipole matrix element, which has the absolute value $\mu = \mu_{12} = \mu_{21}$. The solution of the two coupled differential equations can be found by differentiating (3.65) with respect to time and inserting (3.66). The second order differential equation that emerges can be solved, and one gets

$$\begin{aligned} d_1(t) &= \frac{\hbar}{\mu\mathcal{E}_0}\exp[\mathrm{i}\Delta_\mathrm{d}t/2]\{(\Delta_\mathrm{d}-\Omega_\mathrm{R})C\exp[\mathrm{i}\Omega_\mathrm{R}t/2] \\ &\quad +(\Delta_\mathrm{d}+\Omega_\mathrm{R})D\exp[-\mathrm{i}\Omega_\mathrm{R}t/2]\}, \qquad (3.67)\\ d_2(t) &= \exp[-\mathrm{i}\Delta_\mathrm{d}t/2]\{C\exp[\mathrm{i}\Omega_\mathrm{R}t/2]+D\exp[-\mathrm{i}\Omega_\mathrm{R}t/2]\} \qquad (3.68) \end{aligned}$$

with the Rabi frequency in the time-dependent case

$$\Omega_\mathrm{R} = \sqrt{\Delta_\mathrm{d}^2 + \left(\frac{\mu\mathcal{E}_0}{\hbar}\right)^2}\,. \tag{3.69}$$

The parameters C and D have to be determined from the initial conditions. In the case of non-resonance (corresponding to the non-degenerate case for constant perturbations) the oscillations are again faster than on resonance.

3.7. *Consider a two-level system interacting with a monochromatic laser field.*

(a) *Average the TDSE over times long in comparison to $1/(\omega+\omega_{21})$ in order to motivate neglecting the counter-rotating terms.*
(b) *Using the initial conditions $d_1(0)=1$ and $d_2(0)=0$, give explicit expressions for C and D and for $d_1(t)$ and $d_2(t)$. Depict $|d_2(t)|^2$ for resonance as well as for off-resonance.*

Furthermore, the quality of the RWA depends on the soundness of the approximation of neglecting the counter-rotating terms. The validity of this assumption can be studied explicitly for a specific example in Exercise 3.8.

3.8. *An electron shall move in an inversion symmetric potential $V(x)=V(-x)$ in one spatial dimension.*

(a) *Show that the eigenfunctions of the TISE must fulfill either*

$$\psi_{2n}(x)=\psi_{2n}(-x) \qquad \text{or} \qquad \psi_{2n+1}(x)=-\psi_{2n+1}(-x),$$

and that diagonal dipole matrix elements $\mu_{nn}=\langle\psi_n|e\hat{x}|\psi_n\rangle$ therefore always vanish.
(b) *Calculate the dipole matrix element between the ground and the first excited state of the harmonic oscillator*

$$V(x)=\frac{1}{2}m_\mathrm{e}\omega_\mathrm{e}^2x^2$$

with a frequency in the visible range, $\omega_e = 3.14 \times 10^{15}\,\mathrm{s}^{-1}$. *Determine the Rabi frequency in the resonance case for 3 different field strengths* $\mathcal{E}_0 = 1, 10^6, 10^{10}\,\mathrm{V\,m}^{-1}$. *Is the condition for the applicability of the RWA fulfilled for all field strengths?*

3.2.3.2 Area Theorem

Finally, we consider the case of resonance, $\Delta_d = 0$, in which $\hbar$ times the external frequency equals the level spacing. Furthermore, we assume that the external field shall be of finite duration, i.e., it shall consist of a laser pulse with an envelope, so that we have to replace $\mathcal{E}_0$ by $\mathcal{E}_0 f(t)$ in the time-dependent Schrödinger equation. According to (3.69), a time-dependent Rabi frequency

$$\Omega_R(t) = \frac{\mu \mathcal{E}_0 f(t)}{\hbar} \tag{3.70}$$

emerges, with the help of which the coupled differential equations can be written as

$$\mathrm{i}\dot{d}_1 = \frac{\Omega_R(t)}{2} d_2, \tag{3.71}$$

$$\mathrm{i}\dot{d}_2 = \frac{\Omega_R(t)}{2} d_1. \tag{3.72}$$

For the initial conditions $d_1(0) = 1, d_2(0) = 0$ the solutions are given by

$$d_1(t) = \cos\left(\int_0^t \mathrm{d}t' \frac{\Omega_R(t')}{2}\right), \tag{3.73}$$

$$d_2(t) = -\mathrm{i}\sin\left(\int_0^t \mathrm{d}t' \frac{\Omega_R(t')}{2}\right), \tag{3.74}$$

as can be verified by insertion. In RWA the population transfer in the resonance case does not depend on the specific shape of the pulse, but only on the area below the pulse. This is the so-called area theorem. A π-pulse, for which $\int_0^t \mathrm{d}t' \Omega_R(t') = \pi$, allows for a complete transfer of population.

3.2.4 *Exactly Solvable Time-Dependent Cases*

In very few special cases, also in the case of a time-dependent perturbation an exact analytical solution of the time-dependent two-level Schrödinger equation can be found [16]. As our starting point we use (3.44) in the case of time-dependent ϵ and ν. After elimination of c_1, the second order differential equation

$$\ddot{c}_2 - \frac{\dot{\nu}}{\nu}\dot{c}_2 + \left(\epsilon^2 + \nu^2 - \mathrm{i}\dot{\epsilon} + \mathrm{i}\epsilon\frac{\dot{\nu}}{\nu}\right)c_2 = 0 \tag{3.75}$$

for the other coefficient can be derived, see also Exercise 3.9. Reasons for the time-dependence of the diagonal as well as for the off-diagonal elements of the Hamiltonian may be the coupling to a light field or nuclear motion in a molecule, which will be considered in detail in Chap. 5.

3.2.4.1 The Rosen-Zener Model

In the case of coupling to a pulsed laser field and in RWA[7] we can choose

$$\epsilon = \Delta/2, \qquad \nu(t) = \nu_0 \mathrm{sech}(t/T_\mathrm{p}), \tag{3.76}$$

defining the Rosen-Zener model with a pulse length parameter T_p, see Fig. 3.3. As found by these authors, the solution of the time-dependent Schrödinger equation for this choice can be determined exactly analytically. With the initial condition $c_1(-\infty) = 1$ and for $t \to \infty$ it is given by [17]

$$|c_2(\infty)|^2 = \sin^2(\pi\nu_0 T_\mathrm{p})\mathrm{sech}^2(\pi\Delta T_\mathrm{p}/2). \tag{3.77}$$

For the resonance case, $\Delta = 0$, this solution is proven in Exercise 3.9. In the resonance case it is also rewarding to note that in the argument of the sine, the pulse area $\nu_0\int_{-\infty}^{\infty} \mathrm{d}t\, \mathrm{sech}(t/T_\mathrm{p}) = \nu_0\pi T_\mathrm{p}$ appears. This is yet another manifestation of the area theorem discussed at the end of Sect. 3.2.3.

3.9. *Consider the TDSE for the two-level Rosen-Zener model.*

(a) *Prove the equation for c_2 that can be gained by the elimination of c_1.*
(b) *Transform the independent variable with the help of*

$$z = \frac{1}{2}(\tanh\frac{t}{T_\mathrm{p}} + 1).$$

What is the differential equation for $c_2(z)$?

(c) *Consider the special case $\epsilon = 0$ and determine $c_2(t = \infty)$ for the initial conditions $c_2(t = -\infty) = 0$ and $c_1(-\infty) = 1$.*
Hint: Use the hypergeometric function (see I. S. Gradshteyn and I. M. Rhyzik, Tables of Integrals Series and Products (Academic Press, San Diego, 1994), Sect. 9.1, or http://dlmf.nist.gov/15) and
$F(a, b; c; 1) = \frac{\Gamma(c)\Gamma(c-a-b)}{\Gamma(c-a)\Gamma(c-b)}$, $\Gamma(1-x)\Gamma(x) = \frac{\pi}{\sin(\pi x)}$, $\Gamma(3/2) = \sqrt{\pi}/2$

[7] This is an approximation and therefore the notion of exact solubility refers to the final equation and not the initial problem.

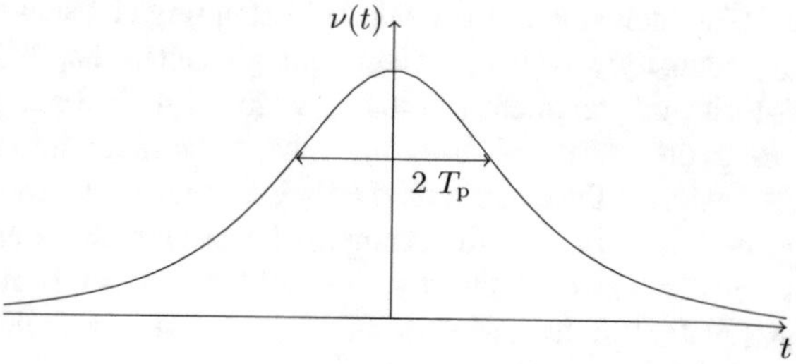

Fig. 3.3 Laser pulse envelope and width parameter T_p for the Rosen-Zener model

3.2.4.2 The Landau-Zener Model

Apart from the Rosen-Zener case, an exact analytic solution is available in the case of a linear time-dependence

$$\epsilon(t) = \lambda t, \qquad \nu(t) = \nu_0 \tag{3.78}$$

of the diagonal terms. This kind of time-dependence can, e.g., be induced in a scattering experiment, by nuclear motion in a molecule, or by time-dependent magnetic fields applied to atoms. The model has been investigated by Landau [18], as well as by Zener [19], Stückelberg [20], and Majorana [21]. An asymptotic solution of (3.75) for the initial condition $c_1(-\infty) = 1$ is given by

$$|c_2(\infty)|^2 = 1 - \exp[-2\pi\gamma], \tag{3.79}$$

where $\gamma = \nu_0^2/|2\lambda|$. The expression for γ can be further specified in molecular theory and leads to the celebrated Landau-Zener formula [22].

We note that for an infinitely fast change of the energy, i.e., $\lambda \to \infty$, it follows that $\gamma \to 0$ and no population will be transferred, i.e., $|c_2(\infty)|^2 \to 0$. On the contrary in the case of $\lambda \to 0$, one finds $\gamma \to \infty$, yielding complete population transfer, i.e., $|c_2(\infty)|^2 \to 1$.

3.3 Notes and Further Reading

Minimal Coupling and Gauge Transformations

Unitary (gauge) transformations of the wavefunction are discussed in depth in [2, 3]. The theory of minimal coupling and the different gauges or frames for field-matter interaction are at least partly covered in the books just mentioned as well as in many other quantum theoretical textbooks, see, e.g. [4]. Schleich's book [3] focuses on the subtleties arising from the inclusion of center of mass motion and contains an appendix, dealing with terms beyond the dipole approximation.

The inclusion of magnetic field coupling is discussed in [23] and in [24]. In [23] an intensity/wavelength diagram is presented that delineates regions, where a fully relativistic treatment (including order $(v/c)^2$ terms) is needed from those, where (magnetic) terms of order v/c have to be taken into account and those, where the inclusion of the electric field only is sufficient. A similar discussion can be found on page 868 of [4]. For the example of a Ti:sapphire laser with 800 nm wavelength, the intensity above which magnetic field effects can become important is 10^{16}W/cm^2. An insightful discussion of the gauge invariant calculation of expectation values and probabilities is given in [25].

Two-Level Systems

Our formulation of the interaction of two-level systems with coherent and incoherent (see Appendix 3.C) light is based on the presentation in Haken's book [26]. A landmark paper in this field is the one by Shirley [27], treating the periodically driven two-level problem in Floquet theory. The Rosen-Zener model is a special case of the first Demkov-Kunike model, which is discussed in the appendix of [28].

The theory of two-level systems interacting with magnetic fields has not been dealt with here but is covered in the book on photon-atom interactions by Weissbluth [29]. This book is also a treasure-house, if one is interested in the effect of damping on the dynamics of a two-level system. The wide field of dissipative quantum systems is usually described in a density matrix formulation (see also Appendix 3.B). More details on that exciting field can be found in the books by Weiss [30] and by Breuer and Petruccione [31].

The laser field is considered to be a classical field throughout the rest of this book. In quantum optics, where the light field is treated quantum mechanically, the RWA can also be performed, and if applied to a driven two-level system, this is known as the Jaynes-Cummings model, which is treated in detail in the book by Schleich [3].

3.A Generalized Parity Transformation

In the case of a symmetric static potential $V(x) = V(-x)$ and in length gauge, with a sinusoidal laser potential of the form $e\mathcal{E}_0 x \sin(\omega t)$, the extended Hamiltonian $\hat{\mathcal{H}}$ in (2.138) is invariant under the generalized parity transformation

$$\hat{\mathcal{P}}: \quad x \to -x, \quad t \to t + \frac{T}{2}\,. \tag{3.80}$$

The Floquet functions thus transform according to

$$\hat{\mathcal{P}}\psi_{\alpha'}(x,t) = \pm\psi_{\alpha'}(x,t)\,, \tag{3.81}$$

i.e., they have either positive or negative generalized parity. With the help of (2.150) it follows that $\psi_{\alpha'}(x,t)$, $\psi_{\beta'}(x,t)$ have the same or different generalized parity, depending on $(\alpha - k) - (\beta - l)$ being even or odd.

As we will see in Chap. 5, exact crossings of the quasienergies as a function of external parameters are of utmost importance for the quantum dynamics of periodically driven systems. For stationary systems, the possibility of exact crossings has been studied in the heyday of quantum theory by von Neumann and Wigner [32]. These authors found that eigenvalues of eigenfunctions with different parity may approach each other arbitrarily closely and may thus cross exactly. This is in contrast to eigenvalues of the same parity, which always have to be at a finite distance, a fact which is sometimes referred to as the non-crossing rule. The corresponding behavior in the spectrum as a function of external parameters is called allowed, respectively avoided crossing. In the Floquet case, the Hamiltonian can also be represented by a Hermitian matrix, see e.g. (2.183), and therefore the same reasoning applies, with parity replaced by generalized parity.

For the investigations to be presented in Sect. 5.5.1 it is decisive if these *exact* crossings are singular events in parameter space or if they can occur by variation of just a single parameter. In [32] it has been shown that for Hermitian matrices (of finite dimension) with complex (real) elements, the variation of three (two) free parameters is necessary in order for two eigenvalues to cross. Using similar arguments, it can be shown that for a real Hermitian matrix with alternatingly empty off-diagonals (as it is e.g., the case for the Floquet matrix of the periodically driven, quartic, symmetric, bistable potential) the variation of a single parameter is enough to make two quasienergies cross.

In the case of *avoided* crossings an interesting behavior of the corresponding eigenfunctions can be observed. There is a continuous change in the structure in position space if one goes through the avoided crossing [33]. Pictorially this is very nicely represented in the example of the driven quantum well, depicted in Fig. 3.4, taken out of [34], where for reasons of better visualization the Husimi transform of the quasi-eigenfunctions as a function of action angle variables (J, Θ) [35] is shown.

3.B Pauli Spin Matrices and the Two-Level Density Matrix

The Pauli spin matrices

$$\boldsymbol{\sigma}_x = \begin{pmatrix} 0 & 1 \\ 1 & 0 \end{pmatrix}; \quad \boldsymbol{\sigma}_y = \begin{pmatrix} 0 & -\mathrm{i} \\ \mathrm{i} & 0 \end{pmatrix}; \quad \boldsymbol{\sigma}_z = \begin{pmatrix} 1 & 0 \\ 0 & -1 \end{pmatrix}, \tag{3.82}$$

together with the 2×2 unit matrix, form a complete basis in the space of complex hermitian 2×2 matrices. In their terms our Hamiltonian (3.45) reads

$$\mathbf{H} = \hbar\nu\boldsymbol{\sigma}_x - \hbar\epsilon\boldsymbol{\sigma}_z. \tag{3.83}$$

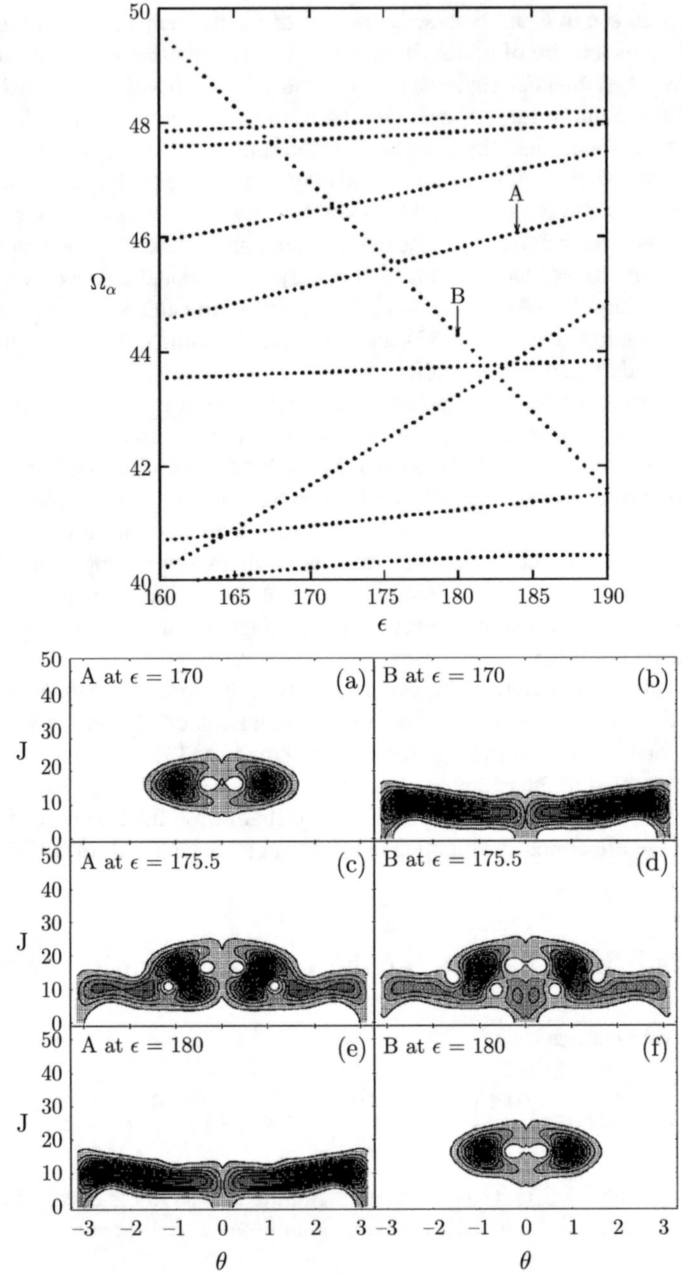

Fig. 3.4 Avoided crossing of Floquet energies (here denoted by Ω_α) as a function of field amplitude (*upper panel*) and associated change of character of the Floquet functions, corresponding to the two levels labelled by A and B in the driven quantum well (*lower panels* (**a**–**f**)); from [34]

Furthermore, a general density operator can be written as

$$\hat{\rho} = \frac{1}{2}\left(\hat{1} + \boldsymbol{r}\cdot\hat{\boldsymbol{\sigma}}\right), \tag{3.84}$$

with a vector $\boldsymbol{r}$ that is of unit length for all times in the case of pure state dynamics, and a vector-operator $\hat{\boldsymbol{\sigma}}$, composed of the Pauli operators. This then allows for a geometrical interpretation of two-level dynamics by going to the Feynman-Vernon-Hellwarth (or Bloch sphere) representation, discussed in the book by Tannor [36].

The pure state density matrix, in the case of a two-level system with energies E_1, E_2, in the basis of the corresponding eigenstates is given by

$$\boldsymbol{\rho} = \begin{pmatrix} |d_1|^2 & d_1 d_2^* \exp\{-\mathrm{i}(E_1 - E_2)t/\hbar\} \\ d_1^* d_2 \exp\{\mathrm{i}(E_1 - E_2)t/\hbar\} & |d_2|^2 \end{pmatrix}, \tag{3.85}$$

with the *populations* of the different energy levels on the diagonal and where the off-diagonal elements are sometimes called *coherences*.

A frequently considered mixed state is the thermal density matrix at temperature T with only diagonal elements

$$\rho_{mn} = \frac{\mathrm{e}^{-\beta E_n}}{Q}\delta_{mn}, \tag{3.86}$$

where $\beta = 1/(k_\mathrm{B}T)$ with Boltzmann constant k_B and where $Q = \sum_{n=1}^{2} \mathrm{e}^{-\beta E_n}$ is the partition function. An initial pure state evolves into a thermal mixed state by relaxation (due to coupling to an environment) which is governed by the time scale for population decay T_1 and the dephasing or coherence decay time scale T_2, which are related via

$$\frac{1}{T_2} = \frac{1}{2T_1} + \frac{1}{T_2^*}, \tag{3.87}$$

with the pure dephasing time T_2^* [36].

3.C Two-Level System in an Incoherent Field

As the starting point of the perturbative treatment of a two-level system in an incoherent external field, we use the Schrödinger equation in the interaction representation (3.60) and (3.61) with the initial conditions $d_1(0) = 1$ and $d_2(0) = 0$. For very small perturbations, the coefficient d_1 is assumed to remain at its initial value, leading to

$$\mathrm{i}\dot{d}_2 = \nu_{21}(t)\exp[\mathrm{i}\omega_{21}t]\,. \tag{3.88}$$

This equation can be integrated immediately to yield

$$d_2(t) = -\mathrm{i}\int_0^t \mathrm{d}t'\nu_{21}(t')\exp[\mathrm{i}\omega_{21}t'] , \tag{3.89}$$

analogous to the first order iteration in (2.28). The field shall consist of a superposition of waves with uniformly distributed, statistically independent phases ϕ_j

$$\mathcal{E}(t) = \frac{1}{2}\sum_{\omega_j>0} \mathcal{E}_j \exp[\mathrm{i}\phi_j - \mathrm{i}\omega_j t] + \mathrm{c.c.} . \tag{3.90}$$

If we insert this into the equation above, we get

$$\begin{aligned} d_2(t) &= -\frac{\mathrm{i}}{2\hbar}\sum_j \boldsymbol{\mathcal{E}}_j \cdot \boldsymbol{\mu}_{21}\exp[\mathrm{i}\phi_j]\int_0^t \mathrm{d}t' \exp[\mathrm{i}(\omega_{21}-\omega_j)t'] \\ &= -\frac{\mathrm{i}}{2\hbar}\sum_j \boldsymbol{\mathcal{E}}_j \cdot \boldsymbol{\mu}_{21}\exp[\mathrm{i}\phi_j]S_j , \end{aligned} \tag{3.91}$$

where the definition

$$S_j = [\mathrm{i}(\omega_{21}-\omega_j)]^{-1}\left\{\exp[\mathrm{i}(\omega_{21}-\omega_j)t] - 1\right\} \tag{3.92}$$

has been introduced. The occupation probability of the second level is then given by the double sum

$$|d_2(t)|^2 = (2\hbar)^{-2}\sum_j\sum_{j'} \exp[\mathrm{i}(\phi_j - \phi_{j'})]\boldsymbol{\mathcal{E}}_j \cdot \boldsymbol{\mu}_{21}\boldsymbol{\mathcal{E}}_{j'} \cdot \boldsymbol{\mu}_{21}^* S_j S_{j'}^* . \tag{3.93}$$

Averaging over the phases is now performed and denoted by $<>$, yielding

$$< \exp[\mathrm{i}(\phi_j - \phi_k)] >= \delta_{jk} . \tag{3.94}$$

One of the sums in (3.93) therefore collapses and

$$< |d_2(t)|^2 >= \left|\frac{\boldsymbol{e}\cdot\boldsymbol{\mu}_{21}}{\hbar}\right|^2 \sum_j |\mathcal{E}_j|^2(\omega_{21}-\omega_j)^{-2}\sin^2[(\omega_{21}-\omega_j)t/2] \tag{3.95}$$

follows for identical polarization, $\boldsymbol{e}$, of the light waves.

Now we have to sum over the distribution of frequencies. To this end we consider the time derivative of the expression above[8]

$$\frac{\mathrm{d}}{\mathrm{d}t} < |d_2(t)|^2 >= \left|\frac{\boldsymbol{e}\cdot\boldsymbol{\mu}_{21}}{\sqrt{2}\hbar}\right|^2 \sum_j |\mathcal{E}_j|^2(\omega_{21}-\omega_j)^{-1}\sin[(\omega_{21}-\omega_j)t]\,. \quad (3.96)$$

With the definition of an energy density per *angular* frequency interval $\rho(\omega_j) = \frac{1}{2}\varepsilon_0|\mathcal{E}_j|^2/\Delta\omega_j$, assuming that the frequencies are distributed continuously, and replacing $\rho(\omega_j)$ by its resonance value $\rho(\omega_{21})$, due to

$$\int_{-\infty}^{\infty} \mathrm{d}\omega \sin(\omega t)/\omega = \pi\ , \quad (3.97)$$

we get

$$\frac{\mathrm{d}}{\mathrm{d}t} < |d_2(t)|^2 >= \frac{\pi}{\varepsilon_0}\left|\frac{\boldsymbol{e}\cdot\boldsymbol{\mu}_{21}}{\hbar}\right|^2 \rho(\omega_{21})\,. \quad (3.98)$$

The right hand side of this expression is a constant and therefore consistent with the assumptions made in the derivation of Planck's radiation law in Chap. 1.

Comparing the equation above with (1.2) for N_1=1 and after switching from the *angular* to the *linear* frequency case [37]

$$B = \frac{2\pi^2}{\varepsilon_0}\left|\frac{\boldsymbol{e}\cdot\boldsymbol{\mu}_{21}}{h}\right|^2 \quad (3.99)$$

is found for Einstein's B coefficient.

References

1. A.D. Bandrauk, in *Molecules in Laser Fields*, ed. by A.D. Bandrauk (Dekker, New York, 1994), Chap. 1, pp. 1–69
2. B. Felsager, *Geometry, Particles and Fields* (Springer, Berlin, 1998)
3. W.P. Schleich, *Quantum Optics in Phase Space* (Wiley-VCH, Berlin, 2001)
4. B.H. Bransden, C.J. Joachain, *Physics of Atoms and Molecules*, 2nd edn. (Pearson Education, Harlow, 2003)
5. M. Göppert-Mayer, Ann. Phys. (Leipzig) **9**, 273 (1931)
6. J.D. Jackson, *Klassische Elektrodynamik*, 2nd edn. (Walter de Gruyter, Berlin, 1983)
7. D. Bauer, D.B. Milôsevíc, W. Becker, Phys. Rev. A **75**, 23415 (2005)
8. F.H.M. Faisal, Phys. Rev. A **75**, 63412 (2007)
9. Y.C. Han, L.B. Madsen, Phys. Rev. A **81**, 63430 (2010)
10. M. Baghery, U. Saalmann, J.M. Rost, Phys. Rev. Lett. **118**, 143202 (2017)
11. L.B. Madsen, Phys. Rev. A **65**, 53417 (2002)

[8]Note that $2\cos(x/2)\sin(x/2) = \sin(x)$.

12. H.A. Kramers, *Collected Scientific Papers* (North Holland, Amsterdam, 1956)
13. W.C. Henneberger, Phys. Rev. Lett. **21**, 838 (1968)
14. F. Grossmann, P. Jung, T. Dittrich, P. Hänggi, Zeitschrift für Physik B **84**, 315 (1991)
15. J.C.A. Barata, W.F. Wreszinski, Phys. Rev. Lett. **84**, 2112 (2000)
16. S. Stenholm, in *Quantum Dynamics of Simple Systems*, ed. by G.L. Oppo, S.M. Barnett, E. Riis, M. Wilkinson (IOP, Bristol, 1996), p. 267
17. N. Rosen, C. Zener, Phys. Rev. **40**, 502 (1932)
18. L.D. Landau, Phys. Z. Soviet Union **2**, 46 (1932)
19. C. Zener, Proc. Roy. Soc. (London) A **137**, 696 (1932)
20. E.C.G. Stückelberg, Hel. Phys. Acta **5**, 369 (1932)
21. E. Majorana, Nuovo Cimento **9**, 43 (1932)
22. D. Coker, in *Computer Simulation in Chemical Physics*, ed. by M.P. Allen, D.J. Tildesley (Kluwer, Amsterdam, 1993)
23. H.R. Reiss, Phys. Rev. A **63**, 13409 (2000)
24. S. Selstø, M. Førre, Phys. Rev. A **76**, 23427 (2007)
25. D.H. Kobe, E.C.T. Wen, J. Phys. A **15**, 787 (1982)
26. H. Haken, *Licht und Materie Bd. 1: Elemente der Quantenoptik* (BI Wissenschaftsverlag, Mannheim, 1989)
27. J.H. Shirley, Phys. Rev. **138**, B979 (1965)
28. B.M. Garraway, K.A. Suominen, Rep. Prog. Phys. **58**, 365 (1995)
29. M. Weissbluth, *Photon-Atom Interactions* (Academic Press, New York, 1989)
30. U. Weiss, *Quantum Dissipative Systems*, 3rd edn. (World Scientific, Singapore, 2008)
31. H.P. Breuer, F. Petruccione, *The Theory of Open Quantum Systems* (Oxford University Press, Oxford, 2002)
32. J. von Neuman, E. Wigner, Phys. Z. **30**, 467 (1929)
33. A.G. Fainshteyn, N.L. Manakov, L.P. Rapoport, J. Phys. B **11**, 2561 (1978)
34. T. Timberlake, L.E. Reichl, Phys. Rev. A **59**, 2886 (1999)
35. H. Goldstein, C. Poole, J. Safko, *Classical Mechanics*, 3rd edn. (Addison Wesley, San Francisco, 2002)
36. D.J. Tannor, *Introduction to Quantum Mechanics: A Time-Dependent Perspective* (University Science Books, Sausalito, 2007)
37. R.C. Hilborn, Am. J. Phys. **50**, 982 (1982)

Chapter 4
Atoms in Strong Laser Fields

In this chapter, modern applications of laser-matter interaction in the realm of atomic physics shall be reviewed. Our focus will be on the interaction of atoms with strong and/or short laser pulses. Due to their availability, a range of new and partly counter-intuitive phenomena can be observed. Some of these are:

- Above Threshold Ionization (ATI)
- Multi-Photon Ionization (MPI)
- Knee structure in double ionization of helium
- Localization of Rydberg atoms by Half-Cycle Pulses (HCP)
- High-order Harmonic Generation (HHG)

These phenomena can in general not be understood in the framework of perturbation theory. It turns out that the time-dependent wavepacket approach of Chap. 2 is the method of choice to describe and understand a lot of the new atomic physics in strong laser fields.

In the following, we will mostly deal with the dynamics of just a *single* electron (or a single active electron), initially bound in a Coulomb potential. The chapter will therefore begin with a short review of the unperturbed hydrogen atom, together with atomic units that will be used subsequently. After a brief introduction to the two-electron helium atom, different aspects of field-induced ionization, as well as of HHG will be reviewed in detail.

4.1 The Hydrogen Atom

At the beginning of this chapter some well-known results from basic quantum mechanics are gathered. These are the eigenvalues and eigenfunctions of the 3D hydrogen atom. Many numerical studies are performed in 1D and therefore also the less familiar eigensolutions in one dimension will be reproduced here.

F. Grossmann, *Theoretical Femtosecond Physics*, Graduate Texts in Physics,
https://doi.org/10.1007/978-3-319-74542-8_4

4.1.1 Hydrogen in 3 Dimensions

The simplest atomic problem is that of the hydrogen atom, where a single electron feels the bare Coulomb potential,

$$V_{\mathrm{C}}(r) \equiv -\frac{1}{4\pi\varepsilon_0}\frac{e^2}{r}, \tag{4.1}$$

due to the proton. This problem is solvable exactly analytically and therefore is at the heart of every basic quantum mechanics course. We assume familiarity of the reader with the solution procedure and only reproduce the final results here.

4.1.1.1 Eigenvalues and Eigenfunctions

Under the approximation of an infinite nuclear mass (i.e. for $M_{\mathrm{p}}/m_{\mathrm{e}} \approx \infty$[1]) the time-independent Schrödinger equation can be solved and the eigenvalues and eigenfunctions for $E < 0$ are given by [1]

$$E_n = -\frac{e^2}{8\pi\varepsilon_0 a_0}\frac{1}{n^2}, \qquad n = 1, 2, \ldots, \tag{4.2}$$

$$\psi_{nlm}(r, \vartheta, \phi) = R_{nl}(r) Y_{lm}(\vartheta, \phi), \tag{4.3}$$

where the Bohr radius

$$a_0 = 4\pi\varepsilon_0\frac{\hbar^2}{m_{\mathrm{e}} e^2} \tag{4.4}$$

has been introduced, spherical coordinates (r, ϑ, ϕ) have been used, and

$$Y_{lm}(\vartheta, \phi) = (-1)^m \sqrt{\frac{(2l+1)(l-m)!}{4\pi(l+m)!}} \mathrm{e}^{\mathrm{i}m\phi} P_{lm}(\cos\vartheta); \quad m \geq 0, \tag{4.5}$$

$$Y_{l-m}(\vartheta, \phi) = (-1)^m Y^*_{lm}(\vartheta, \phi) \tag{4.6}$$

are the spherical harmonics, defined with the help of the associated Legendre functions $P_{lm}(x)$ [2, 3]. In Fig. 4.1, the absolute values of the spherical harmonics up to $l = 2$ are depicted in a polar representation.

The radial function is of the form

$$R_{nl}(r) = N_{nl} \exp(-r'/n) \left(\frac{2r'}{n}\right)^l L^{2l+1}_{n-l-1}\left(\frac{2r'}{n}\right), \tag{4.7}$$

[1] Taking the finiteness of the proton mass into account would lead to the replacement of the electron mass by the reduced mass $\mu = m_{\mathrm{e}} M_{\mathrm{p}}/(m_{\mathrm{e}} + M_{\mathrm{p}})$ in the final solution.

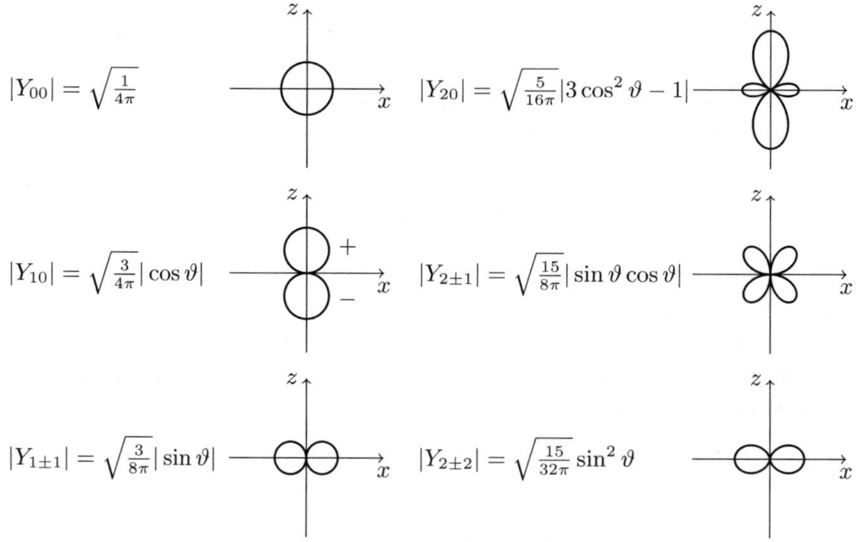

$|Y_{00}| = \sqrt{\frac{1}{4\pi}}$
$|Y_{20}| = \sqrt{\frac{5}{16\pi}}|3\cos^2\vartheta - 1|$
$|Y_{10}| = \sqrt{\frac{3}{4\pi}}|\cos\vartheta|$
+
−
$|Y_{2\pm1}| = \sqrt{\frac{15}{8\pi}}|\sin\vartheta\cos\vartheta|$
$|Y_{1\pm1}| = \sqrt{\frac{3}{8\pi}}|\sin\vartheta|$
$|Y_{2\pm2}| = \sqrt{\frac{15}{32\pi}}\sin^2\vartheta$
z
x

Fig. 4.1 Polar plot of the absolute value of some spherical harmonics as a function of the angle ϑ, measured from the z-axis. The figure is to be imagined rotationally symmetric around the z-axis. The sign of the spherical harmonic with $l = 1$, $m = 0$ is indicated

with the associated Laguerre polynomials[2]

$$L_k^s(x) = \sum_{\nu=0}^{k} \frac{(k+s)!}{(k-\nu)!(s+\nu)!} \frac{(-x)^\nu}{\nu!}, \tag{4.8}$$

the dimensionless radius $r' = r/a_0$, and a normalization factor

$$N_{nl} = \frac{2}{n^2}\sqrt{\frac{(n-l-1)!}{(n+l)!}} a_0^{-3/2}. \tag{4.9}$$

The radial quantum number $n_r = n - l - 1$ gives the number of nodes in the region $r > 0$ of the radial function (4.7), while the angular momentum quantum number fulfills $0 \leq l \leq n - 1$ and the magnetic quantum number m runs between $-l$ and $+l$. In spectroscopic notation, the levels are indicated by two symbols, the principle quantum number n and the angular quantum number l, which is encoded by a letter according to $0 \widehat{=} \text{s}$, $1 \widehat{=} \text{p}$, $2 \widehat{=} \text{d}$, $3 \widehat{=} \text{f}$, to be continued alphabetically.

For future reference we note that the Laguerre polynomials are related to the confluent hypergeometric functions according to [4]

[2]Note that we are using the definition of [4] which leads to a slightly different normalization factor as compared to the one in [1].

$$L_k^s(x) = \binom{k+s}{k} {}_1F_1(-k; s+1; x) . \tag{4.10}$$

For the nodeless ground state wavefunction of hydrogen, which is also referred to as the 1s wavefunction, we use that $L_0^1 = 1$ and $Y_{00} = 1/\sqrt{4\pi}$ in order to arrive at

$$\psi_{100}(r, \vartheta, \phi) = \frac{1}{\sqrt{\pi a_0^3}} \exp\{-r'\}, \tag{4.11}$$

which will be needed again in Chap. 5.

Using the Rayleigh-Ritz variational principle of Sect. 2.1.1 with a Gaussian trial function, an upper bound for the ground state energy can be determined using the variational method of Ritz.

4.1. *Use the trial function*

$$\psi_{\mathrm{var}}(\boldsymbol{r}) = \left(\frac{2\alpha}{\pi}\right)^{3/4} \exp\{-\alpha r^2\} \tag{4.12}$$

with the variational parameter α, to determine an upper bound for the ground state energy of the 3D *hydrogen atom (use atomic units).*
Hint: The radial part of the Laplace operator in spherical coordinates can be written as $\Delta_r = \frac{\partial^2}{\partial r^2} + \frac{2}{r}\frac{\partial}{\partial r}$.

4.1.1.2 Atomic Units

To simplify the notation considerably, from now on we will almost exclusively use atomic units (a.u.). They are defined by

$$\hbar = \mathrm{e} = \mathrm{m_e} = \mathrm{a_0} = 1\mathrm{a.u.} \, . \tag{4.13}$$

Using combinations of powers of these units one can construct atomic units for other observables, as can be seen by looking at Table 4.3 in Appendix 4.A, where SI and atomic units are given for some frequently occurring physical quantities.

From the definition (4.4), the Bohr radius in SI units follows to be $a_0 \approx 0.53 \cdot 10^{-10}$ m. Other SI base units as those of time and current are given by $1\mathrm{s} \approx 4.13 \cdot 10^{16}$ a.u. and $1\mathrm{A} \approx 151$ a.u.. The atomic unit of time therefore is of the order of 24 attoseconds. The relation of atomic units to expectation values of the 3D hydrogen problem is elucidated in Appendix 4.A.

4.1.2 The One-Dimensional Coulomb Problem

After having reviewed the hydrogen atom in full dimensionality, we now concentrate on its one-dimensional analog, which is very frequently used for numerical studies.

4.1.2.1 Exact 1D Coulomb Potential

In atomic units, the exact or "bare" 1D Coulomb potential is given by

$$V(x) = -\frac{1}{|x|} \tag{4.14}$$

and, similar to the 3D potential, has the problem of being singular at the origin. Fortunately, the eigenvalues of the singular potential can be determined analytically [5] and are given by

$$E_n = -\frac{1}{2n^2} \qquad n = 1, 2, 3 \ldots . \tag{4.15}$$

They are converging to the "continuum threshold" $E = 0$ and are equivalent to the eigenvalues of the 3D problem. In addition, however, in the 1D case there is also an eigenvalue $E_0 = -\infty$, with a δ-function type eigenfunction. If the initial state to be propagated is chosen such that it is zero at the origin then the unphysical state [6] is eliminated from the dynamics [7].

The eigenfunctions of the non-singular eigenvalues, which are doubly degenerate are given by [5]

$$\psi_n^\sigma = \sqrt{2/n^3}|x|(\text{sign}(x))^\sigma \exp\left\{-\frac{|x|}{n}\right\} {}_1F_1\left[1-n; 2; \frac{2|x|}{n}\right], \tag{4.16}$$

where ${}_1F_1$ is the confluent hypergeometric function and where the similarity to the radial functions in the 3D-case for $l = 0$ is obvious. For a given n there are two eigenfunctions, one with positive and one with negative parity ($\sigma = 0$: even, $\sigma = 1$: odd). This seems to be contradicting the theorem that there is no degeneracy in 1D quantum spectra. The theorem, however, can only be derived under the assumption that the potential has no singularities, which is not true for the bare Coulomb problem [5, 8].

Furthermore, a short remark on the 1D Coulomb potential restricted to the half space $x > 0$ shall be made. It is given by

$$V(x) = \begin{cases} \infty & x \le 0 \\ -\frac{1}{x} & x > 0 \end{cases} \tag{4.17}$$

and in suitable units describes the problem of a "surface state electron" [9]. The spectrum is again identical to the 3D case. Surface state electrons are hovering above a dielectric surface and are especially interesting in the context of quantum chaos if they are driven by microwave radiation. In [9] this is discussed in great detail.

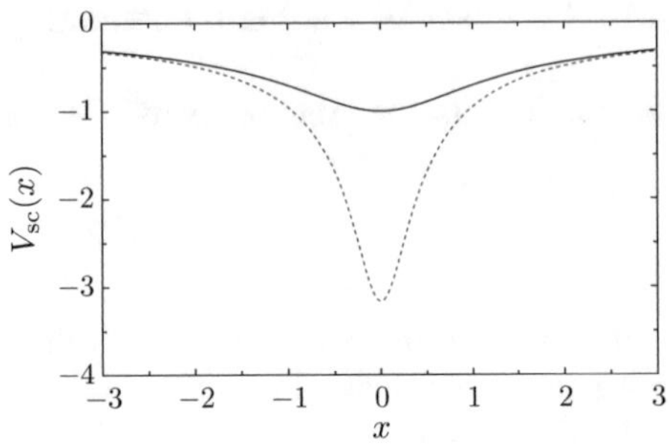

Fig. 4.2 Soft-core Coulomb potential with $a = 1$ (*solid line*) versus $a = 0.1$ (*dashed line*) as a function of x in atomic units

4.1.2.2 Soft-Core Coulomb Potential

For ease of computation, non-singular approximations to the bare (hard-core) Coulomb potential are used frequently. These are created by adding a constant term under the square root in the denominator, leading to

$$V_{sc}(x) \equiv -\frac{1}{\sqrt{x^2 + a}} \tag{4.18}$$

for the "soft-core" Coulomb potential in atomic units, thereby introducing a length scale into the problem already in the potential [10].[3] Frequently, the Bohr radius is used for a in dimensioned units and therefore $a = 1$ in atomic units. Another potential can be gained by choosing $a = 2\,\text{a.u.}$, which leads to a ground state energy of $-0.5\,\text{a.u.}$, equivalent to the bare 3D case. Choosing very small values of the parameter a, the bare Coulomb potential is approached, see Fig. 4.2.

In Sect. 4.3.4, numerical results for the laser-driven dynamics in the soft-core as well as for the bare Coulomb potential will be compared.

4.2 The Helium Atom

In the following, we gather some basic information about the simplest neutral "many electron" atom, the helium atom, consisting of an α particle and two electrons.

4.2.1 Hamiltonian and TISE

For the helium atom with nuclear charge $Z = 2$, we will again consider the approximation of an infinitely heavy nucleus, such that in atomic units the Hamiltonian and the TISE for the undriven 3D problem becomes

---

[3]The length scale a_0 in the bare 3D Coulomb problem appears only in the solution.

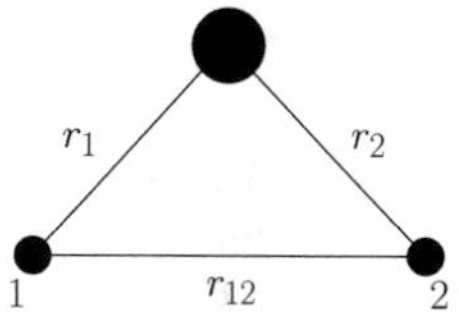

Fig. 4.3 Helium atom consisting of an α particle with $Z = 2$ and two electrons with the distances r_1 and r_2 from the nucleus, and the interelectronic distance r_{12}

$$\hat{H}(\boldsymbol{r}_1, \boldsymbol{r}_2) = \left\{ -\frac{1}{2}\Delta_{\boldsymbol{r}_1} - \frac{1}{2}\Delta_{\boldsymbol{r}_2} - \frac{2}{r_1} - \frac{2}{r_2} + \frac{1}{r_{12}} \right\}, \tag{4.19}$$

$$\hat{H}\psi(\boldsymbol{r}_1, \boldsymbol{r}_2) = E\psi(\boldsymbol{r}_1, \boldsymbol{r}_2), \tag{4.20}$$

where the definitions of the distances displayed in Fig. 4.3 have been used.

Also in the helium case, soft-core potentials are frequently used. In a 1D model with electron coordinates x and y the soft-core Hamiltonian reads

$$\hat{H}_{\rm sc}(x, y) = -\frac{1}{2}\left(\frac{\partial^2}{\partial x^2} + \frac{\partial^2}{\partial y^2}\right) - \frac{2}{\sqrt{x^2 + a}} - \frac{2}{\sqrt{y^2 + a}} + \frac{1}{\sqrt{(x - y)^2 + a}}. \tag{4.21}$$

If the soft-core parameter is chosen as $a = 0.55$, this leads to a ground state energy of -2.897 (in atomic units), which is close to the true ground state energy, -2.902, of 3D helium [11].

In the bare Coulomb case but in 1D, one distinguishes between the eZe and the Zee configuration, with the corresponding Hamiltonians

$$\hat{H}_{\pm}(x, y) = -\frac{1}{2}\left(\frac{\partial^2}{\partial x^2} + \frac{\partial^2}{\partial y^2}\right) - \frac{2}{|x|} - \frac{2}{|y|} + \frac{1}{\sqrt{(x \pm y)^2}}. \tag{4.22}$$

For the eZe configuration, i.e., when the two electrons are on opposite sides of the nucleus, the plus sign holds and in the Zee case, the minus sign holds.

4.2.2 Spin and the Pauli Principle

Although the Hamiltonian (4.19) is not spin-dependent, now the spin of the electrons plays a decisive role. The total wavefunction, which can be written as a direct product

$$\psi(\boldsymbol{r}_1, \boldsymbol{r}_2)\chi(1, 2) \tag{4.23}$$

of a spatial times a spin part, denoted by $\chi(1, 2)$, must be antisymmetric with respect to the interchange of the particle index, due to the Pauli principle.

To achieve this required antisymmetry, the spin part as well as the spatial part must have a definite symmetry with respect to particle exchange. We first briefly consider the spin states of a single electron. Denoting the single particle spin operator for electron one (in atomic units) by

$$\hat{\mathbf{S}}_1 = \frac{1}{2}(\hat{\sigma}_{1,x}, \hat{\sigma}_{1,y}, \hat{\sigma}_{1,z}), \tag{4.24}$$

with the Pauli spin matrices from (3.82) in Appendix 3.B, the eigenvalue equation for the total spin as well as for its z-component read

$$\hat{\mathbf{S}}_1^2|\pm\rangle = S_1(S_1 + 1)|\pm\rangle, \tag{4.25}$$

$$\hat{S}_{1,z}|\pm\rangle = M_{S_1}|\pm\rangle, \tag{4.26}$$

where $S_1 = 1/2$ and M_{S_1} can be either 1/2 or $-1/2$. In the following, we denote the single spin up state $|+\rangle$ with $M_{S_1} = 1/2$ by α and the down state $|-\rangle$ with $M_{S_1} = -1/2$ by β.

Spin states of the composite system that fulfill the requirement of definite symmetry can be classified with respect to their eigenvalues S, which can be either zero or one, and the corresponding M_S, which runs from $-S$ to S, of the total spin operator $\hat{\mathbf{S}} = \hat{\mathbf{S}}_1 + \hat{\mathbf{S}}_2$ of the two-electron system and its z-component

$$\hat{\mathbf{S}}^2\chi = S(S + 1)\chi, \tag{4.27}$$

$$\hat{S}_z\chi = M_S\chi. \tag{4.28}$$

The four possible spin states in the two-electron case are gathered in Table 4.1.

There is only one antisymmetric combination with $S = 0$, denoted as singlet state, whereas one can form a total of three symmetric so-called triplet states with $S = 1$. Now the full wavefunction has to be antisymmetric with respect to particle exchange

Table 4.1 There are one antisymmetric ($S = 0$, singlet) and three symmetric ($S = 1$, triplet) combinations of two single-particle spin states

Two electron spin state χ	S	M_S
$\frac{1}{\sqrt{2}}[\alpha(1)\beta(2) - \beta(1)\alpha(2)]$	0	0
$\alpha(1)\alpha(2)$	1	1
$\frac{1}{\sqrt{2}}[\alpha(1)\beta(2) + \beta(1)\alpha(2)]$	1	0
$\beta(1)\beta(2)$	1	-1

and thus the spin singlet has to be multiplied by a symmetric spatial wavefunction, whereas the spin triplets have to be multiplied by an antisymmetric spatial part.

4.2.3 Semiclassical Determination of Helium Spectra

Provided that spin-orbit interactions can be neglected, radiative transitions between singlet and triplet states are forbidden in the electric dipole approximation. Because of the absence of intercombination lines in the spectroscopy of helium, historically, the atomic "species" with total spin $S = 0$ is called parahelium and the one with $S = 1$ is refered to as orthohelium.

The determination of the spectrum of (both species of) helium is not such an easy exercise as in the hydrogen case. In the heyday of quantum mechanics a lot of unsuccessful efforts have been made to determine the spectrum on the basis of Bohr-Sommerfeld quantization [3]. The main reason for the failure is the fact that a two-electron system is classically unstable, because for almost all initial conditions one of the electrons can approach the nuclueas arbitrarily closely and thereby transfer its energy to the other one, eventually leading to ionization.

Only in 1991 a satisfactory semiclassical determination of the energy levels for collinear helium (electrons are restricted to lie on a straight line on different sides of

Table 4.2 Comparison of semiclassical (Herman-Kluk) and full quantum singlet and triplet state binding energies of the collinear helium (eZe) atom in atomic units, adapted from [13]

State		Singlet		Triplet	
N	n	HK	QM	HK	QM
1	1	3.2102	3.2459	...	...
	2	2.2225	2.2028	2.2393	2.2254
2	2	0.8216	0.8224	...	...
	3	0.6045	0.6098	0.6150	0.6184
3	3	0.3621	0.3662	...	...
	4	0.2883	0.2890	0.2906	0.2935
	5	0.2615	0.2603	0.2616	0.2618
4	4	0.2050	0.2061		
	5	0.1690	0.1695		
	6	0.1528	0.1532		
	7	0.1437	0.1441		
5	5	0.1314	0.1320		
	6	0.1112	0.1117		
	7	0.1006	0.1016		
6	6	0.09127	0.09204		
	7	0.07922	0.07928		

the nucleus, see the Hamiltonian in (4.22) in the eZe case) was achieved by using semiclassical periodic orbit quantization techniques (cycle expansion) based on the dynamical zeta function [12]. Compared to a full numerical solution of the TISE, the results have an accuracy of a few percent of the average level spacing.

The same case has lateron also been treated by using the Herman-Kluk propagator of Sect. 2.3.4 for both the singlet as well as the triplet states with comparable accuracy [13], see Table 4.2. There, the binding energies $-\mathrm{Re}(E_{Nn})$ as a function of the approximate quantum numbers $N, n = N, N+1, \ldots$ for the inner and outer electron are shown for both bound states, i.e. $N = 1$, as well as resonance states $N > 1$. The semiclassical energies have been extracted from autocorrelation functions as indicated in Sect. 2.1.3, whereas the full quantum results have been obtained by using the complex rotation method.

4.3 Field-Induced Ionization

Apart from the excitation of higher lying states, the most prominent effect that an external field can have on the dynamics of an electron initially bound in an atom is ionization. We will deal with relatively strong fields in the following and will thereby focus on ionization phenomena in several different regimes, ranging from the quasi-stationary case to the case of almost δ-function like perturbation by half-cycle pulse fields.

4.3.1 *Tunneling Ionization*

In the case of long wavelengths and relatively strong fields ($I \approx 10^{14}\,\mathrm{W\,cm^{-2}}$), the ionization of an atom induced by a laser can be treated under the assumption of a strong quasi-static field, leading to a relatively low potential barrier. A snapshot of the distorted Coulomb potential

$$V(\boldsymbol{r}, t) = V_{\mathrm{C}}(r) + \boldsymbol{r} \cdot \boldsymbol{\mathcal{E}}_0 \sin(\omega t) \tag{4.29}$$

at $t = -\pi/(2\omega)$ with the electric field polarized in the z-direction is shown in Fig. 4.4. Due to the low barrier, an electron, initially in the ground state, can tunnel out of the region of attraction with a rate given below. For stronger fields, a direct "over the barrier" ionization would even be possible. In early experiments, the experimental realization of tunneling ionization was done with the help of mid-infrared lasers like, e.g., the CO_2-Laser with a rather long wavelength of 10.6 μm [14], leading to a large ponderomotive energy and thus a small Keldysh parameter, to be discussed in Sect. 4.3.3. Nowadays, laser intensities in the near infrared are high enough to generate small Keldysh parameters, but the quasi-static nature of the tunneling barrier becomes more and more questionable in these cases.

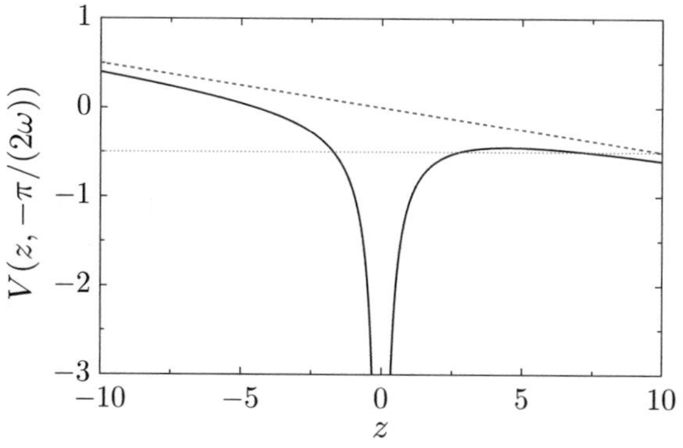

Fig. 4.4 Statically distorted Coulomb potential (*solid line*) along the z-axis and potential induced by an external field of amplitude $\mathcal{E}_0 = 0.05$ a.u., polarized in the z-direction (*dashed line*). The ground state energy of the unperturbed hydrogen atom in atomic units is indicated by the *horizontal dotted line*

For the 3D hydrogen atom, the tunneling ionization rate in the case of $\mathcal{E}_0 \ll 1$ a.u. is given by (see Exercise 1 on p. 283 in [15])

$$\Gamma_{\rm tu} = \frac{4}{\mathcal{E}_0} \exp\left\{-\frac{2}{3\mathcal{E}_0}\right\}. \tag{4.30}$$

A generalization of this formula for arbitrary atoms has been given by Ammosov, Delone and Krainov (ADK) [16]. An easy derivation of an analogous form of the exponential dependence on inverse field strength is done in Exercise 4.2, where we consider tunneling out of a distorted square potential well of zero range, which for better visibility can also be thought of as a finite range potential.

4.2. *Calculate the Gamov factor*

$$\Gamma_{\rm tu} \sim \exp\left[-2^{3/2}\int_0^{r_f} {\rm d}r \sqrt{V(r) - E_{\rm g}}\right]$$

for the rate of tunneling at a negative energy $E_{\rm g}$ out of a bound state of a tilted square well in 1D *(with $V(r{=}0){=}0$) in the presence of a quasistatic external field $\mathcal{E}_0$ (see Fig. 4.5) and compare the result with the exponential factor in the case of* 3D *hydrogen.*

4.3.2 *Multi-Photon Ionization*

Up to now we have considered the quasi-static case of a Coulomb potential, distorted by a constant electric field. In the focus of the presentation below, however, is the

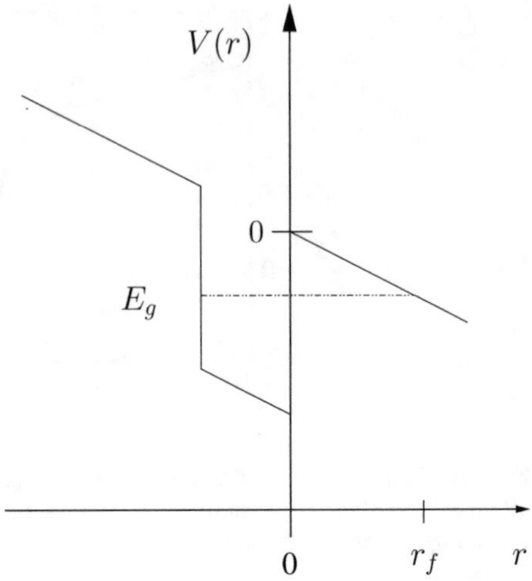

Fig. 4.5 Tunneling out of a square well, supporting a bound state at E_g

question which effect a *time-dependent* external field exerts on an initially bound electron. If the energy of the photon is not enough to overcome the ionization threshold and if tunnel ionization is extremely unlikely, then the transition to the continuum may happen via the absorption of several photons. In this so-called multi-photon ionization (MPI) two principally different scenarios are usually distinguished:

- Resonant MPI (REMPI): Here the spectrum of the unperturbed system has spacings close to the photon frequency, a review of that topic can be found in [17]
- Nonresonant MPI: The frequency of the photons is not in resonance with energy differences of the system

A simple model that allows to understand the basic features of MPI is that of an electron in a Gauß potential given by

$$V_G(x) \equiv -V_0 \exp\{-\sigma x^2\} . \tag{4.31}$$

This choice allows the investigation of resonance phenomena, that occur whenever the average level spacing is equal to the external frequency. Furthermore, the potential has the nice feature that, in contrast to the Coulomb case, it allows only for a finite number of bound states and therefore, the ionization probability can be calculated with great ease, see below and [18]. We will review numerical results for the ionization in the Coulomb case in Sect. 4.3.4 for 1D, as well as in Sect. 5.2.1, where 3D H-atom results are compared to the ones for the hydrogen molecular ion H_2^+.

For the parameters $V_0 = 0.6$ a.u. and $\sigma = 0.025$ a.u., the potential together with its bound state energies at $E_0 = -0.518$, $E_1 = -0.37$, $E_2 = -0.23$, $E_3 = -0.12$,

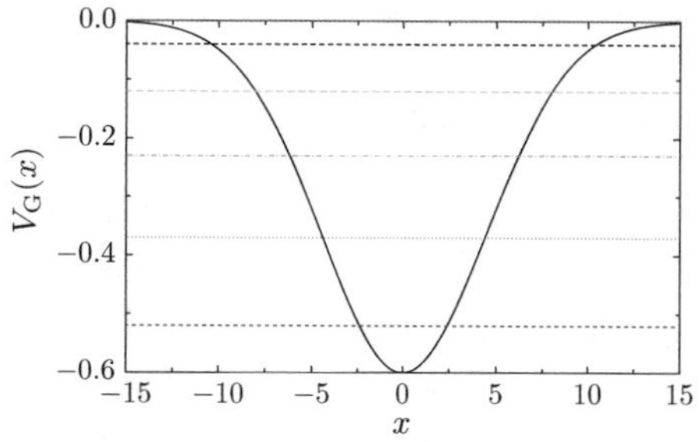

Fig. 4.6 Gauß potential (*solid line*) with the six eigenvalues (*horizontal lines*) for the parameters given in the text

$E_4 = -0.04$, $E_5 = 0$ (all in a.u.) is depicted in Fig. 4.6.[4] The average distance between nearest neighbors is

$$\Delta E = \frac{1}{4}\sum_{n=0}^{3}(E_{n+1} - E_n) \approx 0.12 \text{ a.u.}. \tag{4.33}$$

In order to study ionization, the initial state is taken as the ground state of the system, which is to a good approximation given by a Gaussian wavepacket

$$\Psi_\gamma(x, 0) = \left(\frac{\gamma}{\pi}\right)^{1/4} \exp\left\{-\frac{\gamma}{2}x^2\right\} \tag{4.34}$$

with $\gamma = 0.154$ a.u.. The ionization probability under laser irradiation as a function of time in the case of the Gauß potential can be calculated most easily from the finite(!) sum

$$P_\mathrm{I}(t) = 1 - \sum_{n=0}^{5} |\langle\Psi_\gamma(t)|\psi_n\rangle|^2, \tag{4.35}$$

because it has a finite number of bound states (six). This is the probability *not* to be in a bound state (calculated by solving the TISE numerically) any more. This probability typically increases as a function of time, if the electron can climb up the ladder of energy eigenstates and leave the bound states. In order to calculate it, the solution of the time-dependent Schrödinger equation is needed, and therefore we need to know the explicit form of the laser field. The one that has been used in the investigations of van de Sand [19] is

$$V_\mathrm{L} = \mathcal{E}_0 x \sin^2\left(\frac{\omega t}{16}\right)\sin(\omega t)\ , \tag{4.36}$$

---

[4] Note that each symmetric well potential in 1D has at least one bound state, no matter how shallow it is [15]. Furthermore, by using Wentzel Kramers Brillouin (WKB) quantization

$$W(E) = \oint \mathrm{d}x\, p = 2\pi(n + 1/2) \tag{4.32}$$

with the short action $W(E) = S(t) + Et$, the spectrum is very well reproduced [19].

where the first sine factor is the envelope of the pulse, extending over 8 cycles, $T = 2\pi/\omega$, of the field (thereafter the pulse is assumed to be zero). The field strength is $\mathcal{E}_0 = 0.032$ a.u.. The frequency of the laser has been varied in order to study resonance phenomena. The numerical results have been gained by solving the full quantum problem, using the split-operator FFT method, as well as by using the semiclassical Herman-Kluk propagator (see Sect. 2.3).

(a) Nonresonant case: $\omega = 0.09$ a.u. :
In Fig. 4.7, the ionization probability as well as the occupation probabilities of the states 0–2 are depicted as a function of time. A nonresonant Rabi oscillation (with amplitude much smaller than unity) between the ground and the first excited state can be observed. The fast oscillations (with 2ω) in the occupation probabilities are due to the counter-rotating term, which would not be present if the RWA had been invoked.[5] Furthermore, some probability leaks into the continuum and at the end of the pulse the ionization probability is about 15%. The semiclassical results in panel (b) very nicely reproduce the quantum ones.
(b) Resonance case: $\omega = 0.124$ a.u. :
For the results shown in Fig. 4.8, the same field strength as above has been used. The frequency, however, is now close to the average level spacing. The final ionization then is around 80% and the wavepacket methodology is very well suited to describe this nonperturbative effect of the laser field. From the figure it can be seen that the occupation of the ground state decreases in a quasi monotonous way (and does not rise at the end of the pulse as in case (a)). Probability is being transferred successively into higher states and finally into the continuum $E > 0$. This process is called REMPI (**R**esonantly **E**nhanced **M**ulti **P**hoton **I**onisation). As in case (a), the semiclassical results reproduce the main features of the resonant dynamics very well.

The ionization probability after the laser excitation as a function of external frequency is given in Fig. 4.9. Maxima in $P_{\rm I}(8\,T)$ occur at $\omega = 0.12$ a.u. and at $\omega = 0.14, 0.107$ a.u.. These are again very well reproduced semiclassically and are related to transitions in the spectrum as follows

$$n = 2 \to n = 3 : \qquad \omega = 0.107\,\text{a.u.} \tag{4.37}$$

and

$$n = 0 \to n = 1 \quad \text{and} \quad n = 1 \to n = 2 : \qquad \omega = 0.14\,\text{a.u..} \tag{4.38}$$

In principal, even higher ionization probabilities could be achieved if the frequency of the external field would be allowed to change in the course of time. So-called down-chirps have been investigated in molecular dissociation and will be dealt with in detail in Chap. 5.

[5]A direct comparison with the RWA results of Chap. 3 would ask for an extension of the two-level results to pulsed driving and in addition the dipole matrix element would have to be calculated.

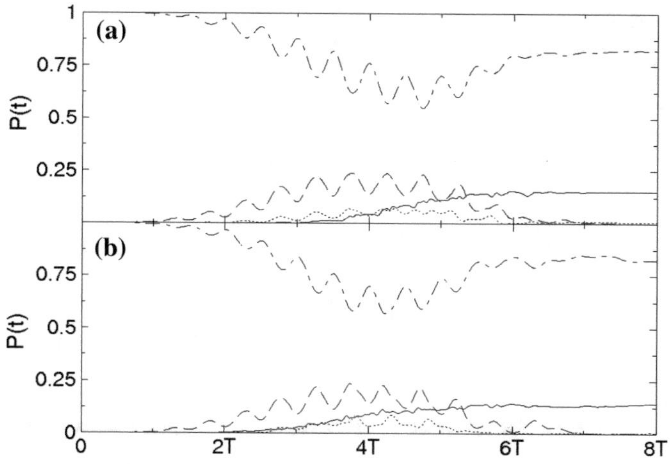

Fig. 4.7 Ionization probability (*solid line*) and occupation probabilities ($n = 0$ *long dash-short dash*, $n = 1$ *dashed*, $n = 2$ *dotted*) as a function of time in the nonresonant case $\omega = 0.09$ a.u. **a** full quantum results, **b** semiclassical results; adapted from [19]

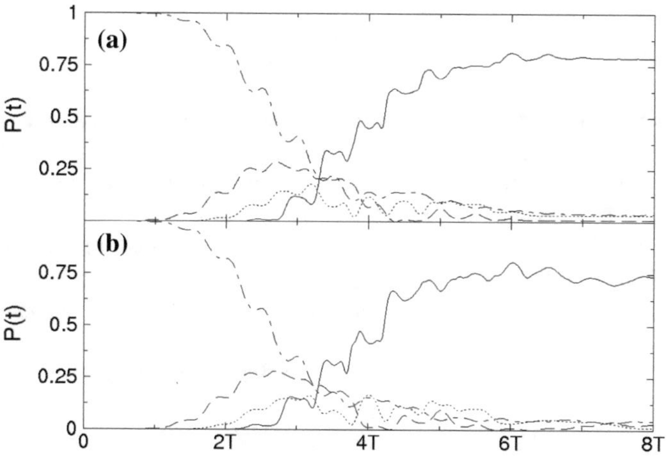

Fig. 4.8 Ionization probability (*solid line*) and occupation probabilities ($n = 0$ *long dash-short dash*, $n = 1$ *dashed*, $n = 2$ *dotted*) as a function of time in the resonant case $\omega = 0.124$ a.u. **a** full quantum results, **b** semiclassical results; adapted from [19]

To finish this section, a short remark on classical trajectory calculations to reproduce the quantum results is in order. These calculations are based on the solution of Hamilton's equations and do *not* take phase information into account. They have been performed by van de Sand, and for the present problem lead to considerably worse results than semiclassics [19].

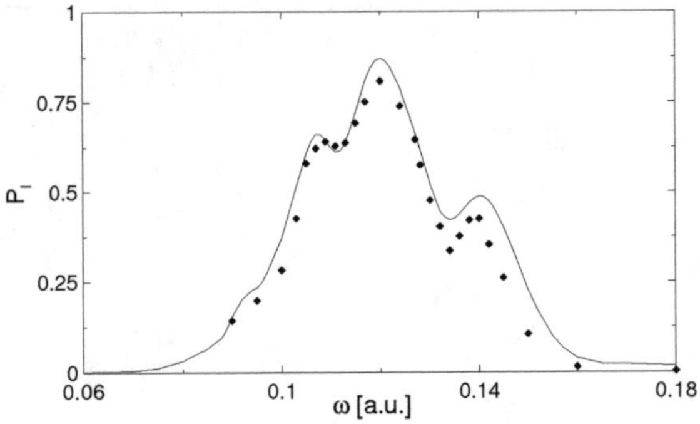

Fig. 4.9 Ionization probability P_I as a function of the laser frequency, semiclassical results are depicted by the *filled diamonds*; from [19]

4.3.3 Keldysh Parameter and Strong-Field Approximation

Of central importance for the understanding of the ionization mechanism is the Keldysh parameter, defined by

$$\gamma_K = \sqrt{\frac{I_p}{2U_p}}, \tag{4.39}$$

where I_p is the ionization potential of an atom and the ponderomotive energy U_p is given by (3.28). γ_K compares the energy of the external field with a typical atomic energy. In the case of the hydrogen atom and in SI units $I_p \approx 13.6\,\mathrm{eV}$ and at a wavelength of 800 nm and intensity of $10^{14}\,\mathrm{W\,cm^{-2}}$, $\gamma_K \approx 1$.

For $\gamma_K \gg 1$ the external field is considered to be a small perturbation and multi-photon ionization of Sect. 4.3.2 is the main ionization mechanism, whereas for $\gamma_K < 1$ the atomic potential plays the role of a perturbation and tunneling ionization of Sect. 4.3.1 starts to become dominant.

With the two different breakups of a system field Hamiltonian

$$\hat{H}(t) = \hat{H}_i + V_L(t) = \hat{T}_k + V_C + V_L(t) = \hat{H}_f(t) + V_C \tag{4.40}$$

in mind, one defines the T-matrix element for the transition probability amplitude from an initial to final state either as

$$f(t) = \langle \Phi_f(t) | \Psi_i^+(t) \rangle, \tag{4.41}$$

with $\Psi_i^+(-\infty) = \Phi_i$ being an eigenstate of $\hat{H}_i$ or as

$$f(t) = \langle \Psi_f^-(t) | \Phi_i(t) \rangle, \tag{4.42}$$

with $\Psi_f^-(\infty) = \Phi_f$ being the Volkov state in length gauge of Sect. 3.1.4 in atomic units.

From the time-dependent Schrödinger equation it can be shown by solving Exercise 4.3 that, up to an irrelevant phase factor,

$$f(t) = \int_{-\infty}^{t} \mathrm{d}t' \langle \Phi_f(t') | V_\mathrm{C} | \Psi_i^+(t') \rangle, \tag{4.43}$$

$$f(t) = \int_{t}^{\infty} \mathrm{d}t' \langle \Psi_f^-(t') | V_\mathrm{L}(t') | \Phi_i(t') \rangle, \tag{4.44}$$

i.e., the transition amplitudes can also be put in an integral form [20].

4.3. *By using the TDSE, show that the transition amplitudes can be written in integral form.*
Hint: use the derivative of a parameter integral $\frac{\mathrm{d}}{\mathrm{d}t}\int_{a(t)}^{b(t)} \mathrm{d}t' f(t') = \frac{\mathrm{d}b(t)}{\mathrm{d}t} f(b(t)) - \frac{\mathrm{d}a(t)}{\mathrm{d}t} f(a(t))$.

Using the above integral forms, one can apply the first-order Born approximation [21], which amounts to the replacements of the solutions of the full time-dependent Schrödinger equation by the one of the corresponding arrangement channel according to (4.40). In the present context this is the so-called strong-field approximation (SFA) or Keldysh-Faisal-Reiss (KFR) approximation [22] and leads to the two equivalent forms

$$f^\mathrm{SFA} = \int_{-\infty}^{\infty} \mathrm{d}t' \langle \Phi_f(t') | V_\mathrm{C} | \Phi_i(t') \rangle, \tag{4.45}$$

$$f^\mathrm{SFA} = \int_{-\infty}^{\infty} \mathrm{d}t' \langle \Phi_f(t') | V_\mathrm{L}(t') | \Phi_i(t') \rangle. \tag{4.46}$$

Therefore in the SFA, one can use either potential to evaluate the transition amplitude. The first form can be viewed as scattering of the electron from the ionic core and subsequent overlap with a Volkov state. In both cases, multiphoton as well as tunneling ionization can be described, depending on the magnitude of γ_K. The second form is interpreted as absorption of a single photon and an overlap with the same Volkov state.

Based on the exact expression (4.41), an investigation of the transition from tunneling ionization to the multiphoton type by varying γ_K has been performed [20], which has corroborated the prediction by Keldysh [23]. The starting point is the time-periodic Ansatz (Fourier expansion, see also (2.173))

$$|\Psi(t)\rangle = \exp[\mathrm{i}I_p t] \sum_j |\psi_j\rangle \mathrm{e}^{-\mathrm{i}j\omega t} \tag{4.47}$$

for the solution of the TDSE. In the case of a circularly polarized external field

$$\boldsymbol{\mathcal{E}}(t) = \mathcal{E}_0 \left[\boldsymbol{e}_x \cos(\omega t) + \boldsymbol{e}_y \sin(\omega t) \right] \tag{4.48}$$

the solution of the inhomogenous equations

$$\left[\hat{H}_0 + (I_p - n\omega)\right] |\psi_n\rangle = \frac{\mathcal{E}_0}{2}(x + \mathrm{i}y)|\psi_{n-1}\rangle \tag{4.49}$$

for off-shell Coulomb wavefunctions (energy is not conserved in the laser driven case) is determined as decribed in Appendix A of [20]. The derivation of (4.49) from the TDSE is very similar to the procedure in Sect. 2.3.1 and a term proportional to $|\psi_{n+1}\rangle$ has been discarded in the spirit of perturbation theory.

Tunneling will then be monitored at the tunnel exit $\boldsymbol{r}_f$, which is calculated from the static potential at maximal field strength (see Fig. 4.5), and by looking at

$$\Psi^+(\boldsymbol{r}_f, t_f) = \int \mathrm{d}^3 r_i K(\boldsymbol{r}_f, t_f; \boldsymbol{r}_i, t_i)\Psi(\boldsymbol{r}_i, t_i). \tag{4.50}$$

Inspired by the Fourier series above, $\Psi_n^+(\boldsymbol{r}_f)$ is defined as a virtual state that has absorbed n photons. For the quantity

$$P_n^{\mathrm{rel}} = \frac{|\Psi_n^+(\boldsymbol{r}_f)|^2}{|\Psi_{\max}^+(\boldsymbol{r}_f)|^2}, \tag{4.51}$$

which is the relative ionization probability (normalized to the maximal value) after absorption of n photons, in the case of the 3D hydrogen atom, the behaviour displayed in Fig. 4.10 is found [20]. For small Keldysh parameter, the highest probabilities are found for ionization without absorption, i.e. ionization from the ground state via tunneling. For larger values of γ_{K} multiphoton ionization becomes more probable.

In the case of higher frequencies, the transition to predominant ground state tunneling proceeds already at larger values of the Keldysh parameter, as can be seen in the right panels of Fig. 4.10.

4.3.4 ATI in the Coulomb Potential

If the photon frequency is of the order of an atomic unit, then already the absorption of a single photon is enough to ionize an atom (the ground state energy of hydrogen is −0.5 a.u.). Intriguingly, even in that case an atom can absorb more than one single photon, however, as we will see in the following. This breed of MPI is called above threshold ionization (ATI). Experimentally, the effect has first been shown to exist in the end of 1970s by using laser driven Xe atoms [24]. Theoretical studies showing the effect have been done for several models of the hydrogen atom by solving the time-dependent Schrödinger equation.

In the following a study using both 1D models, the soft-core and also the bare Coulomb potential from Sect. 4.1.2, will be reviewed. A lot of work in this field has also been done by using the strong-field approximation [25].

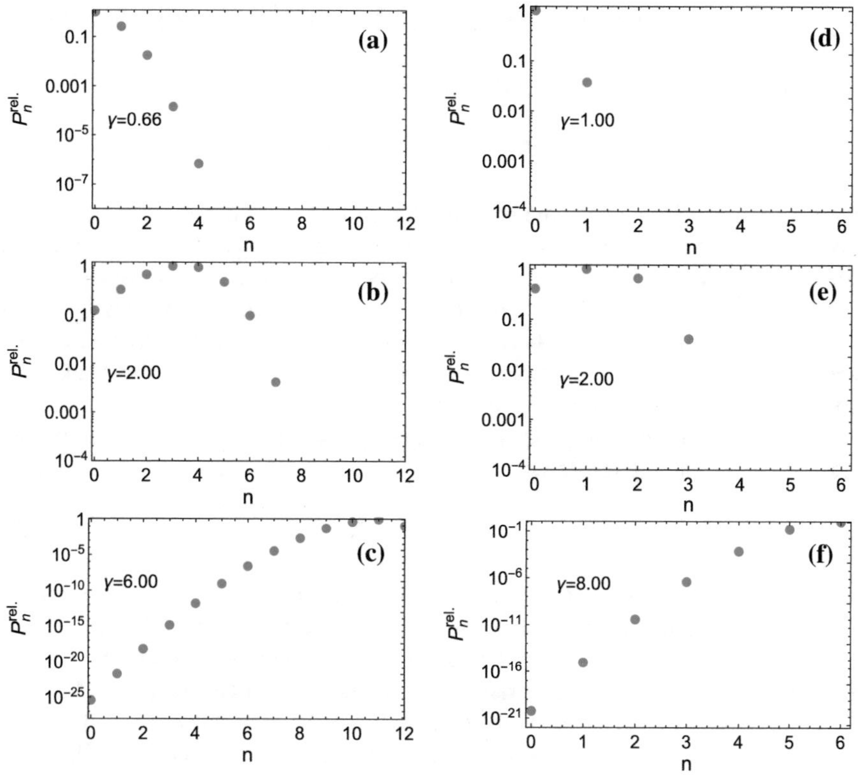

Fig. 4.10 Relative ionization probabilities of laser (circularly polarized) driven hydrogen as a function of the number of absorbed photons, n, for $\omega = 0.025$ (**a**–**c**) and for $\omega = 0.05$ (**d**–**f**), for different values of Keldysh parameter (here denoted by γ) [20]

For the numerical study of ATI an external field of the form

$$\mathcal{E}(t) = \mathcal{E}_0 f(t) \sin(\omega t), \tag{4.52}$$

$$f(t) = \begin{cases} \sin^2[\pi t/(2\tau_1)] & 0 < t < \tau_1 \\ 1 & \tau_1 \le t \le \tau_2 - \tau_1 \\ \cos^2[\pi(t + \tau_1 - \tau_2)/(2\tau_1)] & \tau_2 - \tau_1 < t < \tau_2 \end{cases} \tag{4.53}$$

with the parameters

$$\mathcal{E}_0 = 1\,\text{a.u.}, \tag{4.54}$$

$$\omega = 1\,\text{a.u.} \tag{4.55}$$

has been used in [7]. The ponderomotive energy and the Keldysh parameter then have the values $U_p = 0.25\,\text{a.u.}$, and $\gamma_K = 1$, respectively. The total pulse duration is 25 optical cycles ($\tau_2 = 50\pi/\omega$) and the field is switched on over 5 optical cycles

($\tau_1 = 10\pi/\omega$), is constant for 15 optical cycles, and is switched off in another 5 optical cycles.

Numerical results from [7] for the initial condition

$$\Psi(x,0) = \psi_1^{\sigma=1}(x) = \sqrt{2}x\mathrm{e}^{-x} \tag{4.56}$$

are depicted in Fig. 4.11 for the bare and the soft-core (with $a = 1\,\mathrm{a.u.}$) Coulomb potential. Especially in the case of the bare Coulomb potential a distinct splitting of the wavepacket into different subpackets can be observed. These subpackets correspond to parts of the wavefunction that have absorbed a different number of quanta of radiation energy. Although the field is treated classically, the excitation of the system shows clear peaks around multiple integers of the photon energy, which leads us to speak of the absorption of quanta of energy.

There is a certain probability that an electron has asymptotically the velocity $\Delta x/\Delta t$ corresponding to a kinetic energy

$$E_{\mathrm{kin}} = n\hbar\omega - I_{\mathrm{p}} \tag{4.57}$$

equal to the energy of the ATI peak.[6] For the present field parameters and for the hard-core potential, the first peak in the spectrum is at an energy of 1/2 a.u.. This is the Einstein peak, because it is the peak well known from the photo-electric effect, for the explanation of which Einstein was awarded the 1921 Nobel Prize in Physics. Further ATI peaks well visible in the bare Coulomb case are separated by 1 a.u..

Comparing the results for the bare and the soft-core Coulomb case in Fig. 4.11, we first notice that the Einstein peak is shifted towards lower energies in the second case. This is due to the fact that $a = 1\,\mathrm{a.u.}$ has been chosen for the soft-core parameter here. For $a = 2\,\mathrm{a.u.}$ the peak would be at the same position as in the bare Coulomb case. More importantly, however, the soft-core potential leads to a dramatic underestimation of the first as well as all higher ATI peaks!

A theoretical understanding of ATI can be gained by using the Herman-Kluk propagator, as has been shown by van de Sand [19]. It turns out that interfering trajectories are responsible for the formation of the ATI peaks, which cannot be described by using pure classical mechanics.

Furthermore, for very high intensities, the low energy ATI peaks are usually shifted and suppressed due to the AC Stark shift of the atomic Rydberg levels and the continuum in the presence of the laser field. If the AC Stark shift of the groundstate can be neglected, it turns out that the ionization potential in the presence of a laser with intensity I, leading to a ponderomotive energy U_{p} (see (4.58)), is given by [27]

$$I_{\mathrm{p}}(I) \simeq I_{\mathrm{p}} + U_{\mathrm{p}} \tag{4.58}$$

[6]The increase of I_{p} by U_{p} (see (4.58)) does not play a role here [26].

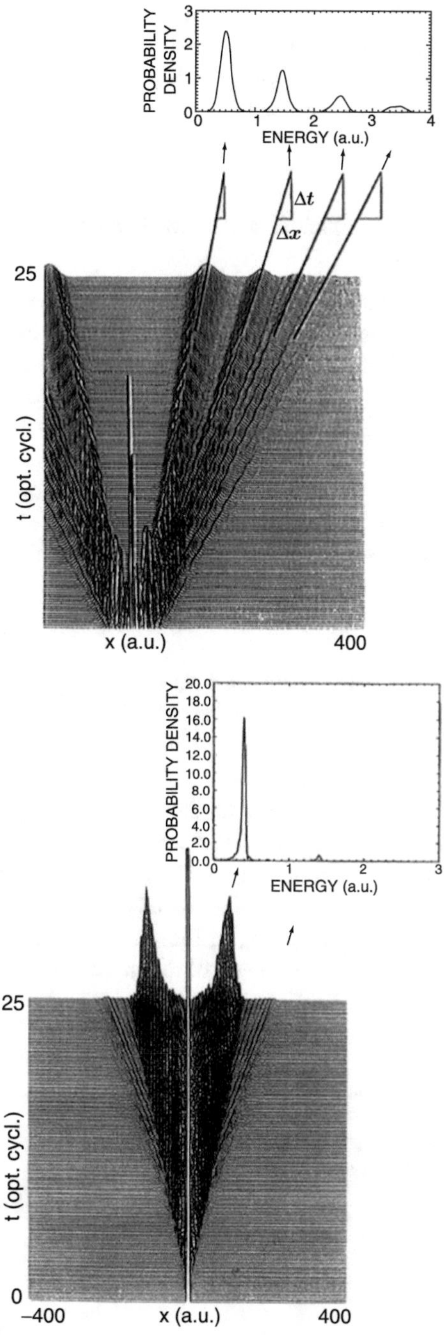

Fig. 4.11 Photo electron spectra and wavefunctions for ATI in the "bare" (*upper panels*) and in the "soft-core" (*lower panels*) Coulomb-Potential, adapted from [7]

and if $n\omega < I_p(I)$ ionization by n photons is forbidden, although this is not a strict argument, because the laser pulse usually has a smoothly varying envelope. In the case of atomic hydrogen, however, also the AC Stark shift of the ground state has be considered, see [28].

4.3.5 Stabilization in Very Strong Fields

A counter-intuitive phenomenon has been predicted theoretically to occur in very strong external fields, i.e., fields that are even stronger than an atomic unit. It has been shown that a continued increase of the field strength will eventually lead to a decrease of photoionization. Frequently used techniques to study this phenomenon are, as in the ATI case, the strong-field or Keldysh-Faisal-Reiss approximation discussed in Sect. 4.3.3.

Due to the fact that most of the features of stabilization can be understood classically, in the following we focus on a classical mechanics study of a 1D soft-core Coulomb potential, driven by a "monochromatic" electric field

$$\mathcal{E}(t) = \mathcal{E}_0 f(t) \sin(\omega t) \ . \tag{4.59}$$

If the field is switched on over 5 periods ($\omega = 0.8$ a.u.) and then oscillates for 50 periods with the constant amplitude $\mathcal{E}_0$, the behavior shown in Fig. 4.12 is observed [29]. For this figure 5000 different bound ($E < 0$) initial conditions have been chosen and the solution of Newton's equation has been calculated numerically. As a function of the field amplitude, the fraction of electrons that have positive energy after the pulse (the fraction of ionized electrons) is then plotted. For fields larger than an atomic unit, ionization as a function of field strength on average drops monotonically.

This phenomenon, which is referred to as stabilization against ionization, can be explained by going to the Kramers-Henneberger frame of Chap. 3. The classical soft-core Hamiltonian of the driven system then is given by

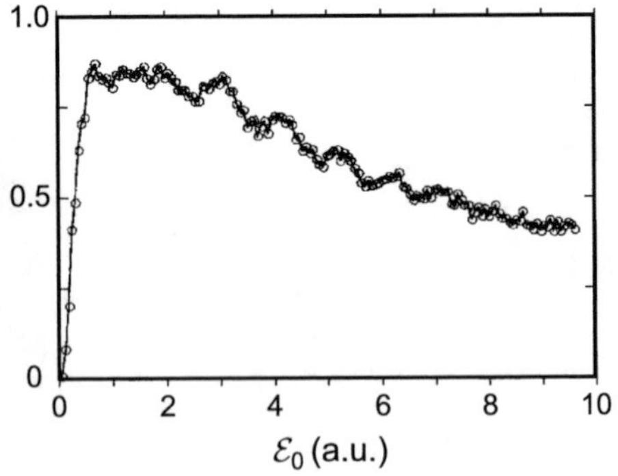

Fig. 4.12 Fraction of ionized electrons in an almost monochromatic laser pulse as a function of field strength [29]

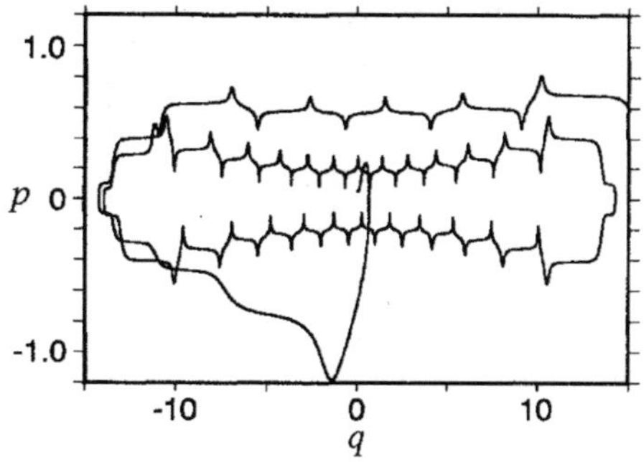

Fig. 4.13 Phase space orbit of an electron for $\omega = 0.8$, $\alpha_0 = 11.7$ and $q(0) = p(0) = 0$, all in atomic units, adapted from [29]

$$H(p, q) = \frac{p^2}{2} - \frac{1}{\sqrt{1 + [q + \alpha(t)]^2}}, \tag{4.60}$$

in atomic units (and in 1D with q denoting the position in phase space) and

$$\dot{\alpha}(t) = -A(t), \tag{4.61}$$

and thus

$$\ddot{\alpha}(t) = \mathcal{E}(t) \tag{4.62}$$

hold. The solution of this differential equation after the field is switched on is given by

$$\alpha(t) = -\alpha_0 \sin(\omega t), \tag{4.63}$$

with the quiver amplitude $\alpha_0 = \mathcal{E}_0/\omega^2$. In the KH frame, the nucleus can be thought of as oscillating with a typical velocity of $\omega\alpha_0$ between the turning points $\pm\alpha_0$.

For the electron, moving in the field of the nucleus, the behavior shown in Fig. 4.13 is typically found. Starting from the origin of phase space, after an initial acceleration, the electron is drifting into a certain direction for a long time and is passed by the nucleus frequently (twice in an optical cycle). At each encounter (a spike in the phase space orbit) it is shortly pulled either to the left and then to the right (or the other way round). The net effect is very small because the two pulls almost cancel each other. If the nucleus is at a turning point, however, its influence on the electron motion can be very strong. First of all, the nucleus is very slow at the turning point and can act for a long time on the electron and secondly, both pulls can have the same effect, because the nuclear motion may have changed its direction in-between both events. The electron then also turns around [29] and acquires energy until it has enough to leave the nucleus.

Stabilization can now be explained by the fact that for strong fields, the electron stays for longer and longer times in regions where no energy gain is possible. These regions increase in size with growing α_0.

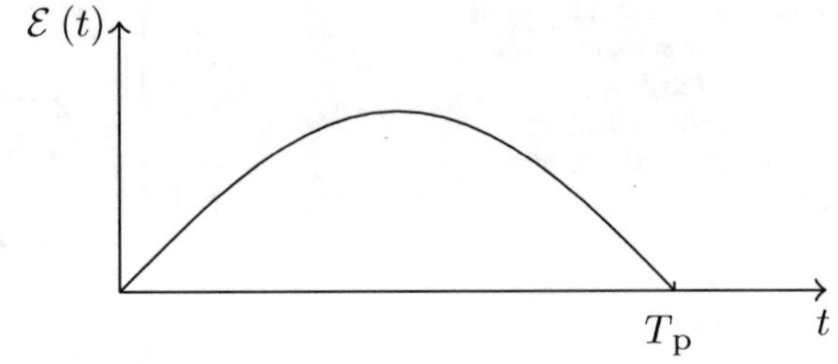

Fig. 4.14 Schematic plot of a half-cycle pulse of length T_p

4.3.6 Atoms Driven by Half-Cycle Pulses

The external driving considered so far was either static or (almost) monochromatic. What happens if a pulse, which is very short and therefore very polychromatic, hits an atomic system? The ultimate form of such a short pulse is a half-cycle pulse (HCP) schematically depicted in Fig. 4.14. It is assumed to be nonzero only for half an oscillation period. The generation of such pulses is the subject of a book on its own. Depending on the shortness required, they can be generated by irradiating semiconductors with pulsed optical lasers [30, 31], or by applying voltage pulses on capacitor electrodes [32].

We begin this section by considering a hydrogen atom initially in its ground state, given in (4.11), in the field of a single pulse and then turn to the case of a Rydberg atom in a periodic train of half-cycle pulses.

4.3.6.1 Ground State Hydrogen Under a Single HCP

A hydrogen atom coupled to an electric field in length gauge is governed by the Hamiltonian

$$\hat{H}(t) = \hat{H}_0 + \hat{\boldsymbol{r}} \cdot \boldsymbol{\mathcal{E}}(t), \tag{4.64}$$

with the unperturbed 3D Coulomb Hamiltonian $\hat{H}_0$. How can the time-dependent Schrödinger equation be solved for half-cycle pulse driving? The answer to this question is given by an application of the Magnus expansion of Sect. 2.2.5. When the length T_p of the pulse is very short compared to a typical orbital time, like $T_\mathrm{e} = 2\pi$ in the case of ground state dynamics in the hydrogen atom, and the field amplitude is very high, the first order Magnus approximation

$$\begin{aligned} |\Psi(t)\rangle &\approx \exp\left(-\mathrm{i}\int_0^t \mathrm{d}t' \hat{H}(t')\right)|\Psi(0)\rangle \\ &\approx \exp\left[-\mathrm{i}\left(\int_0^t \mathrm{d}t' \boldsymbol{\mathcal{E}}(t')\right)\cdot\hat{\boldsymbol{r}}\right]\exp[-\mathrm{i}\hat{H}_0 t]|\Psi(0)\rangle \end{aligned} \tag{4.65}$$

for the wavefunction can be used [33]. Analogous to the procedure in the interaction picture presented in Sect. 2.2.5, it can be derived from the exact time-evolution operator in the Schrödinger picture by neglecting the time-ordering operator. In addition, also the non-commutativity of the perturbation with the unperturbed Hamiltonian is neglected. Both approximations improve, the shorter the time span of the perturbation is.

The first exponent in the second line of the solution above, by comparison to a plane wave $e^{i\boldsymbol{k}\cdot\boldsymbol{r}}$, leads to the definition of the momentum that is transferred to the atomic system

$$\Delta \boldsymbol{p} := -\int_0^t \mathrm{d}t' \boldsymbol{\mathcal{E}}(t') . \tag{4.66}$$

If the initial state is an eigenstate of $\hat{H}_0$, then apart from the phase due to the application of the unperturbed Hamiltonian, the momentum change is the main effect of the total Hamiltonian. Due to the fact that only the pulse area[7] appears in the expression above, the HCP could also have been a δ-pulse with the appropriate strength

$$\boldsymbol{\mathcal{E}}(t) = -\Delta \boldsymbol{p}\delta(t - t_1), \qquad t_1 \in (0, T_\mathrm{p}) . \tag{4.67}$$

In Fig. 4.15, taken from [33], it is shown that in the case of the hydrogen atom starting from the ground state, for a half-cycle pulse with $T_\mathrm{p} = 0.3$ a.u. and for the absolute value of the momentum transfer $q = -\Delta p > 3$ a.u. the atom finally is ionized with almost certainty. Furthermore, full DVR calculations show that no matter if the pulse is rectangular, with $\boldsymbol{\mathcal{E}}_0 = \boldsymbol{q}/T_\mathrm{p}$ or sinusoidal, with $\boldsymbol{\mathcal{E}}_0 = \pi\boldsymbol{q}/(2T_\mathrm{p})$, the outcome is identical to the first order Magnus results.

4.4. *Calculate the excitation probability by a half cycle pulse from the 1s ground state of hydrogen into the 2p state as well as into the 2s state as a function of the (negative) momentum transfer q in first order Magnus approximation.*

So far the length gauge was used for the calculation of the ionization probability. It is worthwhile to note that including terms of third order in the pulse length (in the length gauge this means that the third order Magnus expansion has to be employed) the gauge invariance of the results for excitation and/or ionization probabilities has been shown [34]. By solving Exercise 4.5, analogous findings can be rederived for the velocity gauge and in the Kramers-Henneberger frame.

4.5. *Consider the time-evolution operator (TEO) $\hat{U}(T_\mathrm{p}, 0)$ for the velocity gauge Hamiltonian*

$$\hat{H}_\mathrm{v} = \frac{[\hat{\boldsymbol{p}} + \boldsymbol{A}(t)]^2}{2} + V(\boldsymbol{r})$$

as well for the Kramers-Henneberger (KH) Hamiltonian

[7] This is another occurrence of the area theorem (here assumed to be positive) of Sect. 3.2.

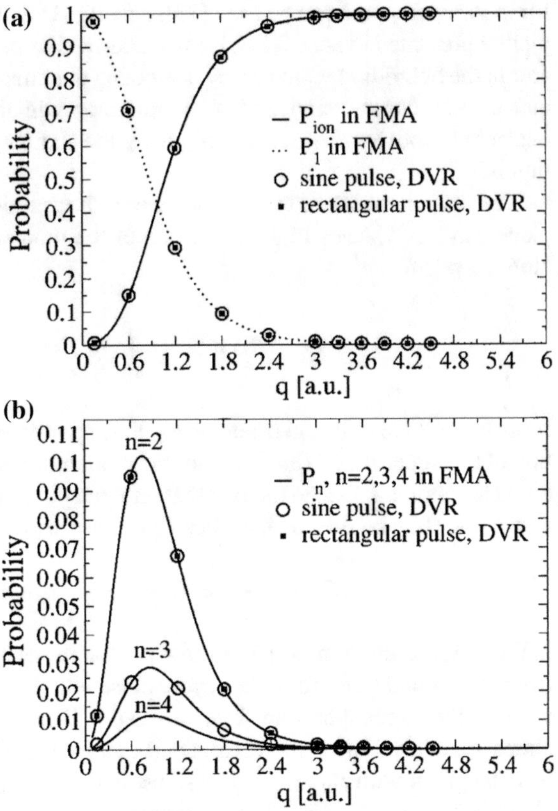

Fig. 4.15 **a** Ionization probability and occupation probability of the ground state, **b** some occupation probabilities for excited states of the hydrogen atom in very intense HCPs as a function of momentum transfer q [33]

$$\hat{H}_{\mathrm{a}} = \frac{\hat{\boldsymbol{p}}^2}{2} + V[\boldsymbol{r} + \boldsymbol{\alpha}(t)]$$

in the Schrödinger picture.

(a) *Use the second order Magnus expansion and the Zassenhaus formula to derive the TEO in the velocity gauge up to third order in the pulse length* T_{p}.
(b) *Use the first order Magnus expansion and the Zassenhaus formula to derive the TEO in the KH frame up to third order in the pulse length* T_{p}.
(c) *Show that both results amount to the same result for excitation and/or ionization probabilities*

$$P = |\langle \Psi(T_{\mathrm{p}})|\hat{U}(\tau, 0)|\psi_{100}\rangle|^2$$

from the ground state.

This section shall be closed by a brief remark on the orders of magnitude that are needed to experimentally realize the findings shown above. In order for the first order Magnus approximation to be applicable $T_{\mathrm{p}} \lesssim 1$ a.u. and thus the pulse length has to be of the order of attoseconds (10^{-18} s). Furthermore, the momentum transfer for total ionization has to be larger than one atomic unit and therefore, the corresponding field strengths have to be of the order of several atomic units and the intensities should be of the order of $10^{18}\,\mathrm{W\,cm^{-2}}$.

4.3.6.2 Rydberg States in Periodic Trains of HCP

If the initial state with energy E_n is a highly excited Rydberg state with a relatively long Kepler period of

$$T_n = T_{\mathrm{e}} n^3 = 2\pi n^3 \,, \tag{4.68}$$

in atomic units then the absolute length of the half-cycle pulse may be very much longer than in the case studied in the previous section and can still be considered to deliver a δ-function like kick to the atom. For Rydberg states of principal quantum number around $n = 400$, Kepler periods are in the range of nanoseconds and for n around 60 they still are in the picosecond range.

An investigation of the effects of a single pulse on the dynamics of Rydberg atoms has been done in [35]. Many interesting effects do emerge, however, if a periodic train of HCPs with period T is applied to a Rydberg atom. A combined experimental and theoretical study of such a system is reported in [32].

In the following, we consider in some detail the *bare* 1D Coulomb potential with a periodic kicking term that leads to the classical Hamiltonian

$$H^{\mathrm{1D}}(p, q, t) = \frac{p^2}{2} - \frac{1}{q} - q\Delta p \sum_{k=1}^{N} \delta(t - k/\nu_T) \tag{4.69}$$

in terms of phase space variables (q, p) with $\nu_T = 1/T$. Here we also assume that the motion is restricted to the positive half space, which can be motivated by the presence of a centrifugal barrier due to a quasi angular momentum Λ, which we let go to zero [32], and which can also be realized experimentally by photo excitation of selected Stark states [36]. Due to the instantaneous kicking, the stroboscopic classical dynamics (over one period) can be factorized according to

$$(q_k, p_k) = M_{\mathrm{Coul}} M_{\Delta p}(q_{k-1}, p_{k-1}) \tag{4.70}$$

into the kick contribution and an unperturbed Coulomb term analogous to the second line in (4.65).

Like in quantum mechanics, the kicking term's action is to change the momentum according to

$$(q_{k-1}, p_{k-1} + \Delta p) = M_{\Delta p}(q_{k-1}, p_{k-1}) \,, \tag{4.71}$$

leading to the new energy

$$E_k = \frac{\Delta p^2}{2} + p_{k-1}\Delta p + E_{k-1} \,. \tag{4.72}$$

The energy conserving Coulomb dynamics over one period of the driving can be extracted from the parametric form of the orbit, which for negative energies is given by [37]

$$q = n_k^2(1 - \epsilon \cos \xi), \qquad \xi - \epsilon \sin \xi = n_k^{-3} t, \tag{4.73}$$

with the eccentricity ϵ, which for vanishing quasi-angular momentum considered here is equal to unity, and $n_k = (2|E_k|)^{1/2}$. Due to the fact that the Coulomb potential is a homogenous function, depending only on a single power of the position variable, the scaling transformations with the initial quantum number n

$$q = n^2 q_0, \tag{4.74}$$
$$p = n^{-1} p_0, \tag{4.75}$$
$$t = 2\pi n^3 t_0 \tag{4.76}$$

can be used to extract the factor n^{-2} from the Hamiltonian. Analogously, for the frequency

$$\nu = \frac{\nu_0}{2\pi n^3} \tag{4.77}$$

applies.

Two different types of classical dynamics can be distinguished. For $\Delta p_0 > 0$, i.e., a kicking force that acts away from the center, one finds that even for very weak driving the phase space plot is completely chaotic, see, e.g., Fig. 4.16a. For a force that acts towards the center of Coulomb attraction ($\Delta p_0 < 0$), however, classical dynamics with a mixed phase space emerges, which is depicted in Fig. 4.16b, where regular and chaotic regions are both present.

Let us concentrate on kicks towards the singularity ($\Delta p_0 < 0$) in the following. The scaled frequency can be chosen either such that the phase space contour for the initial energy is in a regular area (see the left Poincaré section in Fig. 4.17, where the microcanonical initial state is indicated by a red line), or it can be chosen such that the red line is in a chaotic area (see the Poincaré section on the right). In the

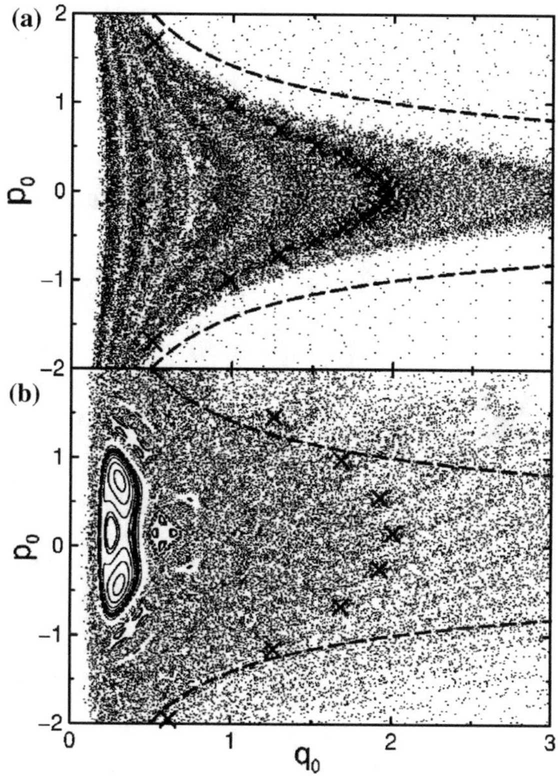

Fig. 4.16 Stroboscopic phase space plot with $n = 50$ and **a** $\Delta p_0 = 0.01$, $\nu_0 = 16.8$ and **b** $\Delta p_0 = -0.3$, $\nu_0 = 15.9$, all in atomic units. The *crosses* are periodic orbits and the *dashed line* shows the location of the ionization threshold energy $E = 0$ [38]

first case by comparing classical and quantum ionization dynamics a stabilization (or localization) of the classical results for the survival probability in a bound state (i.e., $1 - P_I$)

$$P(t) = \begin{cases} \sum_n |\langle \psi_n | \Psi(t) \rangle|^2 & \text{quantum} \\ \int_{E<0} \mathrm{d}p\, \mathrm{d}q\, f(q, p, t) & \text{classical} \end{cases} \tag{4.78}$$

is found. In the quantum result, the summation runs over all bound eigenstates and f is the phase space function corresponding to the quantum mechanical wavefunction. In the second case, if the initial state is in the chaotic range, the opposite effect is observed. Now quantum mechanically the results are localized. This effect remains if the number of kicks is increasing as can be seen in Fig. 4.18.

A semiclassical understanding of the observed effects would be a real breakthrough. First indication of semiclassical localization has been given in a study of the kicked rotor [40]. Right at the onset of localization, however, it is becoming extremely hard to converge the numerical results. In the case of HCP driven Rydberg

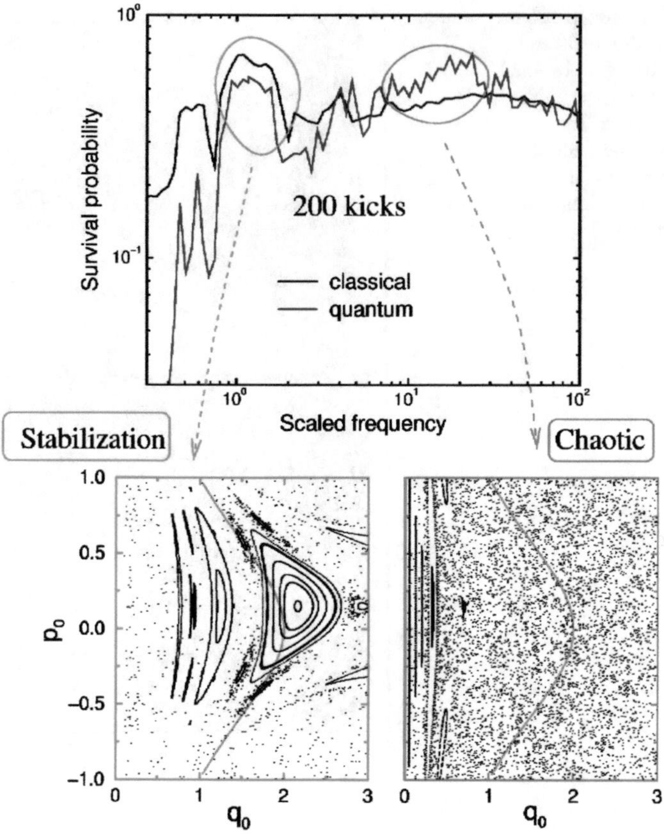

Fig. 4.17 Survival probability after 200 kicks as a function of the scaled frequency ($n = 60$) for $\Delta p_0 = -0.3$ a.u. (*upper panel*), and Poincaré sections for different scaled frequencies (*two lower panels*); red lines indicate the initial state [39]

atoms, as can be seen in Fig. 4.18, application of the Herman-Kluk propagator shows the localization up to around 100 kicks for the chosen parameters. For a larger number of kicks, the results tend towards the classical ones, however. Also here, convergence is a problem due to the fact that at longer times, the Herman-Kluk prefactors assume values of 10^{20} [38].

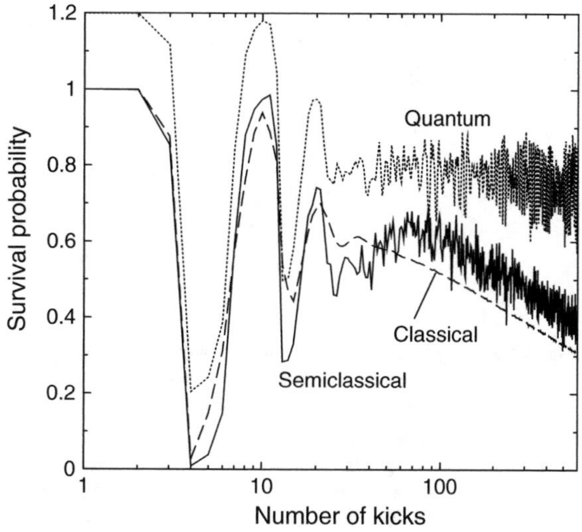

Fig. 4.18 Quantum (*dotted line*), semiclassical (*solid line*), and classical (*dashed line*) survival probability as a function of time for $\Delta p_0 = -0.3$ a.u. and $\nu_0 = 15.9$. Quantum result is shifted by 0.2 for better visibility [38]

4.4 Sundry Topics

We will come back to the discussion of ionization phenomena in sinusoidally driven dynamics of electrons and will discuss some surprising effects that have emerged recently, due to the advent of more powerful computers as well as more elaborate experimental techniques.

Firstly, a more thorough discussion of above threshold ionization, going beyond the 1D model of Sect. 4.3.4 is presented and secondly, the ionization of atoms (and molecules) in mid infrared fields and at low energies of the emitted electrons will be discussed. Thirdly, in the last subsection, surprisingly large double ionization probabilities of the helium atom will be in the focus.

4.4.1 Three-Step Model and ATI Rings

An electron, initially bound in an atom, can undergo different kinds of dynamics, when it is driven by an external laser field as we have learned already. Depending on the value of the Keldysh parameter this can be multi-photon ionization or tunneling ionization.

Some more subtle effects to be discussed below can be well understood by considering the three-step model [41], displayed in Fig. 4.19. After the initial ionization step (i) and subsequent acceleration in the laser field (ii), the possibility of recollision (iii) leads to the interesting effect of High-order Harmonic Generation (HHG) if the electron recombines with the atom and will be discussed in the next section. Accompanying this process is the generation of photo-electrons. In terms of a classical

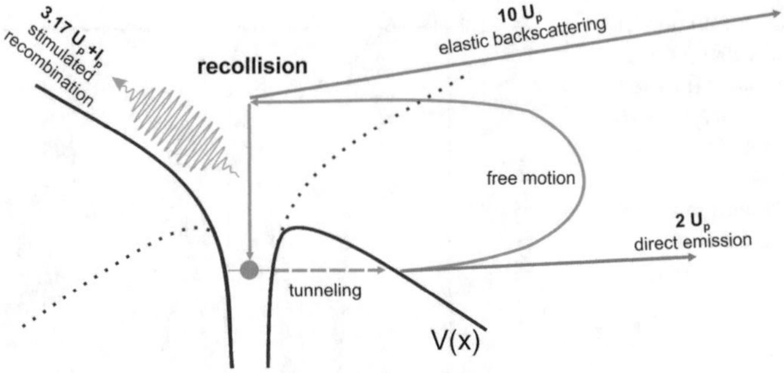

Fig. 4.19 Different scenarios in a laser driven atom based on the three-step model: (i) tunneling of the electron, (ii) acceleration in the laser field (iii) recollision of the electron with the ionic core. The final step can result either in stimulated recombination of the ejected part of the electron with a remaining part at the ionic core under photo emission (this then leads to high-order harmonic generation, discussed in the next section) or to the emission of photo-electrons. The maximal energy of the single photon created and the maximal energies of emitted photo-electrons with or without backscattering are also indicated (courtesy of Thomas Fennel)

description with electron trajectories, the exact timing of the initial ionization event with respect to the laser field is discriminating the possible outcomes (HHG or final photo emission).

In a 2D study of the soft-core hydrogen atom, initially in its groundstate, under a 12 cycle 532 nm pulse ($\omega \approx 0.0857$ a.u.) with intensity of 5×10^{13} W/cm^2, linearly polarized in the y direction, the ATI rings displayed in Fig. 4.20 have been found [42]. They are emerging in the Fourier transformed unbound part of the propagated wavefunction of the electron after the pulse is over. The radius of the rings

$$K_r = \sqrt{K_x^2 + K_y^2} \tag{4.79}$$

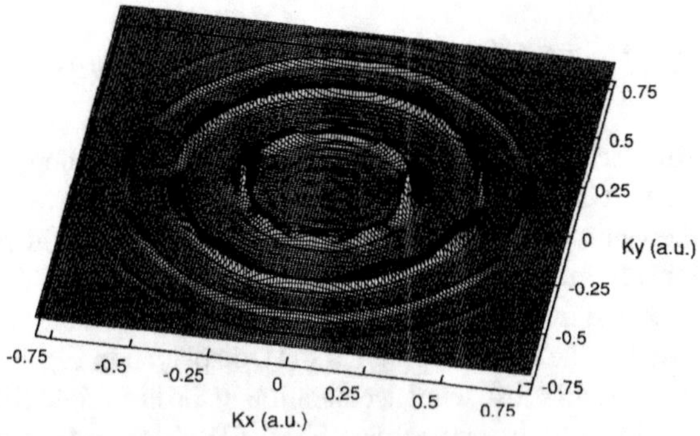

Fig. 4.20 ATI rings for a 12 cycle laser driven 2D hydrogen atom with laser wavelength $\lambda = 532$ nm, intensity $I = 5 \times 10^{13}$ W/cm^2, linearly polarized in the y direction [42]

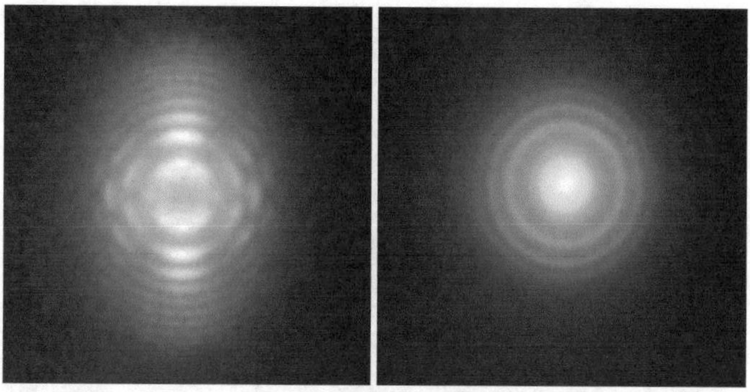

Fig. 4.21 Photo-electron raw image for ionization of xenon with vertical laser polarization parallel to the page (*on the left*) and with laser polarization perpendicular to the page (*on the right*) [45]

gives the photoelectron momentum and if n is the index of the ring, then (in atomic units)

$$\frac{1}{2}[K_r^2(n) - K_r^2(n-1)] = \omega, \tag{4.80}$$

in close analogy to the 1D study of the hydrogen atom discussed in Sect. 4.3.4.

Experimentally, similar rings have first been observed by Yang et al. [43]. The rings would display a higher symmetry, if the laser polarization was oriented perpendicular to the (x, y)-plane, see Fig. 4.21. A formula for the differential cross section for ATI depending on the angle between the laser polarization axis and the direction of emission of the electron has been given in strong-field approximation by Reiss [44].

In a 2D study of the helium atom within a single active electron description and under a two-color laser field

$$\mathcal{E}(t) = f(t)\mathcal{E}_0\left[\boldsymbol{e}_x\cos(\omega t) + 0.1\boldsymbol{e}_y\cos(2\omega t + \Phi)\right], \tag{4.81}$$

with a strong field with 800 nm wavelength ($\omega \approx 0.057$ a.u.) with $\mathcal{E}_0 = 0.107$ a.u. in the x-direction and a weaker field with 400 nm wavelength in the y-direction, with a phase delay Φ, the photoelectron emission probabilities displayed in Fig. 4.22 have been found [46]. Also in this case, ATI rings are visible inside the circle with radius $2U_\mathrm{p}$. The delay between the two fields is barely influencing the slow electrons with maximal ponderomotive energy of $2U_\mathrm{p}$, whereas a strong delay dependence is observed for fast electrons (with up to $10U_\mathrm{p}$) that have undergone at least one recollision. The circles of the fast electrons (which were rescattered after a different number of recollisions) are also centered around nonzero momentum in the x direction, due to the fact that they are accelerated by the field.

Furthermore, it has been shown how ionization times can be extracted from the two-color TDSE results [46].

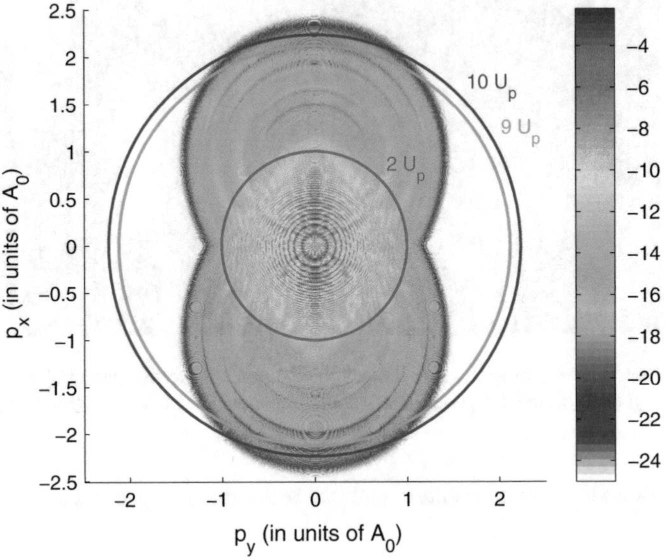

Fig. 4.22 Photo-electron momentum distribution in a single active electron model of helium under a two-color laser field with two-color delay $\Phi = 0.96\pi$, giving the largest fast electron yield (courtesy of Jost Henkel)

4.4.2 Low-Energy Structure

In an experimental study of electron emission on rare gas atoms, using 40 fs laser pulses with 630 nm central wavelength and peak intensities up to 4.4×10^{14} W/cm^2, the results displayed in Fig. 4.23 have been found. The sharp ATI peaks are blurred into a continuous distribution at this high laser intensity and a plateau in the generation of the photo-electrons builds up (see inset in Fig. 4.23), which is most clearly visible for Ar and Xe gases [47]. It was lateron shown that the cutoff of this so-called ATI plateau is at around $10U_\mathrm{p}$ [48], because this is the maximal energy that an electron can acquire after recollision (see also Fig. 4.22).

A surprising detail in the photo-electron emission was measured in the forward direction along the laser polarization axis for atoms (and also molecules) irradiated by long wavelength lasers ($\lambda > 1\,\mu$m) such that the Keldysh parameter is well below unity [49]. On top of the prediction of the strong-field KFR approximation, a maximum is located at low energies, see Fig. 4.23. This maximum has been dubbed "low-energy structure" (LES).

For a theoretical understanding of the LES we now follow closely the review of [50] in [51]. There it is shown that the position of the maximum of the peak depends algebraically on the Keldysh parameter and can be determined by a minimal extension of what we have learned in Sect. 3.1.4 in the 1D case. In the spirit of the three-step model, we concentrate on step two, i.e., the classical motion in the electric field only,

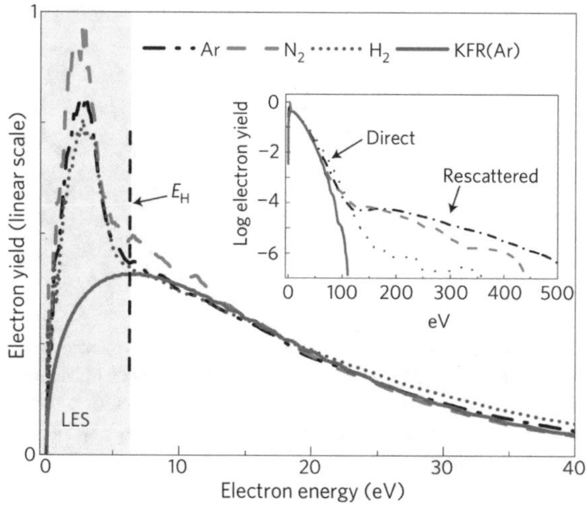

Fig. 4.23 Low-energy structure (LES) in photo-electron yield under 150 TW/cm^2, 2 μm pulses with $\gamma_K \approx 0.4$ on top of the strong field (KFR) prediction in the photo-electron spectrum of atomic argon and two molecular species. The inset shows a wider energy range with the recollision induced ATI plateau [49]. E_H is the high energy limit at which a break in the slope of the yield curve exists

after the initial tunneling step. Furthermore, the solutions of the classical equations of motion are generalized to the case of an arbitrary initial time t_0 and it is assumed that at the initial time the initial position as well as the initial velocity of the electron are both zero. Then the solution of the classical equations of motion in 1D and for a cw-laser polarized in the z-direction is given by

$$\dot{z}(t) = \frac{\mathcal{E}_0}{\omega}\left[\sin(\omega t_0) - \sin(\omega t)\right], \tag{4.82}$$

$$z(t) = \frac{\mathcal{E}_0}{\omega^2}\left[(\omega t - \omega t_0)\sin(\omega t_0) + \cos(\omega t) - \cos(\omega t_0)\right]. \tag{4.83}$$

This is the atomic unit version of (3.26) and (3.27) for variable initial time t_0 and initial conditions $z(t_0) = 0$, $\dot{z}(t_0) = 0$. The LES is due to a recollision at low energy at a time t_1, so we investigate in the following the "final" conditions[8]

$$\dot{z}(t_1) = 0 \tag{4.84}$$

$$z(t_1) = 0. \tag{4.85}$$

Under the well founded assumptions $\cos(\omega t_0) = 1$, $\sin(\omega t_0) = \omega t_0$ this then leads to the condition

$$t_0 = \frac{1}{(n + \frac{1}{2})\pi\omega} \tag{4.86}$$

[8]In 3D, a finite distance of the recolliding electron to the core is observed, see the discussion below.

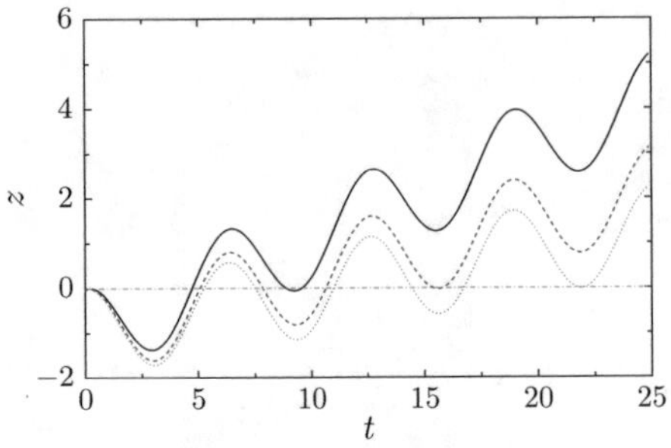

Fig. 4.24 Drifting trajectories according to (4.83) for $\omega = 1$ and $\mathcal{E}_0 = 1$ and different values of t_0, adapted from [51]. Recollision takes place where both $z = 0$ (*green dashed-dotted line*) and the slope of the curve is zero, i.e., once for every trajectory. For the *solid black curve* with $n = 1$ this is at $t_1 = 3\pi$, for the *dashed blue curve* with $n = 2$, we find $t_1 = 5\pi$ and for the *dotted red curve* with $n = 3$, we get $t_1 = 7\pi$

for the initial time, which in turns yields the recollision time

$$t_1 = (2n + 1)\frac{\pi}{\omega}. \tag{4.87}$$

This means that the LES trajectories form a series with the recollision taking place after an odd number of half cycles. The drift momenta (i.e., the slope of the cycle-averaged trajectory) of the corresponding trajectories (see Fig. 4.24) are

$$p_z = \frac{\mathcal{E}_0}{\omega}\frac{1}{(n + \frac{1}{2})\pi}. \tag{4.88}$$

As the order of the recollision increases, the ionization takes place closer to the field maximum, giving a smaller drift momentum. The energies of the (multitude of) peaks corresponding to the drift momenta are proportial to the ponderomotive energy

$$E_{\mathrm{LES}} \sim U_{\mathrm{p}} \sim \gamma_{\mathrm{K}}^{-2}, \tag{4.89}$$

which has also been observed experimentally.

The generalization of the presentation above to pulsed laser driving is discussed in [51]. Furthermore, in [50], it shown in a 3D simulation by using cylindrical coordinates (z, ρ) that the recollisions responsible for LES formation are soft, i.e., they occur at finite values of ρ in contrast to the head-on collisions, responsible for HHG and ATI, see also [52] and Chap. 4 of [51].

4.4.3 Double Ionization of Helium

In this final section on ionization, we will come back to the helium atom introduced in Sect. 4.2 and discuss different ionization scenarios by strong laser fields. There are several ionization possibilities due to the fact that helium contains two electrons:

- single ionization: $He \to He^{+} + e^{-}$
- sequential double ionization: $He \to He^{+} + e^{-} \to He^{2+} + 2e^{-}$
- nonsequential double ionization: $He \to He^{2+} + 2e^{-}$

A knee shape (see Fig. 4.25) is present in the experimental results for the double-ionization probability out of the singlet ground state spatial wavefunction under irradiation with linearly polarized light of 780 nm wavelength as a function of intensity [53]. The single ionization can be treated very effectively by a single active electron model, as has been has shown in [53]. The increased double ionization rate at low intensities has been a topic of intense study and debate, being one of the most prominent and clear electron correlation effects that can be observed experimentally.

It turns out that a classical description on the basis of Newton's equation is adequate to reproduce the knee formation in the double ionization [54]. There it has

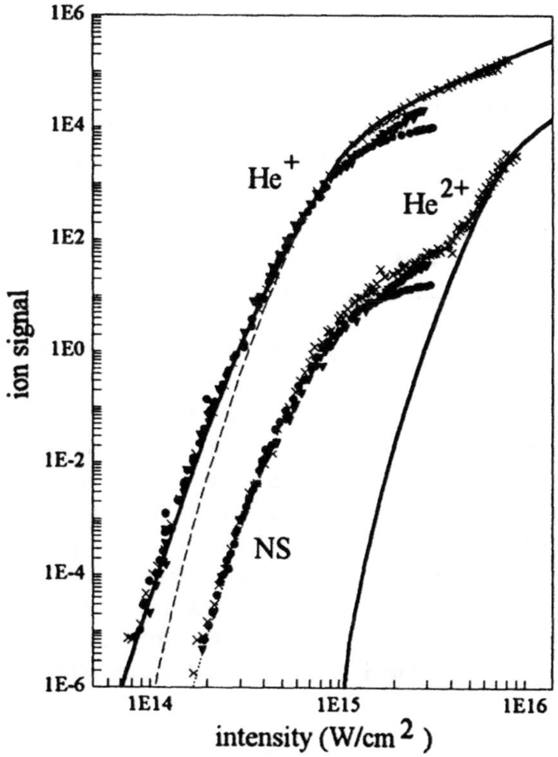

Fig. 4.25 Single and double ionization probabilities of helium for linearly polarized light with a wavelength of 780 nm and a pulse length of 100 fs. The *dashed curve* is the ADK result for single ionization of Helium and the *solid curve* on the left is from a single active electron calculation. The *solid curve* on the right is the corresponding result for He^{+} [53]

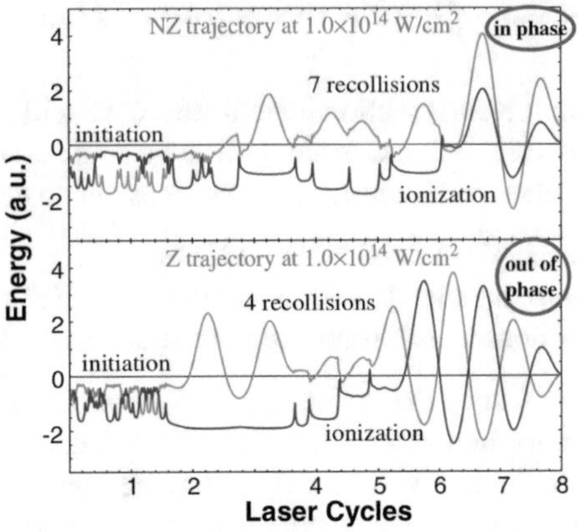

Fig. 4.26 Total energies of the two electrons in the laser driven case (wavelength 780 nm, intensity 10^{14}W/cm^2) as a function of time [54]

also been shown by a 1D soft-core simulation using Newton's equation that a recollision of an initially ionized electron (orange trajectory) with the remaining ion finally leads to the simultaneous liberation of both electrons, i.e., nonsequential double ionization, see Fig. 4.26. Two different cases can be distinguished. Either the final electronic quiver (jitter) motion of the liberated electrons is in phase, or it is out of phase. These differences in the phase relation of the electrons are related to the final ion momentum distribution, which can either be around nonzero (NZ) or around zero (Z) values [54].

In the sequel, we will go into some more detail by reviewing the closely related work of Mauger and coauthors [55], who discuss all three different ionization scenarios listed above in a unified manner. These authors have also used the classical analog of the 1D model Hamiltonian in (4.21) with a soft-core parameter $a = 1$ and an additional laser potential in the form of a socalled 2-4-2 pulse, i.e., a linear ramp turn-on for 2 periods of the field followed by a 4 period constant envelope function and a 2 period linear ramp turn-off. The frequency of the sinusoidal oscillation was chosen as $\omega = 0.0584$, corresponding to the experimental wavelength of 780 nm.

The discussion of the electron dynamics in [55] begins with a dynamical systems theoretical study of the undriven problem. It was found that periodic orbits allow to distinguish an inner from an outer electron. The inner one is denoted by the y-coordinate and the outer one by x. The initial conditions for the classical dynamics were taken from a microcanonical distribution at the quantum mechanical ground state energy of helium, which in the present 1D case and for the chosen soft-core parameter is at -2.24 a.u.. 3000 trajectories were randomly chosen in the accessible phase space. The ionization criterion was defined by a distance of the classical particle(s) of more than 30 a.u. away from the nucleus.

Single ionization happens during the linear ramp turn on of the field. It can be described by the Hamiltonian of the outer electron

$$H_1 = \frac{p_x^2}{2} + \mathcal{E}_0 f(t) x \sin(\omega t) \tag{4.90}$$

with $f(t) = \omega t/(4\pi)$. An approximate solution for the dynamics is given by

$$x(t) = x_0 + p_0 t - \frac{\mathcal{E}_0}{4\pi\omega^2}[\omega t \sin(\omega t) + 2\cos(\omega t) - 2]. \tag{4.91}$$

For the description of the effective dynamics of the inner electron, one can neglect the effect of the outer electron as well as that of the laser field (which is still weak during turn on) and finds a bound motion with a period of $2\pi/\sqrt{2}$.

Sequential double ionization after one electron has been ionized can be understood by looking at the Hamiltonian

$$H_2 = \frac{p_y^2}{2} - \frac{2}{\sqrt{y^2+1}} + \mathcal{E}_0 y \sin(\omega t). \tag{4.92}$$

In the (p_y, y) phase space, a core region corresponding to inner electrons that are not ionized (single ionization in total) can be distinguished from an outer region in which the y electrons are captured by the field and driven away from the nucleus (the soft-core potential plays no role for them) if the intensity is large enough. Complete sequential double ionization in this reduced model is expected at an intensity of $I_c = 1.86 \times 10^{16}$ W/cm^2 [55], which is also corroborated by the results in Fig. 4.27.

For the case of nonsequential double ionization another intensity $I^{(c)} = 4.58 \times 10^{14}$W/cm^2 has been derived in the case of $\omega = 0.0584$. This intensity is indicated by the green line in Fig. 4.27 and is the intensity for which *each* outer electron returns back to the nucleus and has enough energy to ionize the inner one and to stay ionized itsself (see also Fig. 4.26 for a special realization). In essence, this scenario is again a three-step model. To derive the above intensity it was used that the maximum kinetic

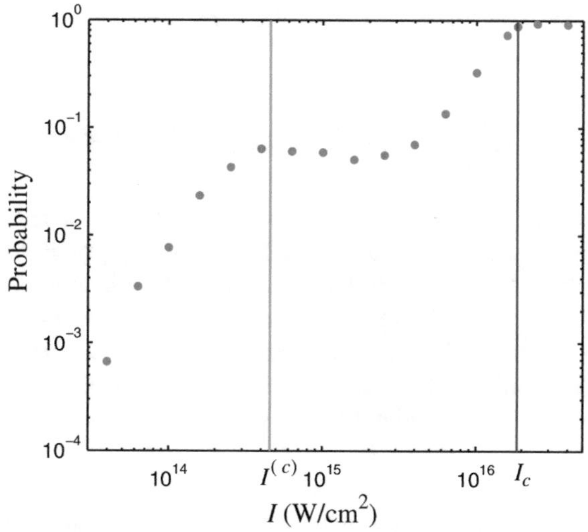

Fig. 4.27 Double ionization probability for laser ($\omega = 0.0584$) driven helium (*red dots*) [56]. Vertical lines indicate intensities for which complete nonsequential and sequential double ionization are expected, which are $I^{(c)} = 4.57 \times 10^{14}$ W/cm^2 (*green line*), respectively $I_c = 1.86 \times 10^{16}$ W/cm^2 (*blue line*)

energy of the returning electron is 3.17 U_p and an implicit equation for the value of the field at which maximum nonsequential double ionization occurs was derived and solved [55].

Full grid based quantum calculations of the dynamics of helium under a laser pulse in 3D are at the forefront of what is computationally possible [57, 58] and the main correlation effect can already be reproduced classically. Nevertheless, alternative methods, more quantum in nature, have been applied to the problem and we finish the discussion of double ionization by briefly commenting on some of them. An early success in reproducing the experimental result of knee formation in the double ionization probability theoretically was based on the intense-field many-body S-matrix theory of Becker and Faisal [59]. Lateron, using a 1D model, the rescattering mechanism (three-step model) has been found to be responsible for double ionization, and shake-off [60] as well as collective tunneling mechanisms have been ruled out [61]. In addition, in a time-dependent density functional theory (TDDFT) study by de Wijn and coauthors [62] it was shown that functionals with a derivative discontinuity are able to reproduce the knee structure in the double ionization at the experimentally used wavelength of 780 nm. The question how (double) ionization probabilities can be calculated from the density via the Runge-Gross theorem is elucidated in [63]. Two approximations have to be made in a corresponding numerical approach. Firstly, the exchange correlation functional is known only approximately and secondly, the functional dependence of the pair-correlation function (necessary for double ionization) on the density is also only approximately known.

Finally, in the work by Kirrander and Shalashilin [64], it could be shown that the experimental results of Walker et al. [53] can be reproduced by the CCS method, which is close in spirit to the semiclassical Herman-Kluk working horse, used frequently throughout this book. CCS goes beyond a pure semiclassical method, however, as the coherent states are coupled, leading, in principle to a full numerical solution of the TDSE. In the present context, the method had to be adapted to the fermionic character of the electrons.

4.5 High-Order Harmonic Generation

Apart from being ionized, an atom that is irradiated by laser light of a certain frequency ω can be radiating itself by emitting photons at odd harmonic frequencies $(2n+1)\omega$ of the original radiation. In the following, we will put our focus on this effect, a schematic representation of which is given in Fig. 4.28. High-order harmonic generation (HHG) via laser excitation has been demonstrated experimentally with harmonics as high as 500 times the fundamental frequency[9] having been reported, and it is reviewed in [26].

[9]The creation of radiation in the soft X-ray regime (with photon energies around 1keV) is thus feasible [65].

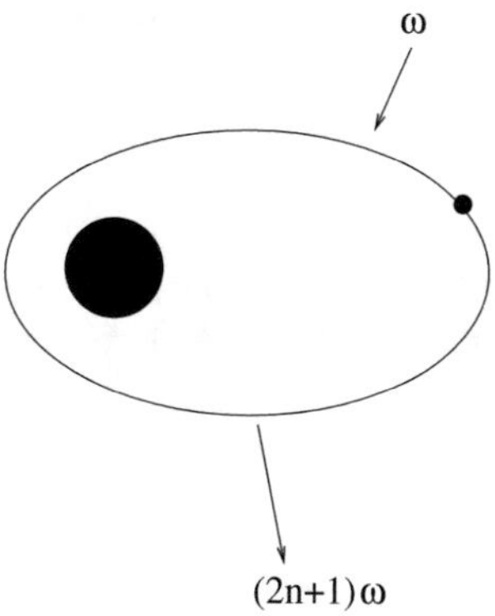

Fig. 4.28 Schematic of HHG for a single electron atom: After irradiation by a field of frequency ω, odd harmonics are emitted

4.5.1 Three-Step Model of HHG

A simple explanation of HHG in a system with a single active electron can be given by considering the three-step model [41], that was already mentioned in Sect. 4.4, and which is repeated here for the case of HHG in a bit more detail:

- The laser field distorts the Coulomb potential, such that (part of) the electron can tunnel out of the range of attraction (see Fig. 4.19)
- The (almost) free electron, born with zero velocity, is accelerated in the laser field
- When the laser field turns around, depending on the phase of the field at its birth, the "free" electron may recombine at the ion and the energy is released in the form of a single photon

It is important to note that the answer of the electron to the applied field is not instantaneous like in the case of low field nonlinear optics but there is a time delay between the stimulus and the radiation of the high harmonics. Furthermore, part of the electron wavefunction has to remain in the ground state, leading to stimulated recombination [66].

In order to describe HHG theoretically, a measurable quantity has to be defined. This is the harmonic spectrum which can be expressed in terms of the Fourier transformation of the dipole acceleration, given by

$$\begin{aligned} a(t) &= \frac{\mathrm{d}^2}{\mathrm{d}t^2}\langle\hat{x}\rangle(t) = \frac{\mathrm{d}^2}{\mathrm{d}t^2}\langle\Psi(t)|\hat{x}|\Psi(t)\rangle \\ &= \frac{\mathrm{d}}{\mathrm{d}t}\langle\Psi(t)|\hat{p}|\Psi(t)\rangle \\ &= -\left\langle\Psi(t)\left|\frac{\mathrm{d}\hat{V}}{\mathrm{d}\hat{x}}\right|\Psi(t)\right\rangle . \end{aligned} \tag{4.93}$$

In the equations above, the general form of the Ehrenfest theorem

$$\mathrm{i}\frac{\mathrm{d}}{\mathrm{d}t}\langle\hat{A}\rangle = \langle[\hat{A},\hat{H}]\rangle + \mathrm{i}\left\langle\frac{\partial}{\partial t}\hat{A}\right\rangle \tag{4.94}$$

for the time-evolution of expectation values has been applied once for $\hat{A} = \hat{x}$ and once for $\hat{A} = \hat{p}$ with the length gauge Hamiltonian

$$\hat{H} = \frac{\hat{p}^2}{2} + \hat{V}_{\mathrm{C}} + \hat{V}_{\mathrm{L}}\,. \tag{4.95}$$

The term due to the laser potential in (4.93) leads only to a contribution at the fundamental frequency and is therefore usually subtracted. In the expectation value then only the term $\frac{\mathrm{d}\hat{V}_{\mathrm{C}}}{d\hat{x}}$ remains. The spectrum is finally given by the Fourier transform

$$\begin{aligned}\sigma(\Omega) &= \frac{1}{\sqrt{2\pi}}\int_0^{T_{\mathrm{t}}}\mathrm{d}t\;\mathrm{e}^{-\mathrm{i}\Omega t}\frac{\mathrm{d}^2}{\mathrm{d}t^2}\langle\hat{x}\rangle(t)\\ &= \frac{1}{\sqrt{2\pi}}\left[\mathrm{e}^{-\mathrm{i}\Omega T_{\mathrm{t}}}\frac{\mathrm{d}}{\mathrm{d}t}\langle\hat{x}\rangle(T_{\mathrm{t}}) + \mathrm{i}\Omega\mathrm{e}^{-\mathrm{i}\Omega T_{\mathrm{t}}}\langle\hat{x}\rangle(T_{\mathrm{t}}) - \Omega^2\int_0^{T_{\mathrm{t}}}\mathrm{d}t\;\mathrm{e}^{-\mathrm{i}\Omega t}\langle\hat{x}\rangle\right].\end{aligned} \tag{4.96}$$

In going to the second line using integration by parts, it has been assumed that $\langle\hat{x}\rangle(0) = \frac{\mathrm{d}}{\mathrm{d}t}\langle\hat{x}\rangle(0) = 0$. In some early literature also the Fourier transformation of the dipole expectation value, i.e., only the last term in the second line of the equation above is used to calculate the harmonic spectrum. This is generally not correct, due to the fact that the boundary terms in the partial integration do not necessarily vanish [67]. The unraveling of the harmonic spectrum when the length T_{t} of the integration interval is increased is displayed in the full quantum calculations shown in Fig. 4.29.

As expected by using perturbation theory, the intensity of the harmonics initially decreases with higher order. Unexpectedly, however, a long *plateau* region can be observed in Fig. 4.29, which is ending with a sharp so-called *cutoff*. These main features are once more depicted schematically in Fig. 4.30. The following questions regarding HHG await an answer:

(a) Why is there a sharp cutoff and what is its value?
(b) Why are harmonics only observed at *odd* multiples of the fundamental frequency?
(c) Why is there a long plateau in the intensity of the harmonics?

By a simple argument using classical mechanics, the answer to question (a) can be given. The explanation of questions (b) and (c) are a little bit more involved. Before explaining the different features, a further remark on the non-instantaneous nature of HHG shall be made, however. It has been shown by numerical simulations of an experiment with Ne atoms that the highest harmonics are created in very short time

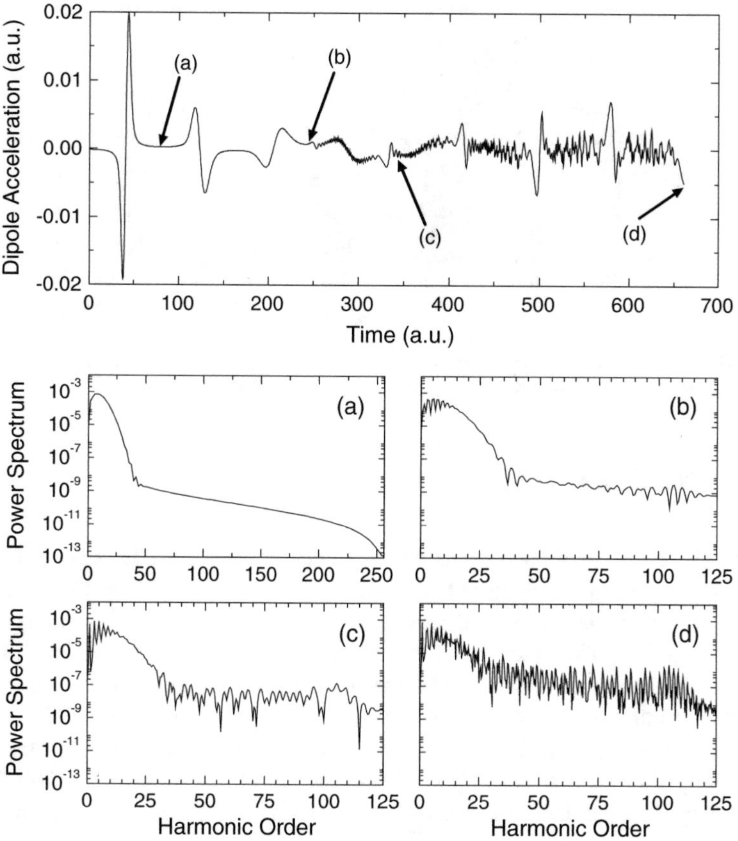

Fig. 4.29 Features of HHG emerging by increasing the time interval T_t in the Fourier transform (4.96) [68]

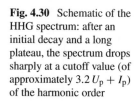

Fig. 4.30 Schematic of the HHG spectrum: after an initial decay and a long plateau, the spectrum drops sharply at a cutoff value (of approximately $3.2\,U_p + I_p$) of the harmonic order

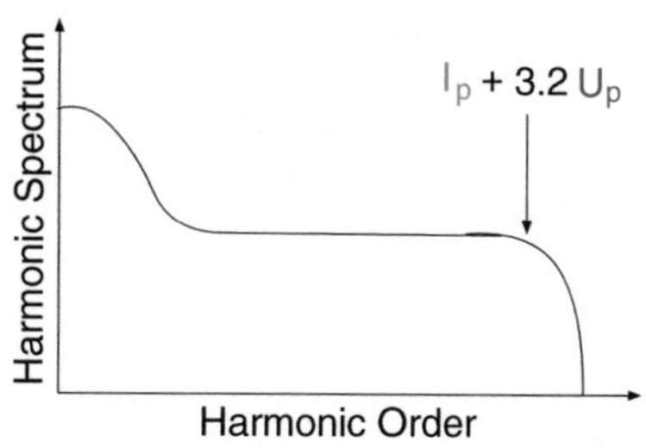

intervals of the length of attoseconds [69]. Figure 4.31 shows the Husimi transform, defined in Chap. 1, of the dipole acceleration. This observation also opens a road to the generation of attosecond pulses of electromagnetic radiation.

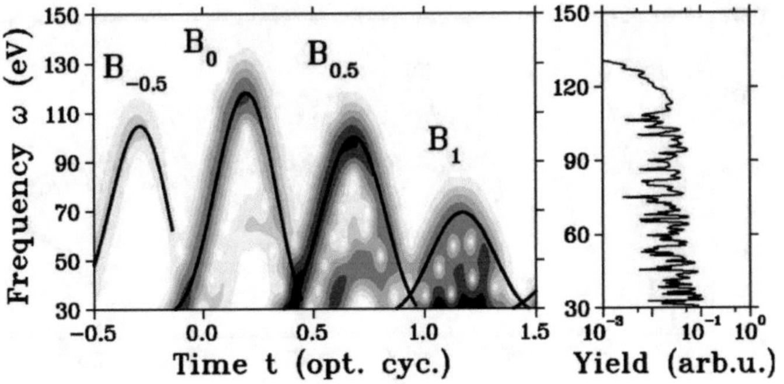

Fig. 4.31 Husimi plot of the dipole acceleration, simulating an experiment with Ne in a 5 fs laser pulse at 800 nm wavelength and with an intensity of $5 \times 10^{14}\,\mathrm{W\,cm^{-2}}$ [69]

4.5.2 The Cutoff

The cutoff can be explained by using the three-step model. The decisive question is how much energy an electron can acquire from a monochromatic laser field. To answer this question, numerical studies based on classical mechanics have been performed [41].[10] The outcome of these calculations is depicted in Fig. 4.32. It is shown there that an electron that is initially bound in an atom can acquire a kinetic energy of maximally $3.17\,U_\mathrm{p}$ with the ponderomotive energy from (3.28).

The maximal energy that can be released upon recombination with the ion is therefore given by $E_C \approx 3.17\,U_\mathrm{p} + I_\mathrm{p}$ with the ionization potential I_p. This finding is robust and independent of the form of the potential and it explains the sharp cutoff observed in the harmonic spectrum.[11]

Fig. 4.32 Energy distribution of electrons at the first encounter with the ion in the case of helium and with $I = 5 \times 10^{14}\,\mathrm{W\,cm^{-2}}$, $\lambda = 800\,\mathrm{nm}$ [41]

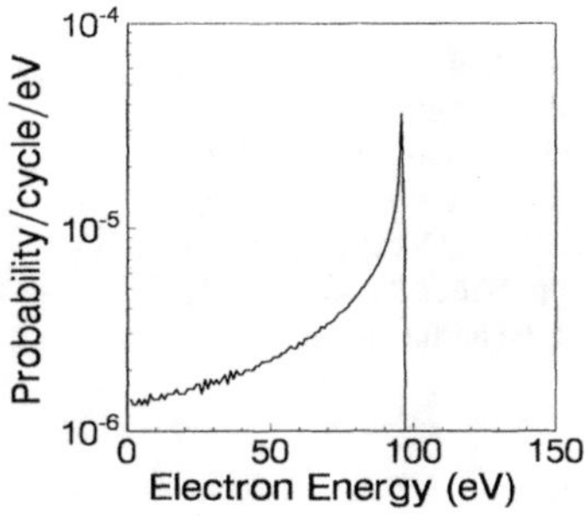

[10] A more involved quantum mechanical reasoning, based on the strong-field approximation, leads to similar results [70].

[11] Without the initial tunnel ionization, as we saw in Chap. 3, the maximal kinetic energy that can be gained in the field is $2\,U_\mathrm{p}$, and the cutoff for scattering initial conditions has to be adjusted accordingly.

4.5.3 Odd Harmonics Rule

To explain the peaks of the HHG spectrum at odd harmonics, we have to invoke some quantum mechanical reasoning. In the case of a periodically driven quantum system the Floquet theorem of Sect. 2.2.8 (in a.u.)

$$\Psi_n(x,t) = \exp\{-\mathrm{i}\epsilon_n t\}\,\psi_n(x,t) \tag{4.97}$$

$$\psi_n(x,t) = \psi_n(x,t+T) \tag{4.98}$$

holds. In the case of a symmetric potential, due to symmetry under the generalized parity transformation

$$\mathcal{P} : x \to -x, \qquad t \to t + T/2\,, \tag{4.99}$$

reviewed in Appendix 3.A, one can conclude that the Floquet functions have to obey the additional condition

$$\psi_n(-x, t + T/2) = \pm\psi_n(x,t)\,. \tag{4.100}$$

Using this fact and by considering the Fourier transformation of the dipole matrix element $\langle\psi_n|\hat{x}|\psi_n\rangle$, it can be shown by solving Exercise 4.6 (neglecting the boundary terms in (4.96)) that the HHG spectrum contains only odd harmonics. An alternative, semiclassical explanation will be given in Sect. 4.5.5.

4.6. *Show that the dipole expectation value between Floquet states has only odd Fourier components.*

In the previous as well as the following numerical results, the theorem of odd harmonics is not fulfilled exactly, however. This is due to the fact that a pulsed driving laser is used and therefore, Floquet theory cannot be applied strictly.

Furthermore, the odd harmonics rule can be violated in *molecules* also due to the breakdown of the Born-Oppenheimer approximation, as has been shown in [71]. The Born-Oppenheimer approximation will be discussed in detail in Chap. 5.

4.5.4 Semiclassical Explanation of the Plateau

Understanding the plateau formation in HHG is a formidable task and has been undertaken by Gerd van de Sand during his PhD thesis work [19, 72]. The main working horses used in these studies are the semiclassical Herman-Kluk propagator and a comparison with full quantum as well as purely classical calculations.

In order to perform all three types of calculation the Hamiltonian

$$H = \frac{p^2}{2} + V_{\mathrm{sc}} + V_{\mathrm{L}} \tag{4.101}$$

with the 1D soft-core Coulomb potential (4.18) and the laser potential in length gauge

$$V_{\mathrm{L}} = f(t)\mathcal{E}_0 x \cos(\omega t) \tag{4.102}$$

has been used in [19]. The parameters appearing in the Hamiltonian have been chosen as follows:

- Soft-core parameter $a = 2$ a.u. leading to an ionization potential of $I_p = 0.5$ a.u. equal to that of 3D hydrogen
- $\mathcal{E}_0 = 0.1$ a.u. and $\omega = 0.0378$ a.u., where $f(t)$ has been chosen such that the laser pulse lasts for 3.5 cycles of the oscillation

The initial wavefunction was assumed to be a Gaussian wavepacket

$$\Psi_\gamma(x, 0) = \left(\frac{\gamma}{\pi}\right)^{1/4} \exp\left\{-\frac{\gamma}{2}(x - q_0)^2\right\} \tag{4.103}$$

with $\gamma = 0.05$ a.u. and $q_0 = 70$ a.u., corresponding to an almost free initial electron being scattered as soon as it reaches the ion under the influence of the laser. This initial state has been chosen due to the fact that the initial tunneling step cannot be described semiclassically, and as we will see below it leads to a cutoff energy that is slightly reduced compared to that of the case of a bound initial state.

The dipole acceleration as well as the corresponding spectra are compared in Figs. 4.33, 4.34 and 4.35. The first figure shows the results of a full quantum mechanical calculation, the second those of a classical Monte-Carlo calculation [73] and the third plot shows semiclassical Herman-Kluk results. In all three cases the dipole acceleration has maxima whenever the electron is close to the nucleus. Due to the broadening of the wavepacket in the course of time the maxima get smeared out. In the quantum, as well as in the semiclassical result, at longer times, fast oscillations of $a(t)$ occur, which are not present in the classical result.

The HHG spectra extracted from the time-domain results are displayed in the lower panels of the respective plots. The full quantum spectrum displays the plateau and a cutoff at 105 times the fundamental frequency. The energy corresponding to this frequency is close to $E_C = 2\,U_{\mathrm{p}} + I_{\mathrm{p}}$, which is the predicted cutoff for a scattering initial condition, differing slightly from the one for a bound state initial condition. The "cutoff" is well represented in the classical, as well as the semiclassical result. The plateau, however, can only be observed in the semiclassical case! The plateau formation therefore seems to be an interference effect.

In order to further investigate the nature of the interference effect, underlying the formation of the plateau, several different classes of classical trajectories have been identified. Apart from "almost free" trajectories, also stranded and trapped trajectories exist, as can be seen in Fig. 4.36. All trajectories in this figure start at the same initial position but with different initial momenta.

Successively neglecting the trapped and stranded trajectories in the semiclassical calculation, the behavior in Fig. 4.37 can be observed: the plateau is vanishing. It is

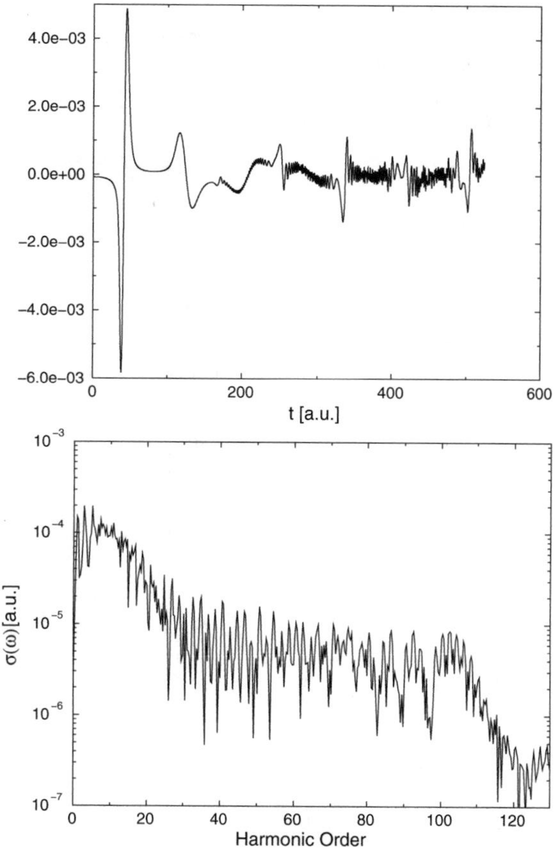

Fig. 4.33 Full quantum mechanical dipole acceleration (*upper panel*) and logarithmic plot of the corresponding spectrum (*lower panel*), adapted from [19]

surprising that neglecting only around 8% of the trajectories is enough to completely suppress the formation of the plateau. These observations led the authors of [72] to conclude that the interference of stranded and trapped trajectories with the free ones is responsible for the plateau formation.

4.5.5 *Cutoff and Odd Harmonics Revisited*

Apart from the explanation of the plateau, also the odd harmonics rule and the cutoff can be understood semiclassically. This is another application of the stationary phase approximation of Sect. 2.2.2 and shall be dealt with below.

Motivated by the fact that the HHG can be explained as an interference effect of two classes of trajectories (almost free and trapped or stranded), the total wavefunction shall be composed of two parts [19]:

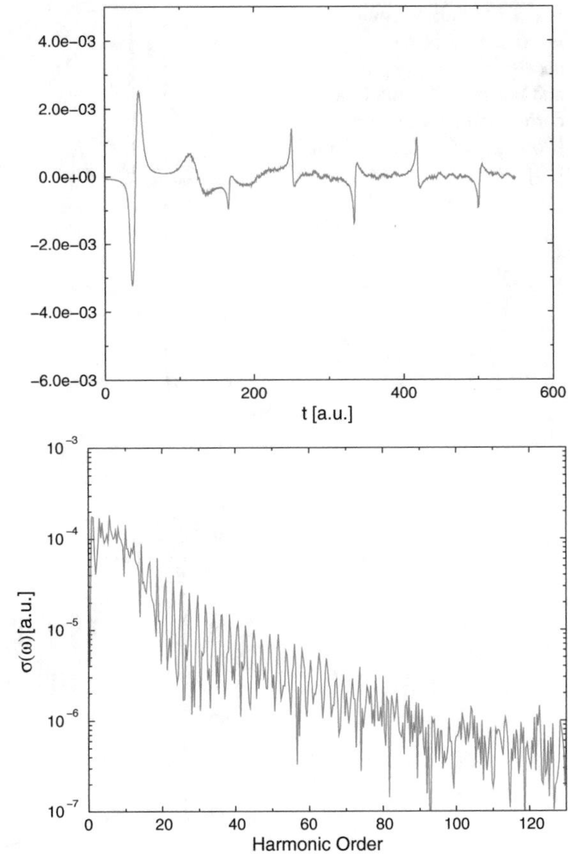

Fig. 4.34 Classical dipole acceleration (*upper panel*) and logarithmic plot of the corresponding spectrum (*lower panel*), adapted from [19]

- An "ionization wavefunction", for the free electron in the laser potential V_{L} with $f(t) = 1$, which is given by the Volkov solution (3.25)
- A "ground state wavefunction", of the electron in the soft-core Coulomb potential without the laser. The phase that this wavefunction acquires in the course of time is determined by the ionization potential via

$$\Phi(t) \approx I_{\mathrm{p}}t. \tag{4.104}$$

It has been shown in [19] that a small admixture of the ground state wavefunction to the Volkov wavepacket leads to a close qualitative agreement with the exact results for the expectation value of the dipole acceleration.

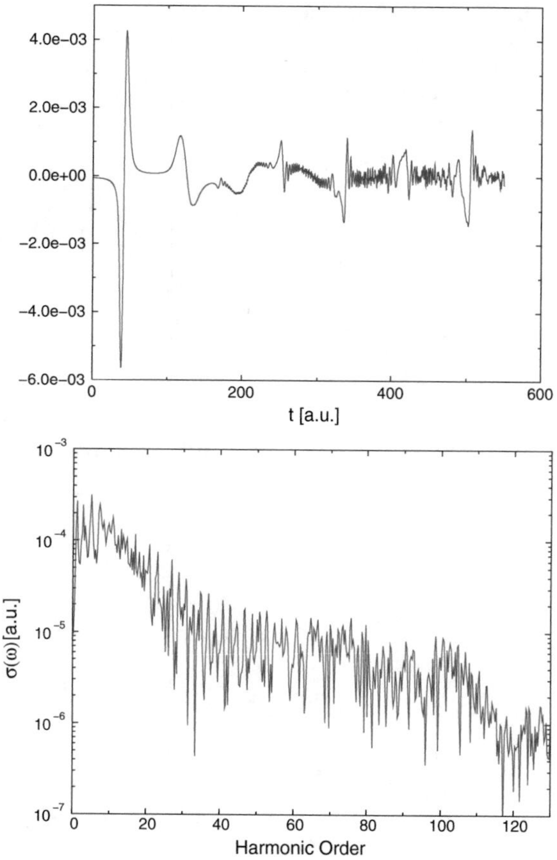

Fig. 4.35 Semiclassical dipole acceleration (*upper panel*) and logarithmic plot of the corresponding spectrum (*lower panel*), adapted from [19]

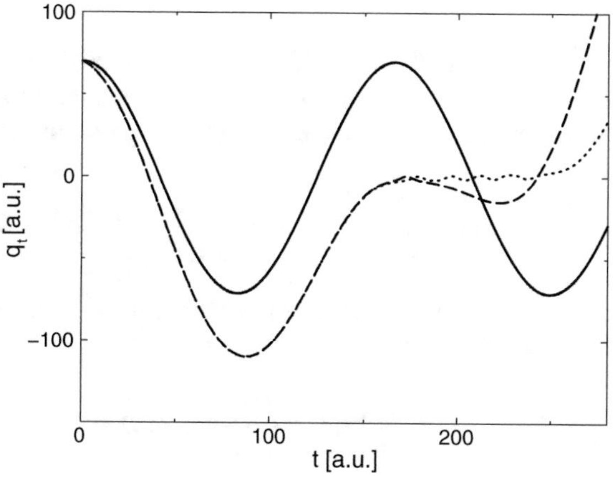

Fig. 4.36 "Almost free" (*solid line*), "trapped" (*dotted line*) and "stranded" (*dashed line*) trajectories, adapted from [19]

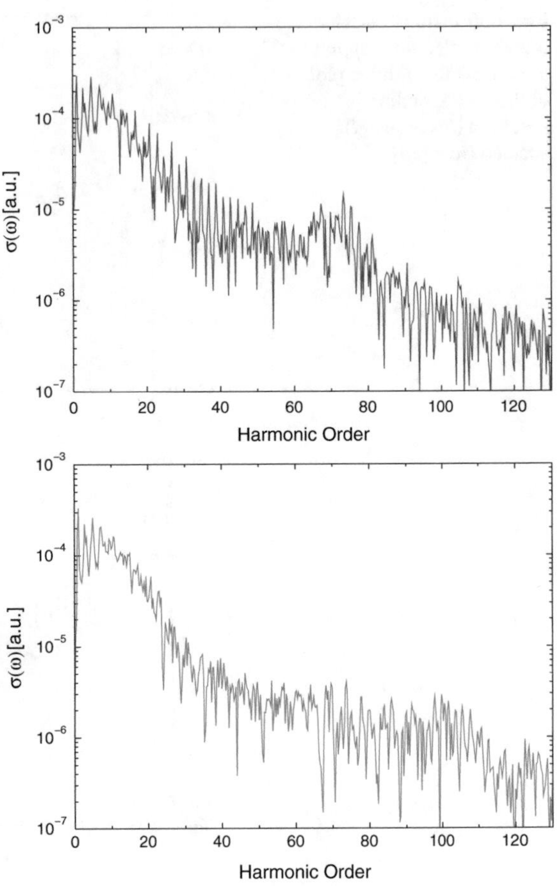

Fig. 4.37 Spectrum without trapped (*upper panel*) and without trapped and stranded trajectories (*lower panel*), adapted from [19]

In order to perform the SPA, the phase in the Fourier integral for the calculation of the HHG spectrum has to be known. This can be extracted from the superposition[12] of the two wavefunctions leading to an expression of the form

$$\sigma(\Omega) = \int \mathrm{d}t\, \mathrm{e}^{-\mathrm{i}\Omega t} \left| \exp\left\{ \mathrm{i}\left[\frac{U_\mathrm{p}}{2\omega} \sin(2\omega t) - U_\mathrm{p} t \right] \right\} + \exp\left\{ \mathrm{i} I_\mathrm{p} t \right\} \right|^2 \quad (4.105)$$

for the spectrum. Calculating the Fourier integral in the stationary phase approximation leads to the condition

$$\frac{\mathrm{d}}{\mathrm{d}t}\left[\Omega t \pm \left(\frac{U_\mathrm{p}}{2\omega} \sin(2\omega t) - U_\mathrm{p} t - I_\mathrm{p} t \right) \right] = 0 \quad (4.106)$$

[12]For reasons of simplicity, we take a superposition with equal weights in the following.

and therefore, for the interference term, the main contributions to the integral are at the (positive) frequencies

$$\Omega = U_{\text{p}}[1 - \cos(2\omega t)] + I_{\text{p}}. \tag{4.107}$$

The maximal value of the expression above follows immediately to be

$$\Omega_{\max} = 2\,U_{\text{p}} + I_{\text{p}}. \tag{4.108}$$

Like in the discussion of the three-step model this frequency is generated by an electron that acquires the maximal kinetic energy in the laser field and transforms it (together with I_{p}) into a single photon in an inverse photo-electric effect.

Also the fact that only odd harmonics are emitted can be derived by using the stationary phase approximation. Again, the interference term

$$\int \mathrm{d}t \exp\left\{-\mathrm{i}\left[(\Omega - U_{\text{p}} - I_{\text{p}})\,t + \frac{U_{\text{p}}}{2\omega}\sin(2\omega t)\right]\right\} \tag{4.109}$$

has to be studied to this end. If the exponent of the sine function is expanded in terms of Bessel functions, using the Jacobi-Anger formula,

$$\exp[\mathrm{i}\alpha\sin(x)] = \sum_{n=-\infty}^{\infty} J_n(\alpha)\exp(\mathrm{i}nx), \tag{4.110}$$

in the exponent of (4.109), this leads to the replacement of the sine by its argument, according to

$$\sum_n J_n\left(\frac{U_{\text{p}}}{2\omega}\right)\int \mathrm{d}t \exp\left\{-\mathrm{i}\left(\Omega - U_{\text{p}} - I_{\text{p}} + 2n\omega\right)t\right\}. \tag{4.111}$$

Along the lines of the SPA, the maxima of this expression occur at

$$\Omega_n = U_{\text{p}} + I_{\text{p}} - 2n\omega \tag{4.112}$$

and are separated by an *even* number of fundamental frequencies 2ω. The fundamental is contained in the spectrum and therefore all the maxima are at *odd* harmonics.

The cutoff can once more be derived by looking at the asymptotic form

$$|n| \gg 1: \qquad J_n(x) \to \frac{(-1)^n}{\sqrt{2\pi|n|}}\left(\frac{\mathrm{e}x}{2|n|}\right)^{|n|} \tag{4.113}$$

of the Bessel function (here $\mathrm{e} \approx 2.72$ is Euler's number). For $|n| > \frac{\mathrm{e}}{2}U_{\text{p}}/(2\omega) \approx U_{\text{p}}/(2\omega)$, the Bessel function goes to zero, leading to an upper bound $\Omega_{\max} = 2\,U_{\text{p}} + I_{\text{p}}$, if for n the minimal value $-U_{\text{p}}/(2\omega)$ is inserted in (4.112).

4.5.6 Dominant Interaction Hamiltonian for HHG

Another possibility to deepen the understanding of the generation of high-oder harmonics is the use of the dominant interaction Hamiltonian (DIH) concept, which has first been applied in the present context in [74].

To understand the basic idea, we split the 1D soft-core Hamiltonian

$$\hat{H}(t) = \hat{H}_i + V_{\mathrm{L}}(t) = \hat{T}_{\mathrm{k}} + V_{\mathrm{sc}} + V_{\mathrm{L}}(t) = \hat{H}_f(t) + V_{\mathrm{sc}} \tag{4.114}$$

analogous to (4.40) and observe that close to the nucleus, the Hamiltonian $\hat{H}_i$ plays the dominant role, while far away, that role is taken by the field Hamiltonian $\hat{H}_f$. In an approach based on classical trajectories, the dynamics can be restricted to evolve under the dominant part of the Hamiltonian only, and using the Herman-Kluk propagator, it has been shown in [74, 75] that this leads to dipole accelerations and spectra that are almost indistinguishable from the full semiclassical and/or full quantum ones. In order to achieve this agreement, a switching criterion for an electron trajectory had been worked out that assumes that the electron is slow and close to the nucleus right before the transition.

In the case of a soft-core parameter $a = 2$ and for the field parameters $\mathcal{E}_0 = 0.1$ a.u. and $\omega = 0.0378$ a.u., used before, a switch from the laser to the soft-core potential occurs, whenever $p_c = 0$ and $|q_c| \le x_c = 3.0083$ are fulfilled and immediately before the switching, the momentum is pointing inwards. For $|q| \le x_c$, the laser potential will never become stronger than the soft-core potential.

It is instructive to investigate the initial conditions of the trajectories which switch and get trapped. It was found that they form bands in the initial phase space. To see that, we use the solution of the equations of motion (3.26) and (3.27) of an electron in a laser field with initial conditions $p(0) = p_0$ and $q(0) = q_0$, which are repeated here in atomic units

$$p(t) = p_0 - \frac{\mathcal{E}_0}{\omega} \sin(\omega t) , \tag{4.115}$$

$$q(t) = q_0 + p_0 t + \frac{\mathcal{E}_0}{\omega^2} [\cos(\omega t) - 1] . \tag{4.116}$$

From the condition $p(t_c) = 0$ one finds $t_c \approx n\pi/\omega$ with $n = 1, 2, \ldots$. Then, the conditions for a switch from V_{L} to V_{C} are

$$\begin{aligned} q_c =& q_0 + \frac{n\pi}{\omega} p_0 + \frac{\mathcal{E}_0}{\omega^2} [\cos(n\pi) - 1] \\ =& q_0 + \frac{n\pi}{\omega} p_0 - \frac{1-(-1)^n}{2} \frac{2\mathcal{E}_0}{\omega^2} . \end{aligned} \tag{4.117}$$

For even n, q_c lies in the interval $[-x_c, 0]$, whereas for odd n, q_c lies in the interval $[0, x_c]$. This implies that the initial phase space points are given by lines $p_0^{(n)}(q_0)$ with a width $\Delta p_0^{(n)} = \omega/(n\pi)x_c$, as illustrated in Fig. 4.38. The explicit expression follows from rearranging (4.117), leading to

$$p_0^{(n)}(q_0) = -\frac{\omega}{n\pi}\left(q_0 - q_c - \left[1 - (-1)^n\right]\frac{\mathcal{E}_0}{\omega^2}\right). \tag{4.118}$$

For the dynamics under the full Hamiltonian, it turns out that the initial conditions for the trajectories which get trapped are located on the same phase space stripes. This has led to the conclusion that the DIH switching condition describes the dynamics relevant for HHG quite well. A comparison of the very similar initial dynamics of two different trajectories (full Hamiltonian versus DIH) in phase space is shown in Fig. 4.39.

A decisive numerical advantage of the DIH approach is that the classical dynamics under the dominant interactions is regular, in contrast to the chaotic dynamics under the complete, nonlinear, driven Hamiltonian. Therefore, the semiclassical DIH calculations could be converged with two orders of magnitude less trajectories as compared to the complete case [76]. In addition, an analytical determination

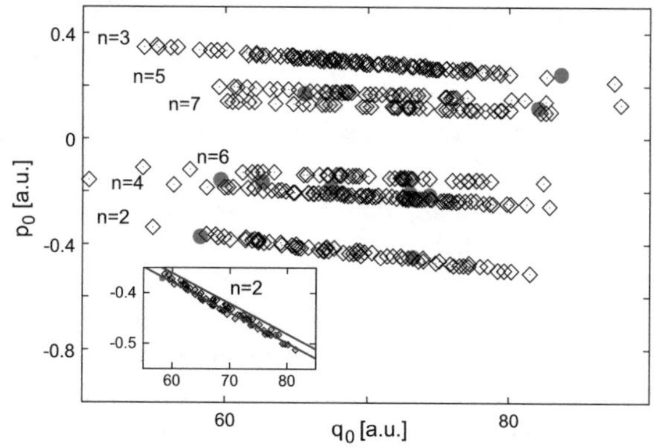

Fig. 4.38 Initial conditions for laser driven trajectories in a soft-core potential with $a = 2$: switched trajectories (*diamonds*), trapped or stranded trajectories in the full potential case (*filled circles*) and analytical conditions (*blue lines* in the inset), see (4.118). The field parameters are $\mathcal{E}_0 = 0.1$ a.u. and $\omega = 0.0378$ a.u. [74]

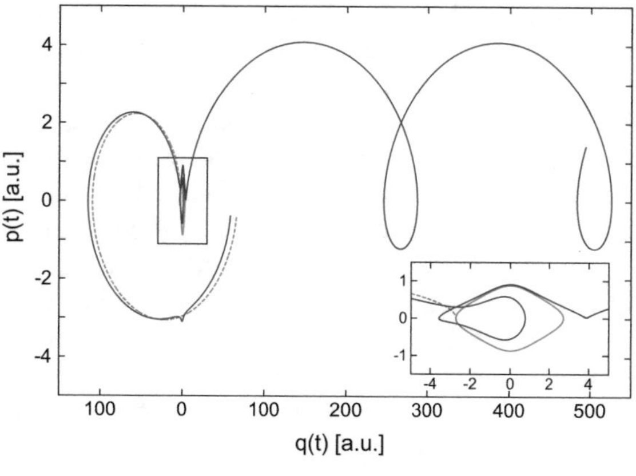

Fig. 4.39 Phase space plot of a DIH trajectory (*dashed red line*) and a stranded trajectory under the full potential (*full blue line*). The switched trajectory ends in a closed orbit, as can be seen in the inset [75]

of the relative weights in the wavefunction mixture that was discussed in detail in Sect. 4.5.5 has been given, using the idea of dominant interaction Hamiltonians [75]. In this publication it was furthermore shown that a linearized semiclassical (truncated Wigner) study of the dipole acceleration does not lead to a plateau formation in the harmonic yield, in line with the classical results presented in [72].

4.6 Notes and Further Reading

Supplementary Books and Basics

The study of the interaction of light fields with atoms has a long history. Classic references are the books "Photon-Atom Interactions" by Weissbluth [77] and "Introduction to the Theory of Laser-Atom Interactions" by Mittleman [78]. The main focus in the first book is on the interaction of atoms with moderately strong fields. Multi-photon and nonlinear processes are treated in the last chapter. The second book has a focus on scattering phenomena in the presence of laser fields and in its second edition also contains multi-photon effects. More recently, the theory of atoms (and also molecules), including their interaction with external (laser) fields, has been covered in the textbook by Bransden and Joachain [27], as well as in the works by Hill and Lee [79], by Joachain, Kylstra and Potvliege [80] and by van der Straten and Metcalf [81].

A book, covering the fundamentals of theoretical atomic physics, which also includes an introductory chapter on quantum theory (with the solution of the Coulomb problem) and a nice collection of exercises with solutions is the one by Friedrich [3]. The book by Feagin contains a lot of material on the hydrogen atom with more information on the properties of spherical harmonics (including a 3D cut-away spherical polar plot) that appear in the solution of the corresponding TISE [82]. The book by Bransden and Joachain [27] contains a chapter on the interaction with strong fields, which covers some topics that we have omitted. These are, e.g., the cooling of atoms with lasers and non-dipole and relativistic effects. Furthermore, we have not considered linear and nonlinear optical spectroscopy in this chapter. Some aspects of nonlinear spectroscopy are dealt with in Chap. 5 and in the textbooks by Weissbluth [77] and Mukamel [83]. More theoretical and experimental information on the topics treated here can be found in the book by Fedorov [84]. Furthermore, chaos in atomic (Rydberg) systems, which can be induced by external laser fields, is covered in the books by Blümel and Reinhard [9], by Reichl [85], and by Bayfield [86].

The AC Stark shift, i.e., the shift of the atomic levels due to time-dependent electric fields is discussed using perturbation theory, e.g., in [27], as well as in [78] and [84]. It is intimately related to the (instantaneous) ponderomotive energy and its dependence on the gauge used is discussed in the supplemental material to [87]. Therein, also a critical discussion of the effect of dynamic interference of photoelectrons in high-frequency laser fields [88] can be found.

Helium

The discussion of the unperturbed helium atom and its semiclassical solution was rather brief. More details on the Pauli principle and the theory of electron spin can be found in standard texts on quantum theory, as well as in the books by Bransden and Joachain [27], Friedrich [3], and van der Straten and Metcalf [81]. Reviews of the early unsuccessful, as well as the successful semiclassical treatment of (collinear) helium [89] can be found in the books by Blümel and Reinhardt [9] and Friedrich [3], and in the article by Tanner, Richter and Rost [90].

Differences between singlet and triplet initial state ionization dynamics in Helium have been investigated in [91]. Corresponding differences between singlet and triplet initial states in the scattering of two electrons have been described semiclassically by using the Herman-Kluk propagator in [92], without invocing artificial Pauli potentials.

The dominant interaction Hamiltonian idea presented in Sect. 4.5.6 has also been used in a study of planar electron-atom scattering in a three-body two-electron system [93].

Photoionization and HHG

Over the years, several authors in addition to the ones already mentioned have worked on the ionization of atoms by (intense) fields in the quasi static regime. This interest started in the heyday of quantum theory [94, 95] and still continues, as demonstrated by recent experimental work on the ionization of Kr atoms by sub-femtosecond laser pulses together with adiabatic tunneling ionization calculations [96].

Photoelectron holography that may reveal structural information of the investigated atomic system is discussed in [97]. Furthermore, recent experimental progress has led to intensive theoretical studies of the photoionization time-delays in atomic systems. In an “attosecond streaking” pump-probe type experiment (see also Chap. 5) a single attosecond XUV pulse ionizes an atom and an infrared laser pulse serves as a clock to measure the delays [98]. In this tutorial introduction and in the more recent review [99] also other related types of experiments are discussed.

The question of how ATI and HHG are related has been answered in [100]. There it is shown that HHG can be understood as ATI followed by electron propagation in the continuum and subsequent stimulated recombination. More information about experimental (and historical) aspects of ATI can be found in Chap. 15 of [27]. Furthermore, recombination and thus creation of high-order harmonics in atoms is suppressed if circularly polarized driving fields are used instead of the linearly polarized ones considered in our 1D studies [101].

Miscellaneous Topics

A Floquet theoretical study of the interaction of atoms with an intense laser pulse is given in [102] and Chap. 4 of [80] is devoted to the same topic. A striking “dichotomy”, i.e., a separation into two parts of the quantum wavefunction (in the

Kramers-Henneberger frame) of the hydrogen atom in superintense high-frequency fields has been unveiled in the Floquet picture in [103].

If generated electromagnetically, due to the properties of the solution of Maxwell's equations in free space, the "half-cycle pulse" (used in the ionization studies of hydrogen in Sect. 4.3.6) should be replaced by an asymmetric mono-cycle pulse. This fact and its consequences are discussed in [104].

Control of quantum dynamics will be discussed at length in the Chap. 5, dealing with molecules in external fields. Also for atoms, however, the concept of control has been discussed, e.g., in the case of high-order harmonic generation, where one is looking for driving fields that enhance the harmonic yield [105].

4.A More on Atomic Units

There is a close connection between atomic units and some expectation values of the hydrogen atom. The Bohr radius, e.g., is related to the expectation value of the position operator in the ground state which is given by

$$\langle \hat{r} \rangle = \frac{3}{2} a_0 \, , \tag{4.119}$$

and in addition the maximum of the probability density for finding the particle in the ground state with a radial separation r from the nucleus is at a_0 [1]. Furthermore, the energy constructed from the atomic base units is twice the absolute value of the ground state energy $E_1 = -\frac{1}{2}$ a.u.≈ -13.6 eV. One could also measure energy in terms of the ionization potential

$$I_{\mathrm{p}} = -E_1 \tag{4.120}$$

of hydrogen, however. The corresponding units are so-called Rydberg units (1 a.u. = 2 Rydberg), in contrast to the atomic or Hartree units.

The atomic unit for velocity follows most easily by considering the quantum mechanical virial theorem [1], which for the hydrogen atom can be stated as

$$\langle \hat{T}_{\mathrm{k}} \rangle = -\frac{1}{2} \langle \hat{V}_{\mathrm{C}} \rangle , \tag{4.121}$$

where the brackets denote expectation values. For the ground state, we find with its help that

$$\langle \hat{T}_{\mathrm{k}} \rangle = -E_1 \tag{4.122}$$

and with the definition

$$\langle \hat{T}_{\mathrm{k}} \rangle =: \frac{m_{\mathrm{e}} v_0^2}{2}, \tag{4.123}$$

the atomic velocity unit $v_0 = \sqrt{2|E_1|/m_{\mathrm{e}}} \approx c\sqrt{27.212\,\mathrm{eV}/0.511\,\mathrm{MeV}}$ is related to the vacuum speed of light via $v_0 = c\alpha = 1\,\mathrm{a.u.}$, with the fine structure constant $\alpha \approx 1/137$. Using v_0 and a_0, the atomic unit of time is formed via $t_0 = a_0/v_0 \approx 24.2\,\mathrm{as}$. The oscillation period of the electron on the Bohr orbit is therefore given by $T_{\mathrm{e}} = 2\pi\,\mathrm{a.u.} \approx 152\,\mathrm{as}$.

The most important unit for this book is the one for the electric field which is given by $\mathcal{E}_{\mathrm{at}} \approx 5.1427 \cdot 10^{11}\,\mathrm{V\,m^{-1}}$. This is the value of the electric field, due to the proton, experienced by the electron at the Bohr radius. The intensity that corresponds to that field is $I_{\mathrm{at}} = \frac{c\varepsilon_0 \mathcal{E}_{\mathrm{at}}^2}{2} \approx 3.5101 \cdot 10^{16}\,\mathrm{W\,cm^{-2}}$. An overview of the units mentioned can be found in Table 4.3. As an example, we use this table to convert frequency into wavelength. For frequencies given in a.u. ($\omega = \mathrm{X}$ a.u.), the wavelength (in SI units) is

$$\lambda = \frac{2\pi c}{\omega} \approx \frac{2\pi\, 137.036}{\mathrm{X}}\,\mathrm{a.u.} \approx \frac{45.5636}{\mathrm{X}}\,\mathrm{nm}\,.$$

Table 4.3 Definition of atomic units and some important quantities in atomic and SI units

	SI (m, kg, s, A)	a.u. ($\mathrm{a_0}$, $\mathrm{m_e}$, ħ, e)
Definitions:		
$\hbar$	$\approx 1.0546 \cdot 10^{-34}\,\mathrm{J\,s}$	$:= 1\,\hbar$
e	$\approx 1.6022 \cdot 10^{-19}\,\mathrm{A\,s}$	$:= 1\,\mathrm{e}$
m_{e}	$\approx 9.1094 \cdot 10^{-31}\,\mathrm{kg}$	$:= 1\,\mathrm{m_e}$
a_0	$\approx 5.2918 \cdot 10^{-11}\,\mathrm{m}$	$:= 1\,\mathrm{a_0}$
Further quantities:		
$4\pi\varepsilon_0 = \frac{10^7}{c^2}\,\mathrm{A\,m\,(V\,s)^{-1}}$	$\approx 1.1126 \cdot 10^{-10}\,\mathrm{A\,s\,(V\,m)^{-1}}$	$= 1\,\frac{\mathrm{a_0\,m_e\,e^2}}{\hbar^2}$
$v_0 \approx \frac{1}{137.036}c$	$\approx 2.1877 \cdot 10^{6}\,\mathrm{ms^{-1}}$	$= 1\,\frac{\hbar}{\mathrm{a_0\,m_e}}$
$2\lvert E_1\rvert = \frac{m_{\mathrm{e}}e^4}{16\pi^2\varepsilon_0^2\hbar^2}$	$\approx 27.212\,\mathrm{eV}$	$= 1\,\frac{\hbar^2}{\mathrm{a_0^2\,m_e}}$
$\boldsymbol{\mathcal{E}}_{\mathrm{at}} = \frac{e\,\boldsymbol{e}_r}{4\pi\varepsilon_0 a_0^2}$	$\approx 5.1427 \cdot 10^{11}\,\mathrm{Vm^{-1}}\,\boldsymbol{e}_r$	$= 1\,\frac{\hbar^2}{\mathrm{a_0}^3\,\mathrm{m_e\,e}}\boldsymbol{e}_r$
$I_{\mathrm{at}} = \frac{c\varepsilon_0 \boldsymbol{\mathcal{E}}_{\mathrm{at}}^2}{2}$	$\approx 3.5101 \cdot 10^{20}\,\mathrm{Wm^{-2}}$	$\approx 5.4556\,\frac{\hbar^3}{\mathrm{a_0^6\,m_e^2}}$
$t_0 = \frac{a_0}{v_0}$	$\approx 2.4189 \cdot 10^{-17}\mathrm{s}$	$= 1\,\frac{\mathrm{m_e\,a_0}^2}{\hbar}$

References

1. E. Merzbacher, *Quantum Mechanics*, 2nd edn. (Wiley, New York, 1970)
2. M. Abramowitz, I.A. Stegun, *Handbook of Mathematical Functions* (Dover Publications, New York, 1965)
3. H. Friedrich, *Theoretical Atomic Physics*, 3rd edn. (Springer, Berlin, 2010)
4. I.S. Gradshteyn, I.M. Ryzhik, *Table of Integrals Series and Products*, 5th edn. (Academic Press, New York, 1994)
5. R. Loudon, Am. J. Phys. **27**, 649 (1959)
6. M. Andrews, Am. J. Phys. **34**, 1194 (1966)
7. U. Schwengelbeck, F.H.M. Faisal, Phys. Rev. A **50**, 632 (1994)
8. M. Andrews, Am. J. Phys. **44**, 1064 (1976)
9. R. Blümel, W.P. Reinhardt, *Chaos in Atomic Physics* (Cambridge University Press, Cambridge, 1997)
10. J.H. Eberly, Phys. Rev. A **42**, 5750 (1990)
11. D. Bauer, Phys. Rev. A **56**, 3028 (1997)
12. G. Ezra, K. Richter, G. Tanner, D. Wintgen, J. Phys. B: At. Mol. Opt. Phys. **24**, L413 (1991)
13. C. Harabati, K.G. Kay, J. Chem. Phys. **127**, 084104 (2007)
14. F.A. Ilkov, J.E. Decker, S.L. Chin, J. Phys. B: At. Mol. Opt. Phys. **25**, 4005 (1992)
15. L.D. Landau, E.M. Lifshitz, *Lehrbuch der Theoretischen Physik*, vol. III: Quantenmechanik, 8th edn. (Akademie-Verlag, Berlin, 1988)
16. M.V. Ammosov, N.B. Delone, V.P. Krainov, Sov. Phys. JETP **64**, 1191 (1986)
17. P. Lambropoulos, X. Tang, Adv. At. Mol. Phys. Suppl. **1**, 335 (1992)
18. G. van de Sand, J.M. Rost, Phys. Rev. A **62**, 053403 (2000)
19. G. van de Sand, *Semiklassische Dynamik in starken Laserfeldern*. Ph.D. thesis, Universität Freiburg, 1999
20. M. Klaiber, J.S. Briggs, Phys. Rev. A **94**, 053405 (2016)
21. J.R. Taylor, *Scattering Theory* (Dover, Mineola, 2006)
22. H.R. Reiss, Opt. Express **8**, 99 (2001)
23. L.V. Keldysh, Sov. Phys. JETP **20**, 1307 (1965)
24. P. Agostini, F. Fabre, G. Mainfray, G. Petite, N.K. Rahman, Phys. Rev. Lett. **42**, 1127 (1979)
25. D.B. Milosevic, G.G. Paulus, D. Bauer, W. Becker, J. Phys. B: At. Mol. Opt. Phys. **39**, R203 (2006)
26. M. Protopapas, C.H. Keitel, P.L. Knight, Rep. Prog. Phys. **60**, 389 (1997). (And refs. therein)
27. B.H. Bransden, C.J. Joachain, *Physics of Atoms and Molecules*, 2nd edn. (Pearson Education, Harlow, 2003)
28. S.I. Chu, J. Cooper, Phys. Rev. A **32**, 2769 (1985)
29. R. Grobe, C.K. Law, Phys. Rev. A **44**, R4114 (1991)
30. R.R. Jones, D. You, P.H. Bucksbaum, Phys. Rev. Lett. **70**, 1236 (1993)
31. D.S. Citrin, Opt. Express **1**, 376 (1997)
32. M.T. Frey, F.B. Dunning, C.O. Reinhold, S. Yoshida, J. Burgdörfer, Phys. Rev. A **59**, 1434 (1999)
33. D. Dimitrovski, E.A. Solov'ev, J.S. Briggs, Phys. Rev. A **72**, 043411 (2005)
34. M. Klaiber, D. Dimitrovski, J.S. Briggs, J. Phys. B: At. Mol. Opt. Phys. **41**, 175002 (2008)
35. R. Gebarowski, J. Phys. B: At. Mol. Opt. Phys. **30**, 2143 (1997)
36. C.L. Stokely, J.C. Lancaster, F.B. Dunning, D.G. Arbó, C.O. Reinhold, J. Burgdörfer, Phys. Rev. A **67**, 013403 (2003)
37. L.D. Landau, E.M. Lifschitz, *Lehrbuch der Theoretischen Physik*, vol. I: Mechanik, 10th edn. (Akademie-Verlag, Berlin, 1981)
38. S. Yoshida, F. Grossmann, E. Persson, J. Burgdörfer, Phys. Rev. A **69**, 043410 (2004)
39. S. Yoshida, *Quantum Chaos in the periodically kicked Rydberg atom*. Ph.D. thesis, University of Tennessee, Knoxville, 1999
40. N.T. Maitra, J. Chem. Phys. **112**, 531 (2000)

41. P.B. Corkum, Phys. Rev. Lett. **71**, 1994 (1993)
42. T. Zuo, A.D. Bandrauk, P.B. Corkum, Chem. Phys. Lett. **259**, 313 (1996)
43. B. Yang, K.J. Schafer, B. Walker, K.C. Kulander, P. Agostini, L.F. DiMauro, Phys. Rev. Lett. **71**, 3770 (1993)
44. H. Reiss, Phys. Rev. A **22**, 1786 (1980)
45. V. Schyja, T. Lang, H. Helm, Phys. Rev. A **57**, 3692 (1998)
46. J. Henkel, M. Lein, Phys. Rev. A **92**, 013422 (2015)
47. G.G. Paulus, W. Nicklich, H. Xu, P. Lambropoulos, H. Walther, Phys. Rev. Lett. **72**, 2851 (1994)
48. G.G. Paulus, W. Becker, W. Nicklich, H. Walther, J. Phys. B: At. Mol. Opt. Phys. **27**, L703 (1994)
49. C.I. Blaga, F. Catoire, P. Colosimo, G.G. Paulus, H.G. Muller, P. Agostini, L.F. DiMauro, Nat. Phys. **5**, 335 (2009)
50. A. Kästner, U. Saalmann, J.M. Rost, Phys. Rev. Lett. **108**, 033201 (2012)
51. E. Diesen, *Low-energy electrons in strong-field ionization*. Ph.D. thesis, Technische Universität Dresden, 2016
52. A. Kästner, U. Saalmann, J.M. Rost, J. Phys. B: At. Mol. Opt. Phys. **45**, 074011 (2012)
53. B. Walker, B. Sheehy, L.F. DiMauro, P. Agostini, K.J. Schafer, K.C. Kulander, Phys. Rev. Lett **73**, 1227 (1994)
54. P.J. Ho, R. Panfili, S.L. Haan, J.H. Eberly, Phys. Rev. Lett **94**, 093002 (2005)
55. F. Mauger, C. Chandre, T. Uzer, J. Phys. B: At. Mol. Opt. Phys. **42**, 165602 (2009)
56. F. Mauger, C. Chandre, T. Uzer, Phys. Rev. Lett. **102**, 173002 (2009)
57. D. Dundas, K.T. Taylor, J.S. Parker, E.S. Smyth, J. Phys. B: At. Mol. Opt. Phys. **32**, L231 (1999)
58. A. Zielinski, V.P. Majety, A. Scrinzi, Phys. Rev. A **93**, 023406 (2016)
59. A. Becker, F.H.M. Faisal, J. Phys. B: At. Mol. Opt. Phys. **29**, L197 (1996)
60. D.N. Fittinghoff, P.R. Bolton, B. Chang, K.C. Kulander, Phys. Rev. Lett. **69**, 2642 (1992)
61. M. Lein, E.K.U. Gross, V. Engel, Phys. Rev. Lett. **85**, 4707 (2000)
62. A.S. de Wijn, M. Lein, S. Kümmel, EPL **84**, 43001 (2008)
63. M. Petersilka, E.K.U. Gross, Laser Phys. **9**, 1 (1999)
64. A. Kirrander, D.V. Shalashilin, Phys. Rev. A **84**, 033406 (2011)
65. J. Seres, E. Seres, A.J. Verhoef, G. Tempea, C. Streli, P. Wobrauschek, V. Yakovlev, A. Scrinzi, C. Spielmann, F. Krausz, Nature **433**, 596 (2005)
66. A. Pukhov, S. Gordienko, T. Baeva, Phys. Rev. Lett. **91**, 0173002 (2003)
67. K. Burnett, V.C. Reed, J. Cooper, P.L. Knight, Phys. Rev. A **45**, 3347 (1992)
68. M. Protopapas, D.G. Lappas, C.H. Keitel, P.L. Knight, Phys. Rev. A **53**, R2933 (1996)
69. V. Yakovlev, A. Scrinzi, Phys. Rev. Lett. **91**, 153901 (2003)
70. M. Lewenstein, P. Balcou, M.Y. Ivanov, A. L'Huillier, P.B. Corkum, Phys. Rev. A **49**, 2117 (1994)
71. T. Kreibich, M. Lein, V. Engel, E.K.U. Gross, Phys. Rev. Lett. **87**, 103901 (2001)
72. G. van de Sand, J.M. Rost, Phys. Rev. Lett. **83**, 524 (1999)
73. R. Abrines, I.C. Percival, Proc. Phys. Soc. **88**, 861 (1966)
74. C. Zagoya, C.M. Goletz, F. Grossmann, J.M. Rost, Phys. Rev. A **85**, 041401(R) (2012)
75. C. Zagoya, C.M. Goletz, F. Grossmann, J.M. Rost, New J. Phys. **14**, 093050 (2012)
76. C. Zagoya, *Semiclassical description of quantum interference in nonlinear dynamics*. Ph.D. thesis, Technische Universität Dresden, 2014
77. M. Weissbluth, *Photon-Atom Interactions* (Academic Press, New York, 1989)
78. M.H. Mittleman, *Introduction to the Theory of Laser-Atom Interactions*, 2nd edn. (Springer, US, New York, 1993)
79. W.T. Hill, C.H. Lee, *Light-Matter Interaction: Atoms and Molecules in External Fields and Nonlinear Optics* (Wiley-VCH, Weinheim, 2007)
80. C.J. Joachain, N.J. Kylstra, R.M. Potvliege, *Atoms in Intense Laser Fields* (Cambridge University Press, Cambridge, 2012)

81. P. van der Straaten, H. Metcalf, *Atoms and molecules interacting with light* (Cambridge University Press, Cambridge, 2016)
82. J.M. Feagin, *Quantum Methods with Mathematica* (Springer Verlag, New York, 1994)
83. S. Mukamel, *Principles of Nonlinear Optical Spectroscopy* (Oxford University Press, New York, 1995)
84. M.V. Fedorov, *Atomic And Free Electrons In A Strong Light Field* (World Scientific, Singapore, 1997). (And Refs. therein)
85. L.E. Reichl, *The Transition to Chaos*, 2nd edn. (Springer, New York, 2004)
86. J.E. Bayfield, *Quantum Evolution: An Introduction to Time-Dependent Quantum Mechanics* (Wiley, New York, 1999)
87. M. Baghery, U. Saalmann, J.M. Rost, Phys. Rev. Lett. **118**, 143202 (2017)
88. P.V. Demekhin, L.S. Cederbaum, Phys. Rev. Lett. **108**, 253001 (2012)
89. D. Wintgen, K. Richter, G. Tanner, Chaos **2**, 19 (1992)
90. G. Tanner, K. Richter, J.M. Rost, Rev. Mod. Phys. **72**, 497 (2000)
91. V. Véniard, R. Taïeb, A. Maquet, J. Phys. B: At. Mol. Opt. Phys. **36**, 4145 (2003)
92. F. Grossmann, M. Buchholz, E. Pollak, M. Nest, Phys. Rev. A **89**, 032104 (2014)
93. M. Gerlach, S. Wüster, J.M. Rost, J. Phys. B: At. Mol. Opt. Phys. **45**, 235204 (2012)
94. J.R. Oppenheimer, Phys. Rev. **13**, 66 (1928)
95. A.M. Perelomov, V.S. Popov, M.V. Terent'ev, Sov. Phys. JETP **50**, 1393 (1966)
96. A. Wirth, M.T. Hassan, I. Grguraš, J. Gagnon, A. Moulet, T.T. Luu, S. Pabst, R. Santra, Z.A. Alahmed, A.M. Azzeer, V.S. Yakovlev, V. Pervak, F. Krausz, E. Goulielmakis, Science **334**, 195 (2011)
97. A. Stodolna, Y. Huismans, A. Rouzée, F. Lépine, M.J.J. Vrakking, J. Phys.: Conf. Ser. **488**, 012007 (2014)
98. J.M. Dahlström, A. L'Huillier, A. Maquet, J. Phys. B: At. Mol. Opt. Phys. **45**, 183001 (2012)
99. U. Thumm, Q. Liao, E.M. Bothschafter, F. Süßmann, M.F. Kling, R. Kienberger, in *Fundamentals of photonics and physics*, ed. by D.L. Andrew (Wiley, New York, 2015). (chap. 13)
100. M.Y. Kuchiev, V.N. Ostrovsky, Phys. Rev. A **60**, 3111 (1999)
101. K.J. Yuan, A.D. Bandrauk, Phys. Rev. Lett. **110**, 023003 (2013)
102. T. Millack, V. Véniard, J. Henkel, Phys. Lett. A **176**, 433 (1993)
103. M. Pont, N.R. Walet, M. Gavrila, C.W. McCurdy, Phys. Rev. Lett. **61**, 939 (1988)
104. C. Wesdorp, F. Robicheaux, L.D. Noordam, Phys. Rev. Lett. **87**, 083001 (2001)
105. C.F. de Morisson Faria, M.L. Du, Phys. Rev. A **64**, 023415 (2001)

Chapter 5
Molecules in Strong Laser Fields

Molecules, compared to atoms, have additional degrees of freedom. Not only do the electrons move around the nuclei, but also the nuclei move relative to each other. This allows for a multitude of new phenomena, already without the action of external fields.

By coupling an external electric field to the molecular dynamics, such diverse topics as femtosecond spectroscopy, control of molecular dynamics, and the realization of quantum logic operations emerge. Before entering that wide field, some basics of molecular theory will be repeated by discussing the simplest molecule, the hydrogen molecular ion. The concept of electronic potential energy surfaces, on which the nuclear dynamics occurs, will thereby be introduced and the analytical Morse potential curve will be reviewed in some detail. The hydrogen molecular ion then serves as an example for the increasing numerical effort that has been taken in the course of time, starting with frozen nuclei and then taking the nuclear dynamics into account.

After the discussion of the Born-Oppenheimer approximation, dynamics on a single as well as on laser-coupled potential energy surfaces will be reviewed. From the field of pump-probe femtosecond spectroscopy three examples, 2D infrared spectroscopy, photoelectron spectroscopy and fluorescence spectroscopy will be dealt with in some detail. The remaining part of this chapter is then devoted to the concept of control of the molecular dynamics by suitably chosen laser fields.

5.1 The Molecular Ion H_2^+

In order to understand laser driven molecular dynamics, we first review the basics of molecular theory. The time-independent Schrödinger equation of the hydrogen molecular ion is approximately as well as exactly [1] solvable. In order to keep the discussion simple, we first review the approximate treatment of H_2^+ in the stationary

F. Grossmann, *Theoretical Femtosecond Physics*, Graduate Texts in Physics,
https://doi.org/10.1007/978-3-319-74542-8_5

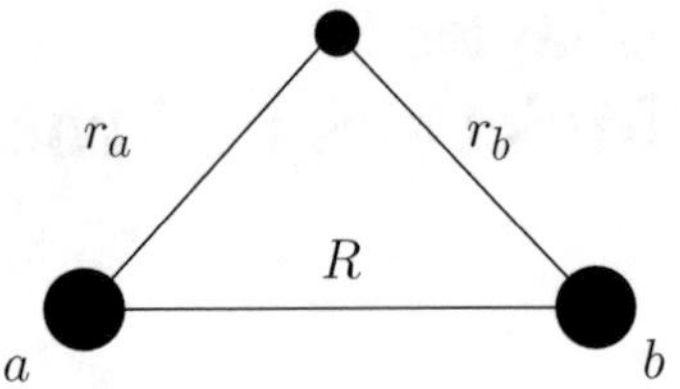

Fig. 5.1 H_2^+ molecular ion consisting of two protons labeled by a and b, with distance R, and a single electron with the distances r_a and r_b from the protons

case, leading to the notion of electronic potential energy surfaces, which will be central for the understanding of almost all the material presented in this chapter.

5.1.1 Electronic Potential Energy Surfaces

The hydrogen molecular ion consists of two protons, apart by a distance of R, and a single electron with the distances r_a and r_b to proton a, respectively proton b. A schematic representation of the molecule is given in Fig. 5.1.

In the following, we try to understand how the energy of the electron is changing as a function of the internuclear distance. It will turn out that there are several possible solutions to this problem, and the outcome has a very intuitive meaning. To make progress, the electron shall first be close to either nucleus a, or nucleus b, while keeping the two nuclei very far apart, i.e., $R \to \infty$. These two limiting cases are simple hydrogen atoms, undisturbed by the other proton, for which the exact solution of the respective time-independent Schrödinger equation

$$\hat{H}_a \psi_a = E_a \psi_a, \tag{5.1}$$

$$\hat{H}_b \psi_b = E_b \psi_b \tag{5.2}$$

is given in Sect. 4.1.1 Furthermore, the two hydrogen atoms are completely equivalent, and therefore the energies are degenerate $E_a = E_b$.

The *full* electronic eigenvalue problem, i.e., the time-independent Schrödinger equation for H_2^+ with a fixed (finite) distance R is given in atomic units by

$$\left\{ -\frac{1}{2}\Delta - \frac{1}{r_a} - \frac{1}{r_b} \right\} \psi_e(r_a, r_b, R) = E(R)\psi_e(r_a, r_b, R) , \tag{5.3}$$

where the Laplace operator can be expressed either by $\nabla_{r_a}^2$ or by $\nabla_{r_b}^2$. The internuclear repulsion leads to an R-dependent shift of the energy scale and shall be neglected for the time being. Due to the fact that we know the solution of the problem for $R \to \infty$, let us calculate the energy $E(R)$ and the corresponding eigenfunction $\psi_e(r_a, r_b, R)$, both depending parametrically on R, as a linear combination of atomic orbitals (LCAO) of the two hydrogen atoms. In the simplest case, we can use just a single hydrogen 1s ground state function from Sect. 4.1

$$\psi_{a,b} = \frac{1}{\sqrt{\pi}} \exp\{-r_{a,b}\} \tag{5.4}$$

per proton. The LCAO Ansatz is then given by

$$\psi_e = c_1\psi_a + c_2\psi_b \,. \tag{5.5}$$

Inserting it into the time-independent Schrödinger equation yields

$$\left(\hat{H}_a - \frac{1}{r_b}\right) c_1\psi_a + \left(\hat{H}_b - \frac{1}{r_a}\right) c_2\psi_b = E(R)(c_1\psi_a + c_2\psi_b). \tag{5.6}$$

The application of the Hamiltonian to the ground state leads to a multiplication of the wavefunction by the ground state energy E_g, and therefore the equation above can be rewritten as

$$\left(E_g - E(R) - \frac{1}{r_b}\right) c_1\psi_a + \left(E_g - E(R) - \frac{1}{r_a}\right) c_2\psi_b = 0 \,. \tag{5.7}$$

This equation can be transformed into a linear system of equations in the usual way by multiplying it from the left with the real eigenfunctions $\psi_{a,b}$ and integration over electronic coordinates. The following definitions are appropriate:

1. overlap integral:

$$\int \mathrm{d}V \psi_a\psi_b =: S(R) = (1 + R + R^2/3)\exp\{-R\} \tag{5.8}$$

2. Coulomb integral:

$$-\int \mathrm{d}V \psi_a \frac{1}{r_b} \psi_a =: C(R) = -(1 - (1+R)\exp\{-2R\})/R \tag{5.9}$$

3. exchange integral (having no classical analog):

$$-\int \mathrm{d}V \psi_a \frac{1}{r_a} \psi_b =: D(R) = -(1+R)\exp\{-R\} \tag{5.10}$$

4. energy difference:

$$\Delta E(R) := E_g - E(R) \tag{5.11}$$

5.1. *Calculate the integrals needed for the LCAO solution procedure for the electronic eigenvalue problem of* H_2^+ *with the help of the hydrogen* 1s *functions. Use prolate spheroidal coordinates (i.e. elliptic coordinates rotated around the focal* *line)* $\mu = (r_a + r_b)/R$, $\nu = (r_a - r_b)/R$, φ, *for which the volume element is given by* $\mathrm{d}V = \frac{1}{8}R^3(\mu^2 - \nu^2)\mathrm{d}\mu\mathrm{d}\nu\mathrm{d}\varphi$, *and where* $1 \le \mu \le \infty$, $-1 \le \nu \le 1$, $0 \le \varphi \le 2\pi$.

The coefficients then fulfill the linear system of equations

$$[\Delta E(R) + C(R)]\, c_1 + [\Delta E(R) S(R) + D(R)]\, c_2 = 0, \tag{5.12}$$

$$[\Delta E(R) S(R) + D(R)]\, c_1 + [\Delta E(R) + C(R)]\, c_2 = 0. \tag{5.13}$$

The eigenvalue problem defined above is a generalized one, due to the fact that the eigenvalues $E(R)$ are also appearing in the off-diagonals because of the non-vanishing overlap integral. The condition for solubility leads to the symmetric so-called $1\sigma_g$ solution, where

$$c_1 = c_2 = \left(\frac{1}{2(1 + S(R))}\right)^{1/2}, \tag{5.14}$$

and the antisymmetric $1\sigma_u$ solution, where

$$c_1 = -c_2 = \left(\frac{1}{2(1 - S(R))}\right)^{1/2}, \tag{5.15}$$

with the corresponding energies ($E_g = 0$)

$$E_\pm(R) = \frac{C(R) \pm D(R)}{1 \pm S(R)}. \tag{5.16}$$

The eigenfunctions displayed in Fig. 5.2 have decisively distinct character. Whereas in the symmetric superposition a sizable part of the wavefunction is located between the two nuclei, the antisymmetric solution $1\sigma_u$ with the minus sign has a node between the nuclei!

For the energies it is important to note that $C(R)$ as well as $D(R)$ are negative, and therefore the energy is reduced in the symmetric case as compared to two infinitely separated nuclei. To discuss the binding character of the solutions, we have to include the nuclear repulsion in our discussion, however, by considering the quantity

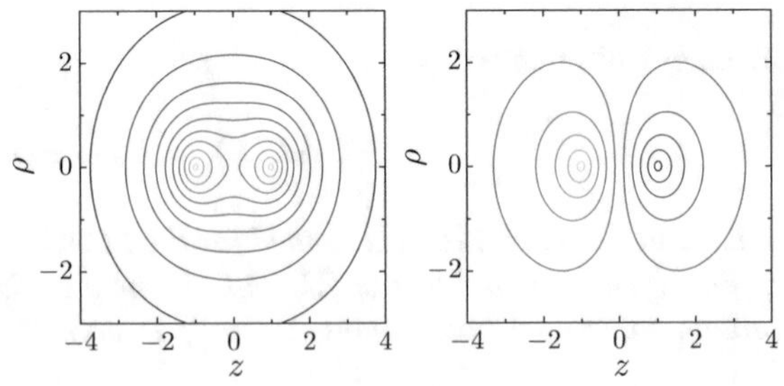

Fig. 5.2 (*Left panel*) Symmetric (binding) and (*right panel*) antisymmetric (anti-binding) LCAO solution for nuclei located at ± 1 a.u. as a function of cylindrical coordinates (ρ, z), with the z-axis along the nuclear axis

$$E_{\pm}^{\text{tot}}(R) = \frac{C(R) \pm D(R)}{1 \pm S(R)} + \frac{1}{R}. \tag{5.17}$$

For $R \to 0$ the nuclear repulsion dominates due to the fact that both C and D are finite in that limit, see Fig. 5.3. Furthermore, as it should be, for $R \to \infty$ both curves have the H + p case with energy $E_g = 0$ as the limiting case. At an intermediate value of R_e, the symmetric solution displays a minimum in the energy curve, with a binding energy (dissociation threshold) of D_e, whereas the antisymmetric one is a continuously decreasing function of R as can be seen in Fig. 5.3. A comparison of experimental and theoretical values for the two parameters of the binding potential can be found in Table 5.1.

We have reviewed an "electronic structure calculation" for a molecule with only a single electron and furthermore, we have used the smallest possible set of basis

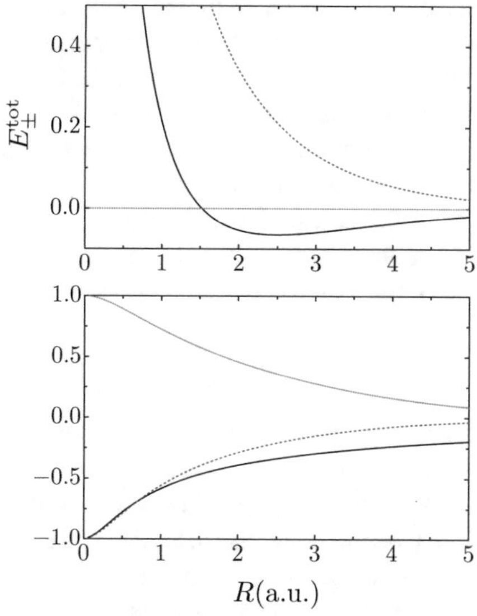

Fig. 5.3 *Upper panel*: LCAO binding (*solid black line*) and anti-binding (*dashed blue line*) electronic potential energy curve of H_2^+ (in atomic units) according to (5.17) as a function of internuclear distance in atomic units; note that at the minimum of the binding curve (around 2.5 a.u.), the binding effect is much smaller than the anti-binding effect. The zero of energy is indicated by a *dotted red line* and corresponds to the H + p case. *Lower panel*: Coulomb (*solid black line*), exchange (*dashed blue*), and overlap integral (*dotted red*)

Table 5.1 Comparison of experimental and theoretical LCAO results for the equilibrium distance, R_e, and the dissociation threshold, D_e, of H_2^+

	Experiment	Theory (LCAO)	Theory (vLCAO)
R_e	106 pm	130 pm	106 pm
D_e	2.79 eV	1.75 eV	2.35 eV

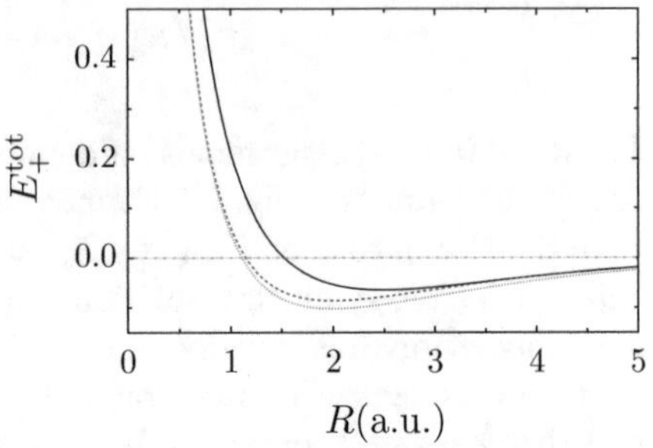

Fig. 5.4 LCAO binding (*solid black line*) electronic potential energy curve of H_2^+ according to (5.17) together with the variational LCAO (*dashed blue line*) and the exact result (*dotted red line*)

functions for this system. Nevertheless, the results are quite reasonable as could be seen by comparing them to experimental results. In general, many electron systems have to be considered, however. A prototypical diatomic example is the hydrogen molecule with two electrons. The methods that are typically used to obtain the electronic wavefunction of that molecule are the molecular orbital method or the Heitler-London method. A discussion of both approaches can be found in [2]. Going beyond the treatment of simple, small systems is done in the field of quantum chemistry, which is dealing with the calculation of the electronic energy curves as a function of internuclear distances in the general case [3].

For our present case, a method to dramatically improve the results by using a slightly modified basis set shall be mentioned. Finkelstein and Horowitz [4] have shown that the variation of the 1s basis functions according to

$$\psi_{a,b} \sim \exp\{-\alpha(R)r_{a,b}\}, \tag{5.18}$$

where $\alpha(R)$ is allowed to vary between the helium value of 2 at $R = 0$ and the hydrogen value of 1 at $R = \infty$ does improve the results tremendously to $R_e = 106\,\text{pm}$ and $D_e = 2.35\,\text{eV}$ (see also Table 5.1). This is another application of the Rayleigh-Ritz variational principle (see also Exercise 4.1), the variational LCAO method. An alternative would be to use many more basis function in the standard LCAO method. The convergence of this approach to the experimental value is rather slow, however. In Fig. 5.4 the variationally improved total energy curve is shown together with the exact one [5]. As expected, the variational curve is much better than standard LCAO but still lies above the exact curve.

5.2. *Prove that the ground state energy of* H_2^+*, calculated with the variational molecular orbital* $[(2\pi/\alpha^3)(1+S)]^{-1/2}(\exp\{-\alpha r_a\}+\exp\{-\alpha r_b\})$*, is given by*

$$E_+ = \alpha^2 F_1(w) + \alpha F_2(w),$$

where the abbreviations $w = \alpha R$ *and*

$$F_1(w) = \frac{1 + e^{-w}(1 + w - w^2/3)}{2 + 2e^{-w}(1 + w + w^2/3)},$$

$$F_2(w) = -\frac{1 + 2e^{-w}(1 + w) - 1/w - (1/w + 1)e^{-2w}}{1 + e^{-w}(1 + w + w^2/3)}$$

have been used.
Now minimize the energy with respect to α at constant R, and plot α as a function of R.

5.1.2 The Morse Potential

Another, arguably more crude, possibility to construct an analytical binding potential energy curve for a diatomic molecule (the two nuclei shall have the masses M_a and M_b) is to use the prototypical Morse potential [6]

$$V_M(R) \equiv D_e[1 - \exp\{-\alpha(R - R_e)\}]^2, \tag{5.19}$$

displayed in Fig. 5.5, and determine the free parameters from experimental values. To do this, we use that the kinetic energy of the relative motion is given by

$$T_R = \frac{M_r}{2}\dot{R}^2, \tag{5.20}$$

with the reduced mass $M_r = M_a M_b/(M_a + M_b)$. Parameters that are available experimentally are, e. g.,

- D_0: dissociation energy (from vibrational ground state)
- ω_e: angular frequency of harmonic oscillations around the minimum

From these, the Morse potential parameters can be extracted according to

- $D_e \approx D_0 + \omega_e/2$: dissociation threshold (from minimum of potential curve)
- $\alpha = \omega_e\sqrt{M_r/(2D_e)}$: range parameter (not to be confused with $\alpha(R)$ in the previous section),

where the α parameter follows from the harmonic approximation to the Morse potential.

The solution of the time-independent Schrödinger equation for the Morse potential can be given analytically, although it should be noted that it is only approximate in character [7]. Defining the anharmonicity constant by $x_e = \omega_e/(4D_e)$ the eigenvalues can be calculated according to

$$E_n = \omega_e(n + 1/2) - x_e\omega_e(n + 1/2)^2, \qquad n = 0, 1, \ldots . \tag{5.21}$$

Table 5.2 Comparison of the exact vibrational eigenvalues of H_2^+ [9] in the electronic ground state and different Morse oscillator eigenvalues in atomic units

n	exact result	$x_e = 1/37$	rel. error (%)	$x_e = 1/39.2$	rel. error (%)
0	0.005495	0.005179	5.747984	0.005183	5.675522
1	0.015479	0.015111	2.376164	0.015147	2.144661
2	0.024884	0.024476	1.638591	0.024576	1.238571
3	0.033729	0.033274	1.348567	0.033469	0.770120
4	0.042029	0.041503	1.249677	0.041826	0.482308
5	0.049798	0.049166	1.269455	0.049647	0.301979
6	0.057046	0.056260	1.378290	0.056933	0.198723
7	0.063782	0.062787	1.560210	0.063683	0.155630
8	0.070010	0.068747	1.804148	0.069897	0.160519
9	0.075730	0.074139	2.101915	0.075576	0.203891
10	0.080942	0.078963	2.445402	0.080719	0.276061
11	0.085640	0.083220	2.826596	0.085326	0.367125
12	0.089816	0.086909	3.236319	0.089397	0.465597
13	0.093454	0.090030	3.663607	0.092933	0.557668
14	0.096538	0.092584	4.095619	0.095933	0.626968
15	0.099044	0.094571	4.516326	0.098397	0.653021
16	0.100942	0.095990	4.906204	0.100326	0.610653
17	0.102200	0.096841	5.243057	0.101719	0.470501
18	0.102798	0.097125	5.518222	0.102576	0.215673
19	0.102902	0.096841	5.890226	0.102897	0.004935

The above definition of the anharmonicity is assuming the exact Morse form of the potential. One could, however, also use the experimental value for x_e which is typically slightly different from the one that is obtained by inserting the experimental values for ω_e and D_e. In the case of H_2^+, the frequency and the dissociation energy are given by $\omega_e \approx 0.0105$ a.u. and $D_e \approx 0.103$ a.u. leading to an anharmonicity of $x_e \approx 1/39.2$ a.u., whereas the direct experimental value as of 1950 is $x_e \approx 1/37.0$ a.u. [8] (the value given in the revised 1979 version of [8] is slightly larger). Choosing $x_e \approx 1/39.2$ a.u. leads to a better agreement of the energies in (5.21) with the exact vibrational eigenvalues of H_2^+ [9], gained without resorting to the Morse potential, as can be seen in Table 5.2. Finally, also the range parameter can be calculated from the anharmonicity according to $\alpha = \sqrt{2M_r\omega_e x_e}$. Again α differs slightly if the direct experimental anharmonicity or the one derived from the Morse potential is used!

The bound eigenvalues of the Morse oscillator are depicted in Fig. 5.5 together with the two lowest eigenfunctions according to [7]. Because of the anharmonicity, the distance between the levels decreases with increasing energy (for the parameters of H_2^+ this is barely visible at low quantum numbers). In contrast to the case of the hydrogen atom, where infinitely many levels below the ionization threshold exist,

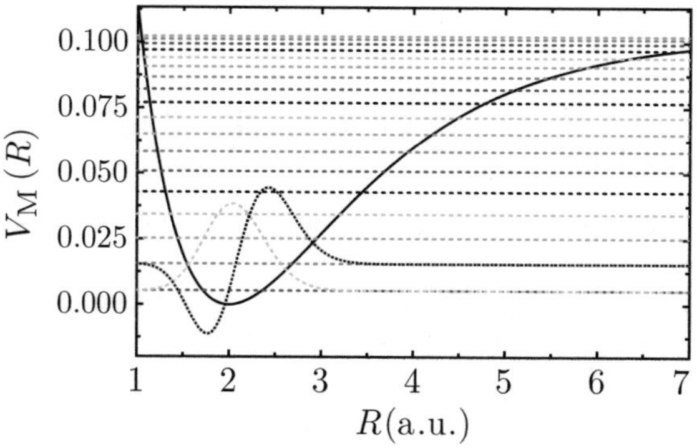

Fig. 5.5 Morse potential in atomic units (*solid line*) with $D_e = 0.103$ a.u. and $R_e = 2.00$ a.u. and $\alpha = 0.72$ a.u., corresponding to the experimental values of H_2^+ [8], together with the bound eigenvalues and the (unnormalized) two lowest eigenfunctions

however, only a finite number of levels lies below the threshold of dissociation. The maximal bound state index in the Morse potential can be determined from (5.21) by setting $E_{max} \leq D_e$ and is given by

$$n_{max} = \text{Int}(1/(2x_e) - 1/2). \tag{5.22}$$

For the direct experimental anharmonicity parameter of H_2^+ this number is 18, corresponding to 19 bound states (the last entry in the third column of Table 5.2 is already larger than $D_e = \omega_e/(4x_e)$), whereas the exact numerical solution displays 20 bound states.

5.2 H_2^+ in a Laser Field

The hydrogen molecular ion is the simplest molecule and therefore it has been the first molecule that has been studied in detail numerically under the influence of an external laser field. Restricting the electronic dynamics to two coupled potential surfaces is allowing the treatment of dissociation via the laser field [10]. Ionization cannot be studied in this framework. We want to focus on both ionization and dissociation phenomena, however. Historically, first the numerical solution of the electron's time-dependent Schrödinger equation with fixed nuclei [11], allowing the investigation of ionization probabilities, has been given. Only after a further increase of computer power, the fully coupled molecular dynamics has been studied [12].

5.2.1 *Frozen Nuclei*

To study the dynamics of a small molecule under the influence of an external electric field, usually several well-founded assumptions are made:

- First, one neglects the translational motion of the center of mass and rotations of the molecule.
- Secondly, the z-axis, along which the nuclei are assumed to be aligned, shall also be the polarization direction of the incident radiation.

The effect of molecular alignment [13] is at the heart of the last assumption. Fortunately, the problem then has a cylindrical symmetry and the time-dependent Schrödinger equation with fixed nuclei in atomic units is given by

$$\mathrm{i}\dot{\Psi}(\rho, z, t) = \left[-\frac{1}{2}\frac{\partial^2}{\partial z^2} + \hat{T}_\rho + V_{\mathrm{CC}}(\rho, z) + z\mathcal{E}(t)\right]\Psi(\rho, z, t), \tag{5.23}$$

where ρ and z are cylindrical coordinates, and the Hamiltonian does not depend on the azimuthal angle φ. Furthermore,

$$\hat{T}_\rho = -\frac{1}{2}\frac{\partial^2}{\partial \rho^2} - \frac{1}{2\rho}\frac{\partial}{\partial \rho} \tag{5.24}$$

stands for the radial part of the kinetic energy and

$$V_{\mathrm{CC}}(\rho, z) = -[\rho^2 + (z - R/2)^2]^{-1/2} - [\rho^2 + (z + R/2)^2]^{-1/2} \tag{5.25}$$

is the Coulomb potential of the electron bi-nuclear interaction.

The singularity of $\hat{T}_\rho$ at $\rho = 0$ as well as of the Coulomb potential at the position of the nuclei can be treated very elegantly, because of the cylindrical symmetry, by expansion in a so-called Fourier-Bessel series [11], see also p. 126 in [14]. If L is the largest distance from the z-axis that is to be described (the wavefunction shall be zero for $\rho \geq L$), then a complete orthonormal system of functions for the expansion of the ρ-dependence of the wavefunction is given by

$$v_n(\rho) = \frac{\sqrt{2}}{L J_1(x_n)} J_0(x_n\rho/L), \tag{5.26}$$

where

$$J_\nu(x) \equiv \left(\frac{x}{2}\right)^\nu \sum_{j=0}^{\infty} \frac{(-1)^j}{j!\Gamma(j+\nu+1)} \left(\frac{x}{2}\right)^{2j} \tag{5.27}$$

are Bessel functions of νth order and the x_n are the zeros of the Bessel function of zeroth order. The basis functions are orthonormal, according to[1]

$$\int_0^L \mathrm{d}\rho\, \rho\, v_n(\rho) v_m(\rho) = \delta_{nm}. \tag{5.28}$$

Application of the radial part of the kinetic energy operator to the basis functions yields

$$\hat{T}_\rho v_n(\rho) = \frac{1}{2}(x_n/L)^2 v_n(\rho), \tag{5.29}$$

i.e., the v_n are eigenfunctions of that operator. This fact can be proven explicitly by using the definition in (5.27).

5.3. *Using the definition of the Bessel function of* 0*th order, show that* v_n *is an eigenfunction of the radial part of the kinetic energy.*

One now expands the wavefunction according to

$$\Psi(\rho, z, t) = \sum_{n=1}^{M} v_n(\rho)\chi_n(z, t). \tag{5.30}$$

After multiplication of the time-dependent Schrödinger equation from the left with v_k and integration over ρ, the system of coupled partial differential equations

$$\mathrm{i}\dot{\boldsymbol{\chi}}(z, t) = \left[-\frac{1}{2}\frac{\partial^2}{\partial z^2} + \mathbf{A}(z) + z\mathcal{E}(t)\right]\boldsymbol{\chi}(z, t) \tag{5.31}$$

is found for the vector of coefficients $\boldsymbol{\chi}(z, t) = (\chi_1(z, t), \ldots, \chi_n(z, t))$ and the non-singular matrix $\mathbf{A}$ (ρV_{CC} is finite at the position of the nuclei) with the elements

$$A_{kn}(z) = \frac{1}{2}(x_n/L)^2\delta_{kn} + \int_0^L \mathrm{d}\rho\, \rho\, v_k(\rho) V_{\mathrm{CC}}(\rho, z) v_n(\rho) \tag{5.32}$$

has been defined.

The time-dependent Schrödinger equation can be solved by again using the split-operator FFT method from Sect. 2.3.2. This leads to the propagated wavefunction vector

$$\boldsymbol{\chi}(z, t + \Delta t) = \exp(-\mathrm{i}\hat{T}_z\Delta t/2)\exp[-\mathrm{i}\mathcal{E}(t')z\Delta t]\exp[-\mathrm{i}\mathbf{A}(z)\Delta t]\exp(-\mathrm{i}\hat{T}_z\Delta t/2)\, \boldsymbol{\chi}(z, t), \tag{5.33}$$

where

[1] The Jacobian determinant, ρ, is essential for the following.

$$\hat{T}_z = -\frac{1}{2}\frac{\partial^2}{\partial z^2} \tag{5.34}$$

is the operator of the kinetic energy in z direction and $t' = t + \Delta t/2$. The only difference to what we have already encountered is the fact that the wavefunction is not a scalar but a vector and correspondingly the non-diagonal matrix $\mathbf{A}(z)$ appears in the exponent. In order to cope with this exponentiated matrix, prior to the propagation, it is diagonalized using a matrix $\mathbf{U}(z)$ for every z and one finally gets

$$\exp[-\mathrm{i}\mathbf{A}(z)\Delta t] = \mathbf{U}(z)\exp[-\mathrm{i}\mathbf{A}_D(z)\Delta t]\mathbf{U}^{\mathrm{T}}(z). \tag{5.35}$$

The matrix is typically of a dimensionality, such that the solution of the eigenvalue problem cannot be done analytically but has to be performed numerically.

5.2.1.1 Numerical Details and Results

It is worthwhile to note some numerical details of the benchmark calculations of the Bandrauk group [11]. First of all, the numerical grid for the remaining z direction was restricted to $|z| < 128$ a.u. Furthermore, it turned out that $L = 8$ and $M = 16$ for the number of Bessel functions was adequate for moderate field intensities around $10^{14}\,\mathrm{W\,cm^{-2}}$. The laser was assumed to be turned on over five cycles of the field with a frequency of $\omega = 0.2$ a.u., corresponding to a wavelength of 228 nm. Finally, the initial state was taken as the electronic ground state $1\sigma_g$. It can be approximated to a good degree by the variational LCAO of the previous section. Alternatively, the LCAO-Ansatz may serve as an initial condition for propagation in imaginary time (see Sect. 2.1.3), distilling the true ground state.

In [11] a measure for ionization was defined with a certain degree of arbitrariness, by introducing

$$P_V(t) = 2\pi \int_{-z_I}^{z_I} \mathrm{d}z \int_0^L \mathrm{d}\rho\, \rho |\Psi(z,\rho,t)|^2, \tag{5.36}$$

with $z_I = 16$ a.u. as the non-ionized part of the probability. Together with the probability to be in the initial state, i.e., the survival probability

$$P_0(t) = |\langle \Psi(0)|\Psi(t)\rangle|^2, \tag{5.37}$$

this quantity is plotted in Fig. 5.6 for two different internuclear distances. In the case $R = 3$ a.u., Rabi oscillations appear for the parameters chosen, which correspond to a one-photon transition between the $1\sigma_g$ and $1\sigma_u$ state, whereas for $R = 2$ a.u., two photons would be needed for the same transition. Although we have plotted only the LCAO result for the surfaces in Fig. 5.3, the drastic decrease of the distance between the two surfaces as a function of R can be observed also there.

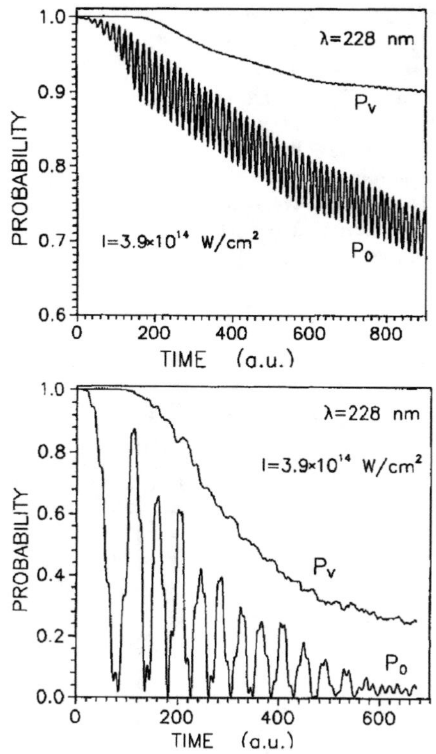

Fig. 5.6 Different probabilities (defined in the text) for different fixed inter-nuclear distances of H_2^+ as a function of time: *upper panel*: $R = 2$ a.u., *lower panel*: $R = 3$ a.u. [11]

5.2.1.2 Charge Resonance Enhanced Ionization

From the quantities defined above, one can extract an ionization rate by using an exponential fit to $P_V(t)$. These rates as a function of intensity can then be compared to the case of the hydrogen atom. In some early work, see, e.g., Fig. 5.7, it turned out that the molecular results tend towards the atomic case for increasing excitation, i.e., increasing R from 2 to 3 a.u. [11].

For even larger distances, however, dramatic rate enhancement far beyond the atom limit was found in [15] for a 1064 nm laser with a five cycle rise. For a fixed intensity this is shown in Fig. 5.8. The explanation of this effect is that a pair of charge resonant states (here the almost degenerate $1\sigma_g$ and $1\sigma_u$ states, which have a similar charge distribution at large R) are strongly coupled to the field at large R, when the dipole moment between them (see also Sect. 5.3) diverges linearly. The effect was therefore termed charge resonance enhanced ionization (CREI).

5.4. *Show that the matrix element of z between the* $1\sigma_g$ *and* $1\sigma_u$ *states of* H_2^+ *diverges linearly with the interatomic distance R.*

The success of the static tunneling picture in the atomic case of Sect. 4.3.1 led the authors of [15] to consider the tunneling out of the statically distorted double-well

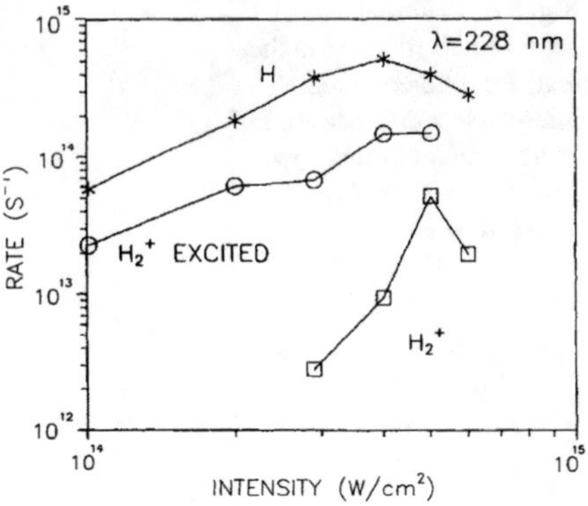

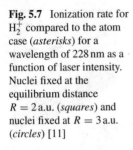

Fig. 5.7 Ionization rate for H_2^+ compared to the atom case (*asterisks*) for a wavelength of 228 nm as a function of laser intensity. Nuclei fixed at the equilibrium distance $R = 2$ a.u. (*squares*) and nuclei fixed at $R = 3$ a.u. (*circles*) [11]

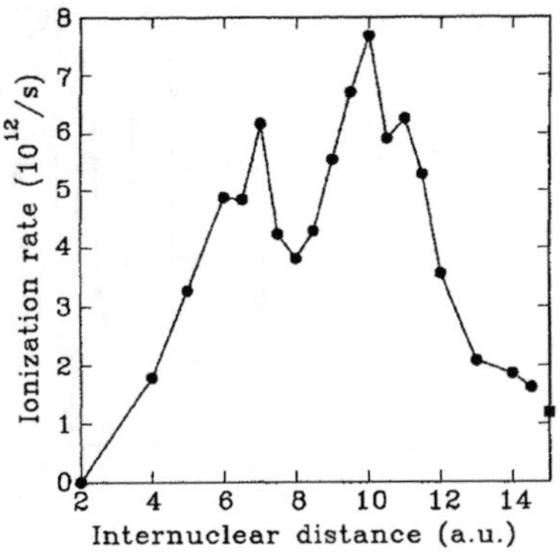

Fig. 5.8 Ionization rate for H_2^+ as a function of internuclear distance for fixed laser intensity of $10^{14}\,\mathrm{W\,cm^{-2}}$ and wavelength of 1,064 nm; hydrogen atom result indicated by a *filled square* [15]

potential that the electron experiences in a H_2^+ molecule with a field induced potential of $\mathcal{E}_0 z$. The results for the energies evolving out of the two lowest, unperturbed electronic eigenstates are displayed in Fig. 5.9. For the distance $R = 10$ a.u., at which a maximum in the ionization rate can be observed, the width of the upper level has a maximum, because it lies just above the inner barrier and the left outer barrier is rather narrow as compared to the $R = 6$ a.u. case. In addition, due to the rapid turn on of the field, the population of the upper level is almost equal to the one of the lower level after the amplitude is constant [15] and therefore the system ionizes to a substantial degree.

5.2.2 *Nuclei in Motion*

Nuclear dynamics in H_2^+ can nowadays be treated on the same level as the electronic dynamics by the solution of the full time-dependent Schrödinger equation.

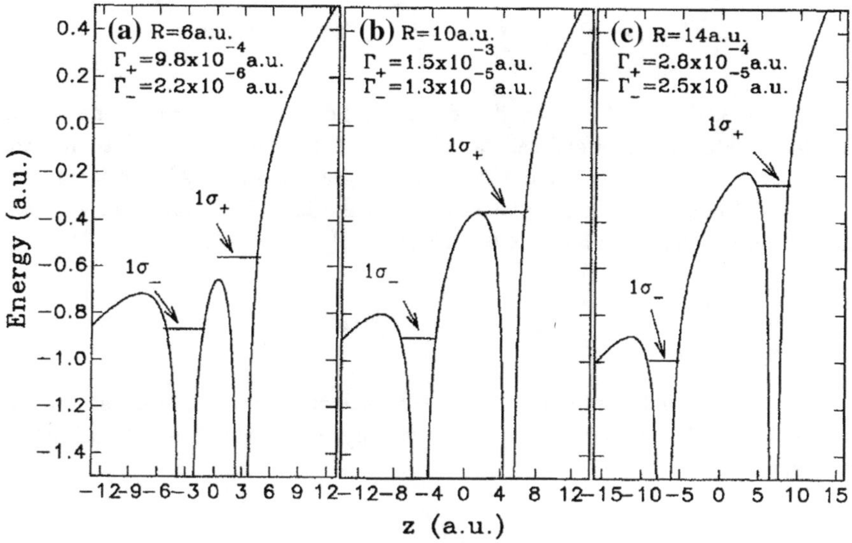

Fig. 5.9 Lowest two static field induced levels and their line widths for H_2^+ for three different internuclear distances: **a** $R = 6$ a.u., **b** $R = 10$ a.u., **c** $R = 14$ a.u. [15]

The wavefunction then depends on the additional degree of freedom R, describing the relative motion of the nuclei. The coupling to the field shall again be given in the length gauge. In this case, the motion of the center of mass motion can be separated away by introducing a center of mass and the relative coordinate between the two nuclei as well as an electron coordinate which is measured with respect to the center of mass of the nuclei. This is a lengthy calculation, however, which is reviewed in Appendix 5.A.

As the final result, it turns out that the Hamiltonian of (5.23) has to be augmented by the kinetic and the potential energy of the nuclei

$$\hat{T}_R + V_R = -\frac{1}{2M_\mathrm{r}}\frac{\partial^2}{\partial R^2} + \frac{1}{R}, \tag{5.38}$$

where $M_\mathrm{r} = M_\mathrm{p}/2 \approx 918$ is the reduced nuclear mass in atomic units. Furthermore, as can be seen in Appendix 5.A, the electron mass (which is unity in atomic units) has to be modified slightly to read $m_i = 2M_\mathrm{p}/(2M_\mathrm{p} + 1)$ and the term with the laser field is to be multiplied by the factor $[1 + 1/(2M_\mathrm{p} + 1)]$. Both modifications are marginal due to the large mass ratio. We stress, however, that for the relative motion of H_2^+, the laser does only couple to the electronic motion and not directly to the nuclei.

The total wavefunction $\Psi(R, \rho, z, t)$ now also depends on R, and distance dependent quantities like

$$f_{\mathrm{I}}(R,t) = 2\pi \int_0^L \mathrm{d}\rho\rho \int_{-z_{\mathrm{I}}}^{z_{\mathrm{I}}} \mathrm{d}z|\Psi(R,\rho,z,t)|^2 \tag{5.39}$$

can be studied. This is the probability density to find the protons a distance R apart and the electron within a cylinder of height $2z_{\mathrm{I}}$, such that H_2^+ is not fully ionized. z_{I} is later on chosen to be 32 a.u. From the R-dependent quantity just defined some integrated quantities can be calculated. These are the dissociation probability without ionization, i.e., the probability for the "reaction" $H_2^+ \to H + H^+$[2]

$$P_{\mathrm{D}}(t) = \int_{R_{\mathrm{D}}}^{R_{\max}} \mathrm{d}R f_{\mathrm{I}}(R,t), \tag{5.40}$$

where again somewhat arbitrarily $R_{\mathrm{D}} = 9.5$ a.u. can be chosen as the onset of dissociation and $R_{\max}$ is the nuclear grid boundary. Furthermore, the probability of ionization

$$P_{\mathrm{I}}(t) = 1 - \int_0^{R_{\max}} \mathrm{d}R f_{\mathrm{I}}(R,t) \tag{5.41}$$

is given by the probability to find the electron outside of a cylinder with $|z| \leq z_{\mathrm{I}}$.

5.2.2.1 Molecular Stabilization

Numerical results for the case of a molecule initially in the electronic $1\sigma_g$ ground state and in an excited vibrational state with quantum number $n = 6$, according to

$$\Psi(R,z,\rho,0) = \phi_{1\sigma_g}(z,\rho,R)\chi_6(R) \tag{5.42}$$

are depicted in Fig. 5.10. Apart from the probabilities defined above, also the probability P_6 to stay in the 6th vibrational state is displayed there. For relatively low intensities (I $= 3.5 \times 10^{13}$ W/cm^2) a stabilization of the initial state, i.e., after a short initial decay, an increase of P_6 is observed. It can be understood due to stimulated emission (Rabi oscillation) from the dissociative $1\sigma_u$ state [12]. For higher intensities (I $= 10^{14}$ W cm^{-2}) this effect vanishes, however, because the system is already ionized to a substantial degree. This is in contrast to the prediction of two-state calculations, which are also displayed in Fig. 5.10 and which show the stabilization effect for both intensities. Furthermore, it is worthwhile to note that the fine oscillations in P_6, well visible for low intensities, occur at twice the laser frequency and are due to the counter-rotating term, which is neglected in the RWA (but not in the full numerical calculations reviewed here).

Further light can be shed on the stabilization (or bond hardening) effect by looking at the nuclear wavefunction at a fixed time. For different field intensities these

[2] We note that there may be an asymmetry as to which proton the electron "belongs to" [16].

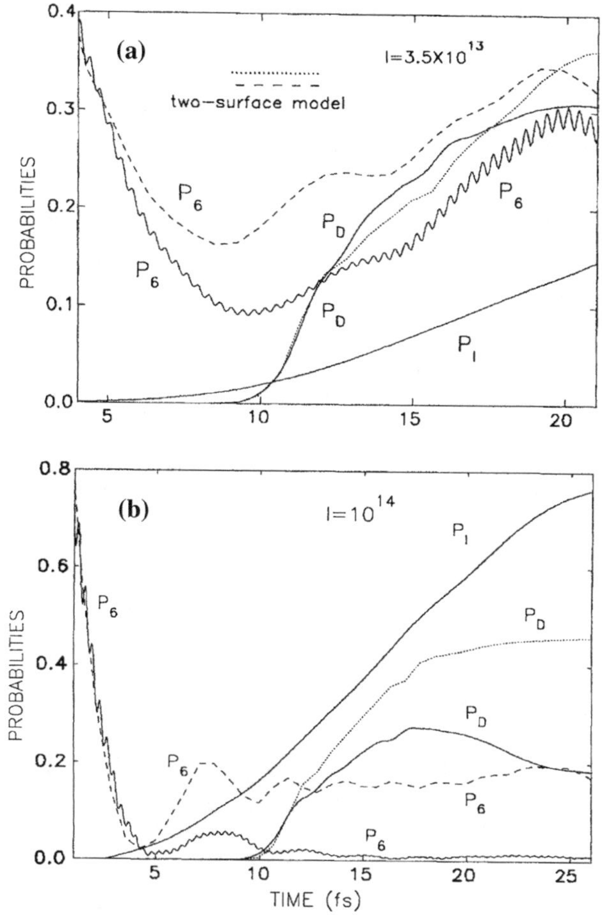

Fig. 5.10 Different probabilities for H_2^+ in a laser field ($\lambda = 212\,\text{nm}$) as a function of time for two different intensities: **a** $I = 3.5 \times 10^{13}\,\text{W cm}^{-2}$, **b** $I = 10^{14}\,\text{W cm}^{-2}$ [12]

results are displayed in Fig. 5.11. Sharp peaks near $R = 3$ and $R = 3.6$ a.u. can be observed. In a two state picture the adiabatic potentials (gained by diagonalization of the Hamilton matrix) corresponding to the $1\sigma_g$ and $1\sigma_u - 1\omega$ surfaces form an avoided crossing[3] [17, 18] and the peaks are at the turning points of the bound motion in the upper adiabatic potential well (dotted red line in Fig. 5.12). For the lower intensity, the shape of the peaks does not vary much as a function of time, whereas for the higher intensity, the peaks decrease considerably due to ionization.

[3]See also Appendix 3.A for avoided crossings of Floquet states.

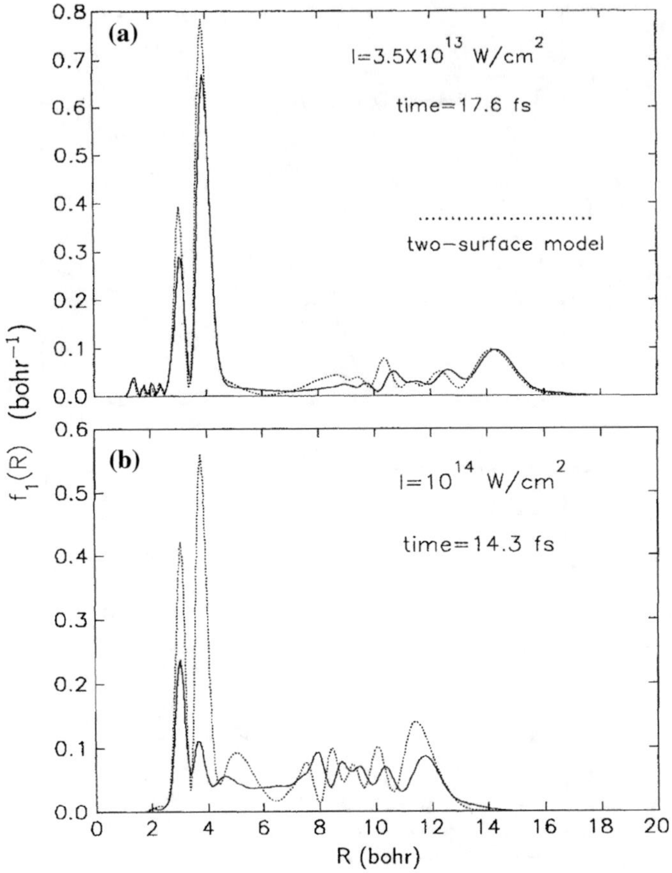

Fig. 5.11 Time evolved nuclear wavefunction of H_2^+ in a laser field of different intensity: **a** $I = 3.5 \times 10^{13}\,\mathrm{W\,cm^{-2}}$, **b** $I = 10^{14}\,\mathrm{W\,cm^{-2}}$ [12]. In these plots also two-surface calculations, which do not account for ionization are displayed

Fig. 5.12 Schematic picture of the formation of an avoided crossing of dressed states emerging from the $1\sigma_g$ (*black line*) and $1\sigma_u - 1\omega$ (*dashed blue line*) surface interacting via a dipole matrix element

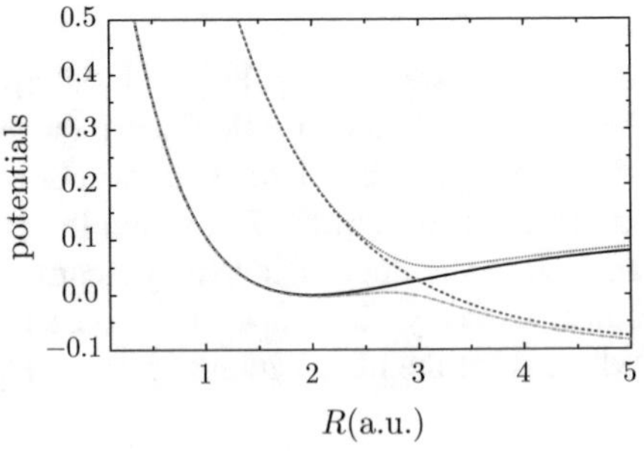

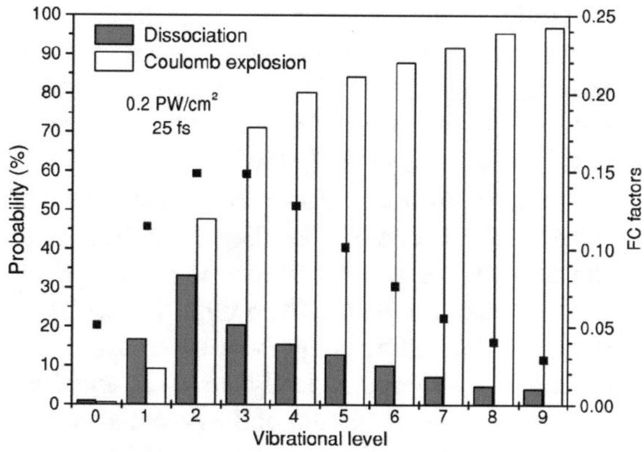

Fig. 5.13 Comparison of the two fragmentation channels, defined in the text, as a function of initial vibrational excitation for a 25 fs laser pulse with an intensity of 0.2 PW cm^{-2} and 800 nm central wavelength. *Black squares* show the overlaps $|\langle H_2^+, n|H_2, n=0\rangle|^2$ [19]

5.2.2.2 Coulomb Explosion Versus Dissociation

In the results that we have discussed so far, the initial vibrational state was fixed to be the $n = 6$ state. What happens for different initial vibrational quantum numbers?

As already discussed, the fragmentation of the molecular ion under the external field can occur via different channels. One channel, in which the electron stays with one (or each) of the two nuclei, is the dissociation channel (without ionization)

$$H_2^+ \rightarrow p + H.$$

The alternative is complete fragmentation, which is our ionization case, given by

$$H_2^+ \rightarrow p + p + e^-.$$

This last channel is also referred to as the Coulomb explosion channel.

The outcome of numerical calculations using a modified 1D soft-core potential, where the softening parameter depends on the internuclear distance [19], and the Crank-Nicolson method has been used for the propagation is displayed in Fig. 5.13. It can be observed that for vibrational levels $n \geq 2$ for the field parameters chosen, the Coulomb explosion channel dominates the dissociation channel. Quantitatively different but qualitatively similar results with up to 60% dissociation probability have been reported for the case of a full 3D hard-core Coulomb interaction [20].

5.3 Adiabatic and Nonadiabatic Nuclear Dynamics

The neutral hydrogen molecule H_2 has an additional electron compared to the hydrogen molecular ion of the previous section. Its dynamics under laser fields has been treated fully quantum mechanically only recently [21]. In order to deal with even more complex systems numerically, approximations and/or restrictions of the dynamics to some relevant degrees of freedom have to be made. An important, if not the most important approximation of molecular theory is the Born-Oppenheimer approximation, to be discussed in the following.

This approximation is closely related to what we have already done for the case of H_2^+ in Sect. 5.1. There the binding potential energy surface $1\sigma_g$ was taken as a sum of the repulsive potential between the nuclei and the attraction due to the electron in-between the two nuclei. If no other (external) force is acting, then the nuclear motion would be attracted by the minimum of the potential. The motion has to be described quantum mechanically, however, and thus a probability distribution with its maximum at the minimum of the potential curve will result. In the following we will realize that the notion above is approximate in nature. Even in the case without an external field the dynamics can, in general, not be restricted to a single electronic surface.

5.3.1 Born-Oppenheimer Approximation

In (5.3) we have neglected the kinetic as well as the potential energy of the nuclei. The total Hamiltonian for a general molecule with M nuclei and N electrons, however, is given by

$$\hat{H}_{\rm mol} = \hat{T}_{\rm N} + \hat{H}_{\rm e} = \hat{T}_{\rm N} + \hat{T}_{\rm e} + V(\boldsymbol{x}, \boldsymbol{X}), \tag{5.43}$$

with the kinetic energies (switching back to SI units for this section)

$$\hat{T}_{\rm N} = \sum_{i=1}^{M} -\frac{\hbar^2}{2M_i}\Delta_i, \tag{5.44}$$

$$\hat{T}_{\rm e} = \sum_{j=1}^{N} -\frac{\hbar^2}{2m_{\rm e}}\Delta_j. \tag{5.45}$$

The potential energy $V = V_{\rm ee} + V_{\rm eN} + V_{\rm NN}$ contains the electron-electron interaction

$$V_{\rm ee}(\boldsymbol{x}) = \frac{e^2}{4\pi\epsilon_0}\frac{1}{2}\sum_{i\neq j}^{N}\frac{1}{|\boldsymbol{r}_i - \boldsymbol{r}_j|}, \tag{5.46}$$

as well as the electron-nucleus (nuclear charge $Z_i e$)

$$V_{\rm eN}(\boldsymbol{x}, \boldsymbol{X}) = -\frac{e^2}{4\pi\epsilon_0}\sum_{j=1}^{N}\sum_{i=1}^{M}\frac{Z_i}{|\boldsymbol{r}_j - \boldsymbol{R}_i|} \tag{5.47}$$

and the internuclear interaction

$$V_{\mathrm{NN}}(\boldsymbol{X}) = \frac{e^2}{4\pi\epsilon_0}\frac{1}{2}\sum_{i\neq j}^{M}\frac{Z_i Z_j}{|\boldsymbol{R}_i - \boldsymbol{R}_j|}. \tag{5.48}$$

All nuclear coordinates are contained in the symbol

$$\boldsymbol{X} = (\boldsymbol{R}_1, \ldots, \boldsymbol{R}_M),$$

whereas the electronic coordinates *including spin* are denoted by

$$\boldsymbol{x} = (\boldsymbol{r}_1, s_1; \ldots; \boldsymbol{r}_N, s_N).$$

As usual, the nuclear coordinates are distinguished from the electronic ones by using capital letters for the former and lower case letters for the latter.

The electronic part of the Hamiltonian commutes with the nuclear coordinates, i.e. $[\hat{H}_\mathrm{e}, \boldsymbol{X}] = 0$. The nuclear coordinates therefore are "good quantum numbers" for the electronic operator and can be viewed as parameters. One now first solves the electronic eigenvalue problem (the time-independent Schrödinger equation of the electrons), which is the generalized analog of (5.3),

$$\hat{H}_\mathrm{e}\phi_\nu = \left[-\sum_j \frac{\hbar^2}{2m_\mathrm{e}}\Delta_j + V(\boldsymbol{x}, \boldsymbol{X})\right]\phi_\nu(\boldsymbol{x}, \boldsymbol{X}) = E_\nu(\boldsymbol{X})\phi_\nu(\boldsymbol{x}, \boldsymbol{X}), \tag{5.49}$$

where the electronic energy, as well as the corresponding wavefunction depend *parametrically*[4] on the nuclear coordinates and ν is a suitable set of quantum numbers of the electronic system. For the electronic functions, orthonormality and completeness relations are assumed to hold,[5] according to

$$\int \mathrm{d}^{4N}x\, \phi_\nu^*(\boldsymbol{x}, \boldsymbol{X})\phi_\mu(\boldsymbol{x}, \boldsymbol{X}) = \delta_{\mu\nu}, \tag{5.50}$$

$$\sum_\nu |\phi_\nu\rangle\langle\phi_\nu| = \hat{1}. \tag{5.51}$$

The *total* wavefunction can thus be expanded by using the electronic states as basis states, according to the Born-Huang expansion

$$\psi(\boldsymbol{x}, \boldsymbol{X}) = \sum_\nu \phi_\nu(\boldsymbol{x}, \boldsymbol{X})\chi_\nu(\boldsymbol{X}), \tag{5.52}$$

[4]Some authors use the notation $\phi_\nu(\boldsymbol{x}|\boldsymbol{X})$ or $\phi_\nu(\boldsymbol{x}; \boldsymbol{X})$ to express this fact, see also Sect. 2.2.7.

[5]This can, e.g., be achieved by putting the molecule in a large box.

with the nuclear functions $\chi_\nu(\boldsymbol{X})$. Inserting this Ansatz into the time-independent Schrödinger equation

$$\left[-\sum_i \frac{\hbar^2}{2M_i}\Delta_i - \sum_j \frac{\hbar^2}{2m_\mathrm{e}}\Delta_j + V(\boldsymbol{x},\boldsymbol{X})\right]\psi(\boldsymbol{x},\boldsymbol{X}) = \epsilon\psi(\boldsymbol{x},\boldsymbol{X}), \quad (5.53)$$

one can now use the electronic Schrödinger equation to replace the electronic Hamiltonian by its eigenvalue and arrives at (written suggestively)

$$\sum_\nu \phi_\nu(\boldsymbol{x},\boldsymbol{X})\left[-\sum_i \frac{\hbar^2}{2M_i}\Delta_i + E_\nu(\boldsymbol{X})\right]\chi_\nu(\boldsymbol{X}) = \sum_\nu \phi_\nu(\boldsymbol{x},\boldsymbol{X})\epsilon\chi_\nu(\boldsymbol{X})$$
$$+\sum_\nu\sum_i \frac{\hbar^2}{2M_i}[2\boldsymbol{\nabla}_i\chi_\nu(\boldsymbol{X})\cdot\boldsymbol{\nabla}_i\phi_\nu(\boldsymbol{x},\boldsymbol{X}) + \chi_\nu(\boldsymbol{X})\Delta_i\phi_\nu(\boldsymbol{x},\boldsymbol{X})]. \quad (5.54)$$

The second line in the equation above follows from the application of the product rule of differentiation and is due to the change of the electronic states induced by the nuclear motion.

If we neglect this second line and multiply the equation from the left by $\phi_\mu^*(\boldsymbol{x},\boldsymbol{X})$, after integration over the electronic coordinates, we get

$$\left[-\sum_i \frac{\hbar^2}{2M_i}\Delta_i + E_\mu(\boldsymbol{X})\right]\chi_\mu(\boldsymbol{X}) \approx \epsilon\chi_\mu(\boldsymbol{X}). \quad (5.55)$$

This is the time-independent Schrödinger equation for the nuclear degrees of freedom in an (adiabatic) potential that is given by the μth eigensolution of the electrons. From the minimum of the ground state energy of the electrons, one can determine the binding energy of the molecule, by comparing that minimum to the sum of the ground state energies of all isolated atoms. We have used this notion already in Sect. 5.1. As we can appreciate now, an approximation has been made along the way, however. The equation above and also its time-dependent analog, which leads to a dynamics on a single potential energy surface, are based on the Born-Oppenheimer approximation that amounts to the complete neglect of the second line in (5.54).

5.5. *Using the Rayleigh-Ritz variational principle, show that the Born-Oppenheimer ground state energy is a lower bound to the exact ground state energy of the full molecular problem. Employ the original Born-Oppenheimer factorization Ansatz* $\psi^{\mathrm{BO}}(\boldsymbol{x},\boldsymbol{X}) = \phi(\boldsymbol{x},\boldsymbol{X})\chi(\boldsymbol{X})$ *to this end.*

The approximation that has been made still needs to be justified. That is we have to argue why the terms in the second line of (5.54) might by small compared to the terms in the first line of that equation. Let us just consider the terms

$$\frac{\hbar^2}{2M_i}\Delta_i\phi_\nu(\boldsymbol{x},\boldsymbol{X}),$$

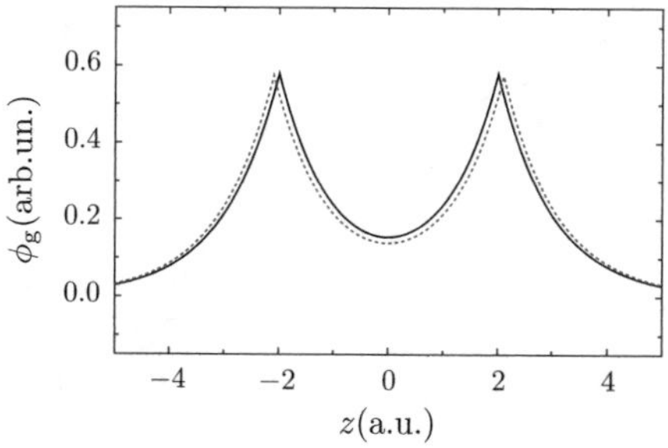

Fig. 5.14 Change of the LCAO electronic ground state of H_2^+ from (5.5) by changing the internuclear distance slightly from $R = 2$ a.u. (*solid line*) to $R = 2.1$ a.u. (*dashed line*). Shown is a cut through the 3D wavefunction (integrated over the azimuthal angle) at $\rho = 0$

containing the second derivative with respect to nuclear variables. The electronic wavefunction depends on $\boldsymbol{X}$ in a similar fashion as it depends on $\boldsymbol{x}$ (depending mainly on the differences $\boldsymbol{R}_i - \boldsymbol{r}_j$). Therefore the respective derivatives are of the same order of magnitude. The corresponding prefactors in (5.53), however, are smaller by almost 3 orders of magnitude due to the large mass ratio M_i/m_e. A similar reasoning is possible for the terms containing the first derivative, see Sect. 11.1.2 of [22]. Furthermore, in many cases there is only a weak dependence of the electronic wavefunction on the nuclear distance. That is why the Born-Oppenheimer approximation is so successful. As an explicit example, in Fig. 5.14, a cut through the electronic ground state in the case of H_2^+ is displayed and the slow variation of this state with the nuclear distance is shown (shown is the LCAO result; full numerical calculations lead to the same conclusion, however).

In general, however, one has to multiply also the second line in (5.54) from the left with $\phi_\mu^*(\boldsymbol{x}, \boldsymbol{X})$ and has to integrate over the electronic coordinates. The terms that emerge then lead to transitions between the different electronic surfaces, the so-called nonadiabatic or non Born-Oppenheimer transitions. For reasons of simplicity, we transform into a coordinate system that moves with the center of mass (see below), and give the corresponding TDSE in the case of a diatom and just two (excited) electronic surfaces that we denote by $V_1^a(R)$ and $V_2^a(R)$ and that depend on the inter-atomic separation R. This leads to

$$\mathrm{i}\hbar \begin{pmatrix} \dot{\chi}_1(R,t) \\ \dot{\chi}_2(R,t) \end{pmatrix} = \left[\begin{pmatrix} \hat{T}_N + U_{11} & Q_{12}\frac{\partial}{\partial R} + U_{12} \\ Q_{21}\frac{\partial}{\partial R} + U_{21} & \hat{T}_N + U_{22} \end{pmatrix} + \begin{pmatrix} V_1^a(R) & 0 \\ 0 & V_2^a(R) \end{pmatrix} \right] \begin{pmatrix} \chi_1(R,t) \\ \chi_2(R,t) \end{pmatrix}, \tag{5.56}$$

where $\hat{T}_N = -\frac{\hbar^2}{2M_r}\frac{\partial^2}{\partial R^2}$ and

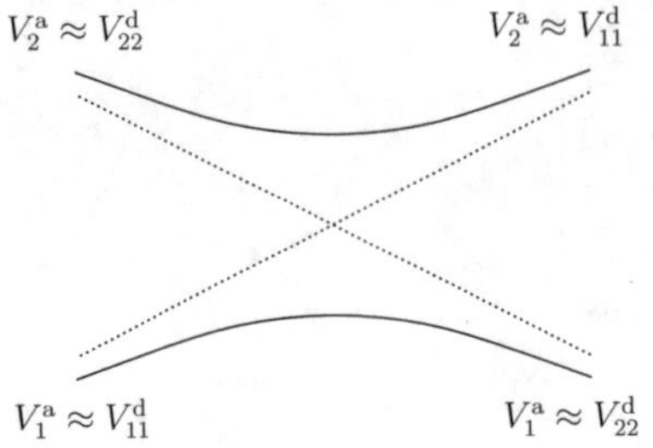

Fig. 5.15 Schematic dependence on internuclear distance of adiabatic (*solid lines*, superscript a) and diabatic (*dotted lines*, superscript d) electronic levels in the vicinity of an avoided crossing of the adiabatic levels (where the physical character of the states can change dramatically). In the diabatic picture, the nonadiabatic transitions are due to non-diagonal coupling matrix elements (not displayed). Realistic adiabatic and diabatic surfaces for the LiH_2 system are displayed in Fig. 12.1 of [24]

$$Q_{\mu\nu} = -\frac{\hbar^2}{M_r}\left\langle \phi_\mu \left| \frac{\partial}{\partial R} \right| \phi_\nu \right\rangle, \tag{5.57}$$

$$U_{\mu\nu} = -\frac{\hbar^2}{2M_r}\left\langle \phi_\mu \left| \frac{\partial^2}{\partial R^2} \right| \phi_\nu \right\rangle, \tag{5.58}$$

where we used Dirac notation for the electronic integration. The transitions between the adiabatic surfaces that occur close to avoided crossings of those surfaces, where the character of the states may change dramatically, are due to the off-diagonal electronic matrix elements of the derivative terms of the second line of (5.54). The first term in the second line of (5.54) has vanishing diagonal elements (which can be proven by assuming that the electronic wavefunctions are real and normalized at all R). Small diagonal contributions stem only from the second term on the second line and are not leading to electronic transitions. The off-diagonal ones are (mainly) coming from the first term (see Exercise 5.6) and still contain a nuclear momentum operator, after performing the integration over electronic coordinates.

The nonadiabatic matrix elements (5.57) and (5.58) are usually quite hard to calculate [23]. Alternatively, nonadiabatic transitions are therefore frequently described in another, so-called diabatic basis, leading to a coupled TDSE of the form

$$i\hbar\begin{pmatrix}\dot{\chi}_1(R,t)\\ \dot{\chi}_2(R,t)\end{pmatrix} = \left[\begin{pmatrix}\hat{T}_N & 0\\ 0 & \hat{T}_N\end{pmatrix} + \begin{pmatrix}V_{11}^d(R) & V_{12}^d(R)\\ V_{21}^d(R) & V_{22}^d(R)\end{pmatrix}\right]\begin{pmatrix}\chi_1(R,t)\\ \chi_2(R,t)\end{pmatrix}. \tag{5.59}$$

A schematic plot of adiabatic and diabatic levels is given in Fig. 5.15. The diabatic levels can cross and their coupling is given by non-diagonal potential terms and not by derivative terms, which might be advantageous computationally. However, the construction of diabatic surfaces is not unique [23].

Before leaving this section, let us come back to the mixed quantum classical methods that we have discussed in Chap. 2. In the framework of the Born-Oppenheimer description the classical dynamics is restricted to a single surface, most frequently, the ground state. For frozen nuclear coordinates the electronic quantum problem is solved for the ground state and the forces which govern the nuclear motion are determined. After a short time step of the nuclear motion the same procedure is repeated. In this way, the forces are calculated "on the fly" and the procedure can also be generalized to the treatment of nonadiabatic dynamics. In the following, however, we assume that the potential and/or forces are at our disposal, either analytically, or numerically on a grid, as the outcome of a quantum chemical calculation.

5.6. *Let R denote a parameter that describes a displacement of the nuclei, such that the electronic Hamiltonian, as well as its (orthonormalized) eigensolution depend on it*

$$\hat{H}_e(q)\phi_\nu(R) = E_\nu(R)\phi_\nu(R).$$

(a) *Proof that the force due to the displacement is given by the Hellmann-Feynman theorem*

$$-F(R) = \frac{\mathrm{d}E_\nu(R)}{\mathrm{d}R} = \left\langle \phi_\nu(R) \left| \frac{\mathrm{d}\hat{H}_e(R)}{\mathrm{d}R} \right| \phi_\nu(R) \right\rangle.$$

(b) *Show that the derivative coupling term $Q_{\mu\nu}$ increases, when the two electronic surfaces come close in energy.*

5.3.1.1 Relative and Center of Mass Coordinates

Similar to the case of H_2^+ (see Appendix 5.A), the treatment of the molecular problem is preferably done in center of mass and relative coordinates. Furthermore, only for the "internal motion", the assumption of normalizable electronic wavefunctions is reasonable [25].

The center of mass is moving freely due to the fact that the potential does not depend on the corresponding coordinate. We therefore concentrate on the description of the relative motion.[6] The relevant masses then are reduced masses. In the diatomic case, this has already been used in Sect. 5.2. For more atoms, matters become complicated rather quickly. Already in the case of collinear motion of three nuclei, cross terms do appear in the kinetic energy.

5.7. *For three collinear masses M_1, M_2, M_3, give relative and center of mass coordinates and write the kinetic energy with the help of the canonically conjugate relative momenta.*

A transition to center of mass and relative coordinates for the complete molecular Hamiltonian (including the electrons) in two steps is given in the supplementary

[6] We will not consider rotational dynamics with the exception of Sect. 5.5.

material to [26]. There, first the transition into the center of mass system of the nuclei is performed and secondly the transition from the nuclear center of mass system to that of the complete system is performed. In this way, the kinetic energies do not include terms mixing electronic and nuclear derivatives.

5.3.1.2 Coupling to a Laser Field

In order to couple a laser field to a molecular system, we will use the length gauge. Generalizing the discussion in Sect. 3.1.2 to the many particle case, this leads us to consider the dipole operator

$$\boldsymbol{d}(\boldsymbol{x}, \boldsymbol{X}) = \sum_i Z_i e \boldsymbol{R}_i - \sum_j e \boldsymbol{r}_j \,. \tag{5.60}$$

After having solved the electronic problem, the dipole matrix element (or (transition) dipole moment)

$$\boldsymbol{\mu}_{ba}(\boldsymbol{X}) = \int \mathrm{d}^{4N}x \, \phi_b^*(\boldsymbol{x}, \boldsymbol{X}) \boldsymbol{d}(\boldsymbol{x}, \boldsymbol{X}) \phi_a(\boldsymbol{x}, \boldsymbol{X}) \tag{5.61}$$

has to be calculated. This is a generalization of the dipole matrix element of Sect. 3.2.1, due to the fact that it still may depend on the internuclear distance.

If we consider different electronic levels, i.e., if $b \neq a$, then due to the orthogonality of the corresponding states only the parts proportional to $\sum_j e\boldsymbol{r}_j$ of the dipole operator survive. If a corresponding transition is not forbidden, then even if one neglects the nonadiabatic terms in the spirit of the Born-Oppenheimer approximation, still a coupled surface time-dependent Schrödinger equation has to be solved due to the presence of the laser.

5.3.2 *Dissociation in a Morse Potential*

Before treating the problem of coupled surfaces, the influence of a laser on the wavepacket dynamics in the electronic ground state shall be studied. Frequently used systems for theoretical calculations are "diatomics" like HF and CH- or OH-groups of larger molecules. *Heteronuclear* systems are chosen due to the fact that symmetric homonuclear molecules as, e.g., H_2 do not have a permanent electric dipole moment in the electronic ground state. If the parameters of a typical infrared laser are chosen appropriately, the diatomic can be driven into dissociation. In contrast to the studies of dissociation of the hydrogen molecular ion of Sect. 5.2, in the following only the nuclear part of the Schrödinger equation will be considered.

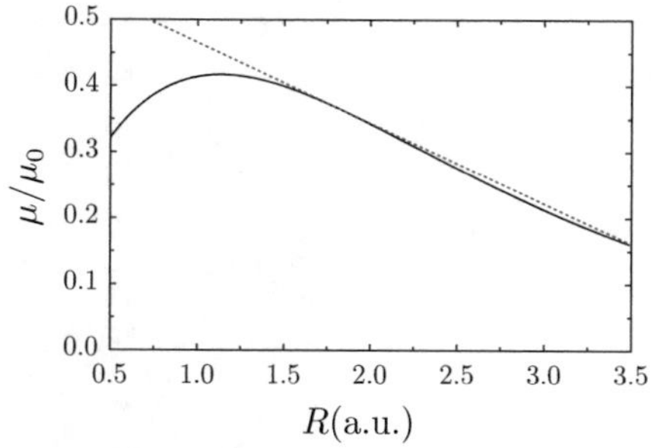

Fig. 5.16 Mecke form for the dipole function (*solid line*) of an X-H diatomic and its linear approximation (*dashed line*) around the minimum of the corresponding Morse potential; parameters used are the ones for the OH stretch in H_2O: $R_e = 1.821$, $R^* = 1.134$ in atomic units [31]

5.8. *Show explicitely that, for the relative motion, the permanent electronic dipole moment of* H_2^+ *in the electronic ground state (approximated by* $1\sigma_g$ *of Sect. 5.1) is zero.*

After the separation of the center of mass motion, the Hamiltonian for the relative motion of the two nuclei in a Morse potential modeling the electronic ground state is given by

$$\hat{H} = -\frac{1}{2M_r}\frac{\partial^2}{\partial R^2} + V_M(R) + \mu(R)\mathcal{E}_0 f(t)\cos(\omega(t)t)\,. \tag{5.62}$$

The potential parameters for HF are $D_e = 0.225$, $R_e = 1.7329$, $\alpha = 1.1741$ in atomic units [27]. In general, one allows for a chirp, see also Sect. 1.3.3, in the laser frequency.

The R-dependent dipole matrix element (or dipole moment) in principle has to be determined by quantum chemical methods and we assume it to be given analytically in the form studied in detail by Mecke [28]

$$\mu(R) = \mu_0 R\, e^{-R/R^*}\,. \tag{5.63}$$

Alternatively, other powers of R than the first may appear in the exponential function. For HF the power of 4 is frequently used [29]. As can be seen from Fig. 5.16, around the minimum R_e of the Morse potential the dipole function in (5.63) can (up to an irrelevant constant) be approximated by a linear function

$$\mu(R) \approx -\mu'(R - R_e). \tag{5.64}$$

The slope of this linear function is referred to as the dipole gradient or effective charge. For HF this quantity is given by $\mu' = 0.297$ in atomic units [30].

What is the reason for allowing a chirp in the laser frequency? As we have seen in Chap. 3, complete population transfer between two levels is only possible in the case

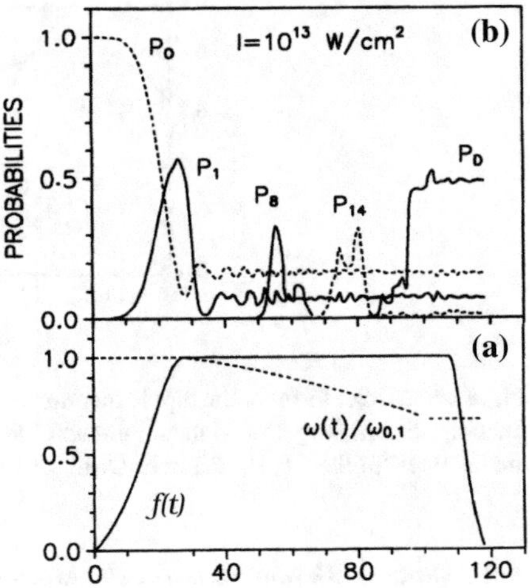

Fig. 5.17 **a** Chirped pulse frequency $\omega(t)$ and envelope function $f(t)$, **b** several probabilities in a driven Morse oscillator, all as a function of time in cycles of the external field; adapted from [30]

of resonance. So if the oscillator is to be excited resonantly on the ladder of energy levels, in the case of the Morse energies (5.21), the frequency has to decrease as a function of time. Furthermore, the excitation "pulse" has to be a π-pulse. Following these arguments, for the HF molecule, the authors of [30] have constructed an analytic form of a pulse that leads to a large final dissociation probability. The envelope of that pulse and its time varying central frequency are depicted in Fig. 5.17 and we stress that the chirp appears in the simple form $\cos[\omega(t)t]$ in the Hamiltonian (not integrated over time like in (1.26)). The frequency decreases from an initial value of ω_{01},[7] equal to the energy difference of the lowest two levels. In the figure also the probabilities for dissociation and the occupation of different vibrational levels of the HF molecule are displayed. The results have been gained by numerically solving the TDSE with the vibrational ground state as the initial state. Apart from the slight generalization of the chirp, this study is completely analogous to the investigation of multi-photon ionization in the Gauss potential of Sect. 4.3.2.

Analogous results have been found by applying a classical mechanics optimization procedure [32] and are reproduced in Fig. 5.18. It is not surprising that also the classical result displays a down chirp of the frequency. Also in classical mechanics, a softening of the bond occurs for higher energies.

The material just presented is already a glimpse of what we will discuss in detail in Sect. 5.5 on the control of quantum systems. There we will, e.g., review the use of optimal control methods to steer a Morse oscillator into a desired excited vibrational state [33].

[7]The corresponding vibrational period of HF is 8.4 fs.

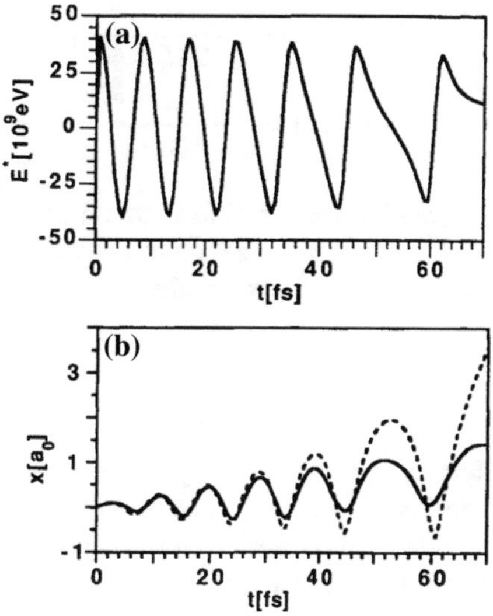

Fig. 5.18 **a** Optimal field and **b** quantum mechanical expectation value of position (*solid line*) and classical trajectory (*dashed line*) for the dissociation of a Morse oscillator [32]

5.3.3 *Coupled Potential Energy Surfaces*

Now we are ready to deal with the problem of nonadiabatic dynamics on coupled potential energy surfaces. The wavefunction that describes the system in this case is a vector and each component evolves on a specific surface. Due to the fact that the equations are coupled the problem is also referred to as a coupled channel problem. In Sect. 5.2, we have encountered such a situation at least formally already. However, the different channels there were the different basis functions in the Fourier-Bessel series expansion.

In this section, we will show how the formalism for the solution of the time-dependent Schrödinger equation has to be augmented to cope with the new situation. First this will be done fully quantum mechanically and then we will deal in some detail with the semiclassical approximation to coupled surface quantum dynamics.

5.3.3.1 Quantum Mechanical Approach

As a simple example, let us start with the case of a homonuclear diatomic molecule and consider two diagonal (diabatic) potential matrix elements $V_{nn}(R), n = 1, 2$. Their coupling shall be given by arbitrary non-diagonal matrix elements $V_{12}(R, t) = V_{21}(R, t)$, which may depend on time.

The corresponding two surface time-dependent Schrödinger equation is given by[8]

---

[8]Note that we are using the symbol for the time-independent nuclear wavefunction also for the time-dependent one.

$$\mathrm{i}\dot{\chi}_1(R,t) = \left[-\frac{1}{2M_\mathrm{r}}\frac{\partial^2}{\partial R^2} + V_{11}(R)\right]\chi_1(R,t) + V_{12}(R,t)\chi_2(R,t), \quad (5.65)$$

$$\mathrm{i}\dot{\chi}_2(R,t) = \left[-\frac{1}{2M_\mathrm{r}}\frac{\partial^2}{\partial R^2} + V_{22}(R)\right]\chi_2(R,t) + V_{12}(R,t)\chi_1(R,t)\,. \quad (5.66)$$

Its solution can be gained by an extension of the split-operator FFT method of Sect. 2.3. A generalization of the known procedure is necessary due to the fact that the potential is a 2×2 matrix

$$\mathbf{V}(R,t) = \begin{pmatrix} V_{11}(R) & V_{12}(R,t) \\ V_{12}(R,t) & V_{22}(R) \end{pmatrix} \quad (5.67)$$

now, in complete analogy to the matrix $\mathbf{A}$ of Sect. 5.2. In order to exponentiate it, one first has to diagonalize it, as in the previous section. In contrast to the previous section, in the case of two levels the diagonalization can be done *exactly analytically*, however, leading to [34]

$$\exp\{-\mathrm{i}\Delta t\mathbf{V}\} = \exp\left\{-\mathrm{i}\Delta t\left(\frac{V_{11}+V_{22}}{2}\right)\right\}\begin{pmatrix} A & B \\ B & A^* \end{pmatrix}. \quad (5.68)$$

Here A and B are complex numbers given by

$$A = \cos\phi - \mathrm{i}\Delta t\lambda\frac{\sin\phi}{\phi}, \qquad B = -\mathrm{i}\Delta t V_{12}\frac{\sin\phi}{\phi}, \quad (5.69)$$

with the phase

$$\phi(R,t) = \Delta t\sqrt{V_{12}^2(R,t) + \lambda^2(R)}, \quad (5.70)$$

and half the potential energy difference

$$\lambda(R) = \frac{V_{11}(R) - V_{22}(R)}{2}. \quad (5.71)$$

The presented approach is still exact if maximally two surfaces are taking part in the dynamics.

5.9. *Prove the explicit formula for the exponentiated two by two matrix* $\exp\{-\mathrm{i}\Delta t\mathbf{V}\}$.

The formalism shall now be applied to the case of coupling between two adiabatic surfaces due to a laser pulse in length gauge using the RWA. The non-diagonal coupling is then given by

$$V_{12}(R,t) = \mu(R)f(t)\mathcal{E}_0\cos(\omega t), \quad (5.72)$$

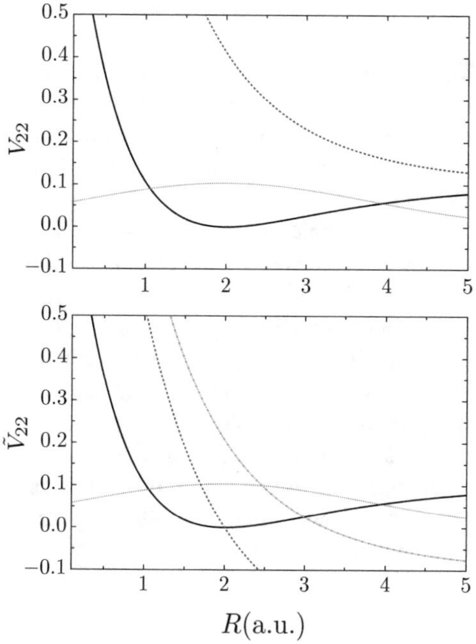

Fig. 5.19 *Upper panel*: Excited electronic surface V_{22} (*dashed blue*). *Lower panel*: the coupling to a field in the resonance case and for non-resonance leads to differently modified excited states $\tilde{V}_{22}$ (*dashed blue*: resonance, *dashed-dotted green*: nonresonance). In both panels also the ground state wavefunction (*dotted red*) and the ground state potential surface V_{11} (*solid black*) are depicted

with the dipole moment μ. To apply the rotating wave approximation of Sect. 3.2.3, the second component of the wavefunction vector is transformed according to

$$\tilde{\chi}_2 = \exp\{\mathrm{i}\omega t\}\chi_2. \tag{5.73}$$

Inserting the transformed wavefunction into the time-dependent Schrödinger equation and neglecting the counter-rotating term, proportional to $\exp\{-2\mathrm{i}\omega t\}$, the only time-dependence that remains is the one due to the envelope. Furthermore, because of the product rule to be used for the time-derivative, the second surface is shifted by $-\omega$ as displayed in Fig. 5.19. The transformed coupled surface equations in RWA thus contain the modified potential matrix elements

$$\tilde{V}_{22}(R) = V_{22}(R) - \omega, \tag{5.74}$$
$$\tilde{V}_{12}(R,t) = \mu(R) f(t)\mathcal{E}_0/2. \tag{5.75}$$

In addition, the Condon approximation can be made. The dipole moment then does not depend on R. Before presenting an application of this formalism, we first discuss a semiclassical approach to the coupled surface problem.

5.3.3.2 Semiclassical Approach: Mapping Hamiltonian

The classical and semiclassical description of the motion on coupled potential surfaces seems to be problematic. A way must be found to let trajectories switch from the motion on one to motion on another surface. A method that is ad hoc in nature, but is rather successful numerically, is the surface hopping technique developed by Tully [35].

Modern semiclassical methods can be *derived* from the underlying quantum mechanics, however, as we will see in the following. These methods have a historical precursor in the so-called "classical electron analog model" by Meier and Miller [36]. We will now review the method by Stock and Thoss, which is based on Schwinger's "mapping formalism" [37]. The N *discrete* electronic levels that shall be taking part in the dynamics are mapped onto N *continuous*, harmonic degrees of freedom in this approach.

The analogy between uncoupled harmonic oscillators and a spin system shall be dealt with in the following for the simple case of $N = 2$. In this case an oscillator of plus type and an oscillator of minus type are defined with the respective annihilation and creation operators

$$\hat{a}_+, \hat{a}_+^\dagger, \qquad \hat{a}_-, \hat{a}_-^\dagger, \tag{5.76}$$

where operators of the same type are fulfilling the standard commutation relations (2.164) and any operators of different type are commuting with each other. Furthermore, occupation number operators

$$\hat{N}_+ = \hat{a}_+^\dagger \hat{a}_+, \quad \hat{N}_- = \hat{a}_-^\dagger \hat{a}_- \tag{5.77}$$

can be defined. Simultaneous eigenkets of $\hat{N}_+$ and $\hat{N}_-$ fulfill the eigenvalue equations

$$\hat{N}_+|n_+, n_-\rangle = n_+|n_+, n_-\rangle, \qquad \hat{N}_-|n_+, n_-\rangle = n_-|n_+, n_-\rangle, \tag{5.78}$$

with the eigenvalues $n_\pm$, and an arbitrary state can be created from the vacuum state by the application of $\hat{a}_+^\dagger$ and $\hat{a}_-^\dagger$, using (2.165), according to

$$|n_+, n_-\rangle = \frac{(\hat{a}_+^\dagger)^{n_+}(\hat{a}_-^\dagger)^{n_-}}{\sqrt{n_+! n_-!}}|0, 0\rangle. \tag{5.79}$$

One can now define the products

$$\hat{J}_+ \equiv \hat{a}_+^\dagger \hat{a}_-, \qquad \hat{J}_- \equiv \hat{a}_-^\dagger \hat{a}_+, \qquad \hat{J}_z = \frac{1}{2}(\hat{a}_+^\dagger \hat{a}_+ - \hat{a}_-^\dagger \hat{a}_-), \tag{5.80}$$

where $\hat{J}_z = \frac{1}{2}(\hat{N}_+ - \hat{N}_-)$. It can be shown that these operators fulfill the angular momentum commutation relations [38]

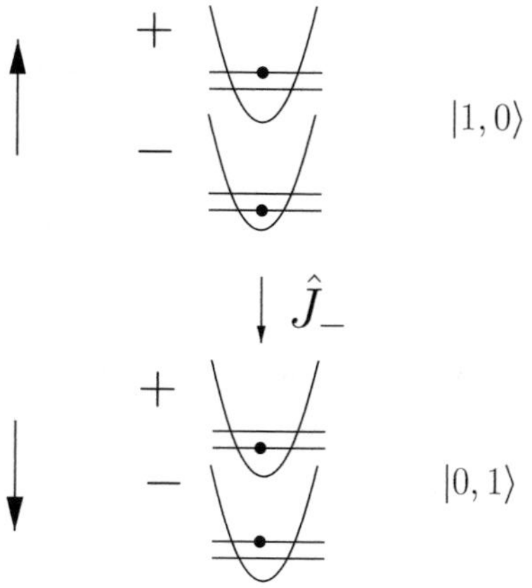

Fig. 5.20 Mapping analogy and occupation number representation

$$[\hat{J}_z, \hat{J}_\pm] = \pm\hat{J}_\pm \tag{5.81}$$

and

$$[\hat{J}_+, \hat{J}_-] = 2\hat{J}_z. \tag{5.82}$$

Furthermore,

$$\hat{\boldsymbol{J}}^2 = \hat{J}_z^2 + \frac{1}{2}(\hat{J}_+\hat{J}_- + \hat{J}_-\hat{J}_+) = \frac{\hat{N}}{2}\left(\frac{\hat{N}}{2} + 1\right), \tag{5.83}$$

with the total occupation number operator $\hat{N} \equiv \hat{N}_+ + \hat{N}_-$holds.

For our purposes the restriction to the subspace $n_+ + n_- = 1$ is appropriate. This is due to the fact that a spin 1/2 particle can be mapped onto two uncoupled oscillators. In case the eigenvalues of the plus and minus oscillator are $n_+ = 1$ and $n_- = 0$, respectively, the particle is in the spin up state. Application of the operator $\hat{J}_-$ leads to $n_+ = 0$ and $n_- = 1$ and the particle is in the spin-down state. The equivalent description of such a system in terms of harmonic oscillators and by using the occupation number representation is depicted in Fig. 5.20.

Now we return to the problem of a Hamilton operator for the coupled dynamics on N surfaces

$$\hat{H} = \sum_{n,m} h_{nm}|\phi_n\rangle\langle\phi_m|. \tag{5.84}$$

Rewriting this Hamiltonian with the help of continuous bosonic variables, one introduces N harmonic degrees of freedom by using the mapping procedure

$$|\phi_n\rangle\langle\phi_m| \to \hat{a}_n^\dagger \hat{a}_m, \tag{5.85}$$

$$|\phi_n\rangle \to |0_1, \dots 1_n, \dots, 0_N\rangle, \tag{5.86}$$

discussed above. This leads to the Hamiltonian

$$\hat{H} = \sum_{n,m} h_{nm} \hat{a}_n^\dagger \hat{a}_m. \tag{5.87}$$

A classical analog of the mapped quantum dynamics can be defined by replacing the position and momentum operators appearing in the creation and annihilation operator

$$\hat{a}_n^\dagger = (\hat{y}_n - \partial/\partial y_n)/\sqrt{2}, \tag{5.88}$$

$$\hat{a}_n = (\hat{y}_n + \partial/\partial y_n)/\sqrt{2} \tag{5.89}$$

by the respective classical variables y_n, p_n. The lower case variables $(\boldsymbol{y}, \boldsymbol{p}) = \{y_n, p_n\}$ with $n = 1, \dots N$ are thus the coordinates and momenta of the auxiliary harmonic oscillators and $(\boldsymbol{R}, \boldsymbol{P})$ are the phase space variables of the relative nuclear motion with the reduced mass M_r. The classical "mapping" Hamiltonian of an N level system is then given by

$$H(\boldsymbol{y}, \boldsymbol{p}, \boldsymbol{R}, \boldsymbol{P}) = \frac{\boldsymbol{P}^2}{2M_\mathrm{r}} + H_\mathrm{e}, \tag{5.90}$$

with the "electronic" Hamiltonian

$$H_\mathrm{e} = \sum_{n=1}^{N} V_{nn}(\boldsymbol{R})\frac{1}{2}(p_n^2 + y_n^2 - 1) + \sum_{n<m=1}^{N} V_{nm}(\boldsymbol{R})(y_n y_m + p_n p_m). \tag{5.91}$$

A semiclassical implementation of the coupled time-dependent Schrödinger equation (5.65) and (5.66) can now be done by using the Herman-Kluk propagator of Sect. 2.3.4. Each of the vectors $\{\boldsymbol{x}, \boldsymbol{p}_i, \boldsymbol{q}_i\}$ in the multi-dimensional formulation of the semiclassical propagator (2.229) contains the nuclear as well as the harmonic degrees of freedom. The initial state is a direct product of, e.g., a Gaussian wavepacket in the nuclear coordinate times the ground state eigenfunction of the initially unoccupied harmonic mode and the first excited state in the occupied harmonic mode [37]. The overlap with the coherent state in (2.229) can again be determined analytically.

5.3.3.3 Application to a Model System

In the following, full quantum as well as semiclassical results of the solution of the coupled surface time-dependent Schrödinger equation (5.65) and (5.66) will be reviewed for a model that has been used in order to study the breakdown of the

Rosen-Zener "approximation" [39]. The dimensionless variables

$$Q \equiv R/R_c \quad \text{and} \quad \tau \equiv t/t_c \tag{5.92}$$

are the same as used there, with

$$R_c = \sqrt{\hbar/(\sqrt{2}M_r\omega_e)} \quad \text{and} \quad t_c = \sqrt{2}/\omega_e\,. \tag{5.93}$$

Here ω_e is the frequency of the harmonic ground electronic surface and the model potentials are given by

$$V_{11}(Q) = Q^2/2, \tag{5.94}$$

$$\tilde{V}_{22}(Q) = -AQ + B, \tag{5.95}$$

whereas the off-diagonal potential, proportional to the envelope of the external field with dimensionless pulse length parameter T_p, is given by

$$\tilde{V}_{12}(\tau) = D\,\mathrm{sech}\left[\frac{\tau - \tau_0}{T_p}\right]. \tag{5.96}$$

For an inverse hyperbolic cosine pulse as above, a driven two-level system can be treated analytically and its Rosen-Zener solution has been reviewed in Sect. 3.2.4. The only difference to the case we study here is the absence of the kinetic energy in the Rosen-Zener model. Therefore, although the problem without kinetic energy is solvable exactly analytically, now this solution is an approximation! With these remarks, it is clear that the Rosen-Zener (RZ) approximation is the exact analytical solution of the approximate coupled time-dependent Schrödinger equation

$$\mathrm{i}\dot{\chi}_1^{\mathrm{RZ}}(\tau) = \lambda(Q)\chi_1^{\mathrm{RZ}}(\tau) + \tilde{V}_{12}(\tau)\tilde{\chi}_2^{\mathrm{RZ}}(\tau), \tag{5.97}$$

$$\mathrm{i}\dot{\tilde{\chi}}_2^{\mathrm{RZ}}(\tau) = \tilde{V}_{12}(\tau)\chi_1^{\mathrm{RZ}}(\tau) - \lambda(Q)\tilde{\chi}_2^{\mathrm{RZ}}(\tau), \tag{5.98}$$

with

$$\lambda(Q) = \frac{\tilde{V}_{22}(Q) - V_{11}(Q)}{2}. \tag{5.99}$$

In the spirit of the so-called Franck-Condon approximation [40, 41], electronic transitions take place at fixed nuclear positions Q. We can therefore use the solution of Rosen and Zener

$$|\chi_1^{\mathrm{RZ}}(\lambda,\tau)|^2 = \left|F\left[DT_p, -DT_p; \frac{1}{2} - \mathrm{i}\lambda T_p; z(\tau)\right]\right|^2, \tag{5.100}$$

with the hypergeometric function F [42] and

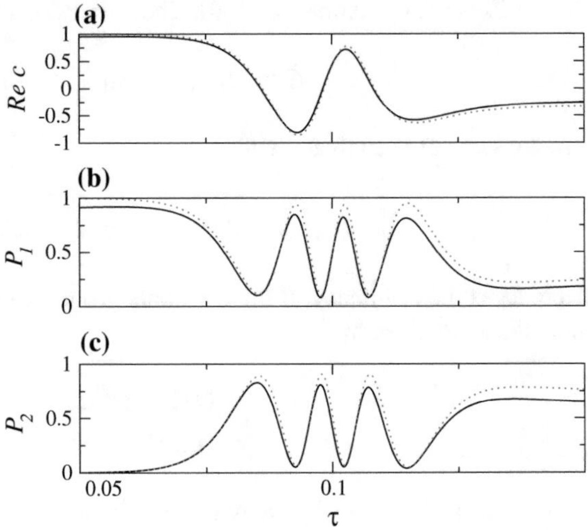

Fig. 5.21 Real part of the auto-correlation function (**a**) and the population of level 1 (**b**) and level 2 (**c**) as a function of time for model I; *solid line*: semiclassical result, *dotted line*: full quantum result; adapted from [43]

$$z(\tau) = \frac{1}{2}[\tanh(\tau/T_\mathrm{p}) + 1]. \tag{5.101}$$

$|\chi_1^{\mathrm{RZ}}|^2$ depends on Q via λ and provides the probability to be in the ground state at a given Q. The total probability to be in the ground state can be gained by multiplying the Rosen-Zener solution with the initial probability density and integrating over position. For the probability to be in the excited state

$$P_2^{\mathrm{RZ}}(\tau) = 1 - \int \mathrm{d}Q|\chi_1(Q, -\infty)|^2|\chi_1^{\mathrm{RZ}}(2\lambda = \Delta V(Q), \tau)|^2 \tag{5.102}$$

then follows.

In Table 5.3, the model parameters for the results to be presented are gathered. In both cases, the initial nuclear wavepacket is the ground state wavefunction of the harmonic surface. The Gaussian part of the 3D wavefunction in the semiclassical case is thus centered around the origin and has the width parameters $(\gamma_{11}, \gamma_{22}, \gamma_{33}) = (2^{-1/2}, 1, 1)$. Quantities of interest are the auto-correlation function

$$c(\tau) = \langle\chi_1(\tau)|\chi_1(0)\rangle + \langle\chi_2(\tau)|\chi_2(0)\rangle\ , \tag{5.103}$$

as well as the occupation probabilities

$$P_{1,2}(\tau) = \langle\chi_{1,2}(\tau)|\chi_{1,2}(\tau)\rangle \tag{5.104}$$

of the different levels.

For model I, a considerable number of Rabi oscillations occurs as can be seen in Fig. 5.21. The quality of the semiclassical results is good. This is so, although

Table 5.3 Dimensionless model parameters for the problem of two coupled surfaces

	A	B	D	τ_0	T_p
Model I	50	10	300	0.1	0.01
Model II	0.1	0.01	2.5	2	0.4

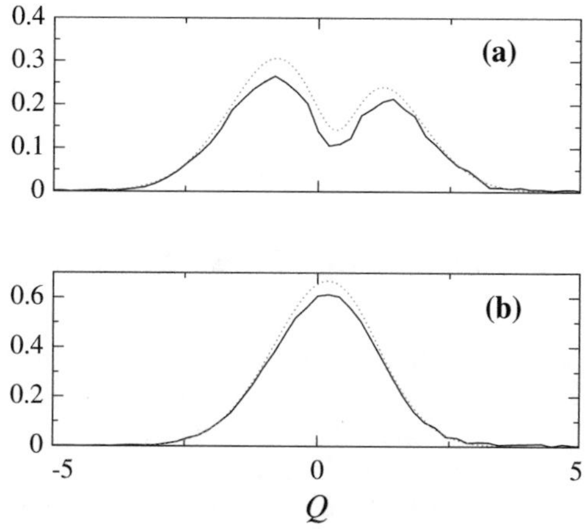

Fig. 5.22 Absolute value of the wavefunction in level 1 (**a**) and level 2 (**b**) at time $\tau = 0.135$ for model I; *solid line*: semiclassical result, *dotted line*: full quantum result [43]

the parameters chosen for model I lead to a substantial nonlinearity of the classical equations of motion.[9] Also the semiclassical wavefunctions at time $\tau = 0.135$ in the different levels in Fig. 5.22 give a good account of the quantum wavefunction. Especially the double humped structure of the wavefunction in the ground state is correctly reproduced.

For model I, the Rosen-Zener approximation (which is not shown) would be very well founded due to the shortness of the pulse and there would be almost no difference compared to the full quantum results. Let us consider a case, where the neglect of the kinetic energy, sometimes referred to as the short time approximation, breaks down, however. This is the case of model II that uses the same wavepacket parameters as model I. In Fig. 5.23 a comparison of three different results is displayed. The semiclassical as well as the full quantum and the approximate Rosen-Zener quantum results for the probability to be in the second level are shown. As in case I above, the semiclassical results are representing the full quantum results quite well. In the Rosen-Zener approximation for longer times, a deviation from the quantum result can be observed, however, which is due to the neglect of the kinetic energy in (5.97) and (5.98).

[9]Both potentials, if uncoupled, would show no nonlinearity in the classical dynamics and would be solvable exactly analytically; when coupled, however, the mapping Hamiltonian is highly anharmonic!

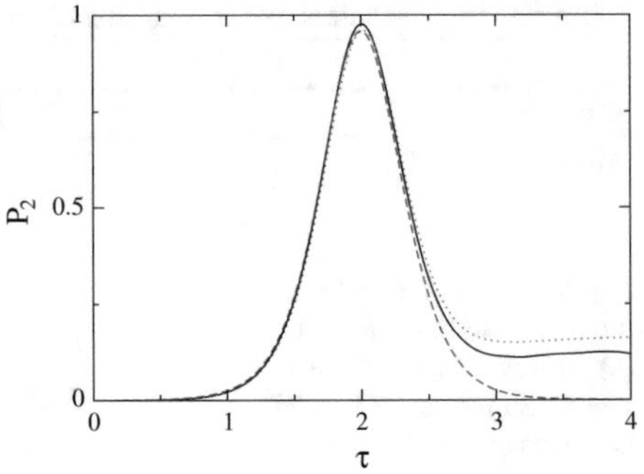

Fig. 5.23 Comparison of the semiclassical (*solid*), the quantum (*long-dashed*) and the Rosen-Zener result (*short dashed*) for the population of level 2 as a function of time for model II [43]

5.4 Femtosecond Pump-Probe Spectroscopy

The study of ultra-fast molecular processes is a rapidly growing research field that has gained considerable attention due to the 1999 Nobel prize in Chemistry for Zewail. Before we concentrate on some theoretical aspects of that field, let us get acquainted with orders of magnitude of time scales in molecular dynamics.

To this end we will convert times into energy (and vice versa) using the formula

$$E/h = 1/T\ . \tag{5.105}$$

In different units this reads

$$E(\mathrm{eV}) = \frac{4.134}{T(\mathrm{fs})}, \qquad E(\mathrm{cm}^{-1}) = \frac{33,368}{T(\mathrm{fs})}. \tag{5.106}$$

From molecular spectra [22] the following ranges for times (periods) in which typical phenomena occur can be extracted[10]:

- Rotation: from 1–100 ps ($\bar{\nu}_{J=0\to J=1}$ from 20 to 0.25 cm^{-1})
- Normal vibrations: from 10–300 fs ($\bar{\nu}_{n=0\to n=1}$ from 4,000 to 100 cm^{-1})
- Vibrational relaxation: 100 fs–100 ps
- Direct photodissociation: up to 100 fs [23]

In order to investigate molecular phenomena on a femtosecond scale, time-resolved measurements, which are frequently referred to as "pump-probe" experiments, are performed. In such experiments a sample is excited by a first, so-called pump pulse. After a variable time delay T_d a second, so-called probe pulse is impinging on the excited system and a signal

$$S(T_\mathrm{d}) = S(\text{with pump}) - S(\text{without pump}) \tag{5.107}$$

[10] $\bar{\nu} = \nu/c$ is the wavenumber.

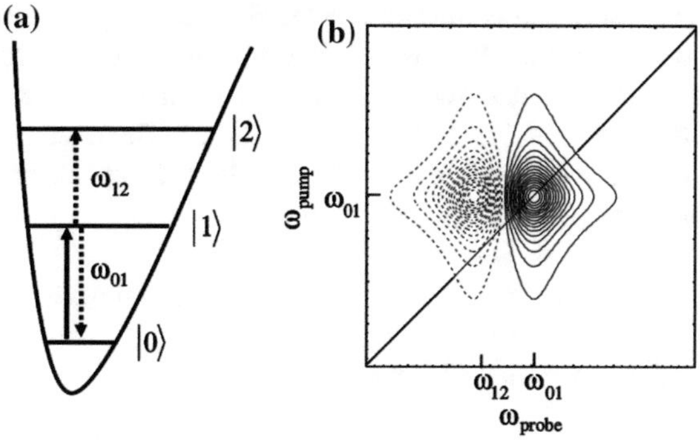

Fig. 5.24 **a** Morse potential with dipole-allowed transitions and **b** absorptive 2D IR spectrum of an anharmonic molecular vibration (*dashed line*: positive, and *solid line*: negative signal) [44]

is measured. This may, e.g., be the absorption of the system, the fluorescence of the system or the probability to emit an electron with a certain energy. The time resolution of the experiment is given by the full width at half maximum of the pulses, which shall be used in the following to characterize their shortness.

5.4.1 2D IR Spectroscopy

In 2D infrared (IR) spectroscopy, one is interested in describing the coupling between vibrational modes. They manifest themselves in the so-called cross peaks in the 2D spectrum. In this section, we concentrate on 2D IR spectroscopy in the frequency domain and follow closely the excellent presentation of the subject by Hamm and Zanni [44]. We start by looking at a single vibrational mode first and then move on to the case of two coupled vibrational modes.

5.4.1.1 Single anharmonic vibrational mode

A single Morse oscillator model for a carbonyl stretch in an acetone molecule will be used as the underlying physical system in the following. A 2D IR spectrum in the frequency domain is displayed in Fig. 5.24 together with the Morse potential, whose energy spectrum is given in (5.21). The anharmonic character of the Morse potential and the corresponding feature of its spectrum is decisive for a nontrivial signal in 2D IR spectroscopy.

One scans the pump frequency $\omega_{\rm pump}$ and the probe frequency $\omega_{\rm probe}$ and uses them as y axis and x axis, respectively. What is then displayed as a contour plot is the difference spectrum, i.e., the absorption with the pump pulse switched on, minus the absorption with the pump pulse switched off. Because the pump pulse depletes the ground state, there is less absorption from the ground state (an effect which is called bleach) and in addition due to the stimulated emission from the excited state the absorption is further decreased. These two effects lead to a negative sign in the 2D spectrum, which is depicted by solid contour lines. In contrast, the absorption from the excited state can only occur of the pump pulse was switched on and is therefore positive (broken contour lines). The negative and positive contribution can be distinguished because the transition from the first to the second excited level occurs at

$$\omega_{12} = \omega_e - 4\omega_e x_e, \tag{5.108}$$

which is red-shifted from

$$\omega_{01} = \omega_e - 2\omega_e x_e. \tag{5.109}$$

In the case of the harmonic oscillator $x_e = 0$ and the positive and negative contributions would occur around the same central frequency and cancel out, because, although there are two signals contributing to the on-diagonal peak, the intensity of the excited state absorption is twice as strong as the 0–1 transition. This is due to the fact that the squared transition dipoles scale as $\mu_{12}^2 = 2\mu_{01}^2$.

5.10. *Show that the transition dipoles for the harmonic oscillator fulfill the scaling relation* $\mu_{12}^2 = 2\mu_{01}^2$.

If the anharmonic shift is larger than the bandwidth of the transition, the anharmonicity constant can be read off directly from the 2D IR spectrum.

5.4.1.2 Two coupled anharmonic vibrational modes

The more prominent reason for the study of 2D IR spectroscopy is the possibility to monitor interactions between coupled modes. We first consider two coupled local modes in an exciton model Hamiltonian (ignoring zero point energy)

$$\hat{H}_h = \omega_1 \hat{a}_1^+ \hat{a}_1 + \omega_2 \hat{a}_2^+ \hat{a}_2 + \beta_{12}(\hat{a}_1^+ \hat{a}_2 + \hat{a}_2^+ \hat{a}_1), \tag{5.110}$$

with the creation and annihilation operators of the local oscillators introduced in Sect. 2.3.1. The non-rotating wave terms $\hat{a}_1 \hat{a}_2$ and $\hat{a}_1^+ \hat{a}_2^+$ that originate from a bilinear coupling term in the mode coordinates have been neglected already. The coupling term can be modeled by a transition dipole coupling of the form

$$\beta_{12} = \frac{\boldsymbol{\mu}_1 \cdot \boldsymbol{\mu}_2}{r_{12}^3} - 3\frac{(\boldsymbol{r}_{12} \cdot \boldsymbol{\mu}_1)(\boldsymbol{r}_{12} \cdot \boldsymbol{\mu}_2)}{r_{12}^5}, \tag{5.111}$$

with the transition dipoles $\boldsymbol{\mu}_i$ of the local modes and the distance vector $\boldsymbol{r}_{12}$ connecting the two sites.

The corresponding Hamilton matrix in a product basis of number states (up to double excitations)

$$\{|00\rangle, |10\rangle, |01\rangle, |20\rangle, |02\rangle, |11\rangle\} \tag{5.112}$$

is given by

$$\mathbf{H}_h = \begin{pmatrix} 0 & 0 & 0 & 0 & 0 & 0 \\ 0 & \omega_1 & \beta_{12} & 0 & 0 & 0 \\ 0 & \beta_{12} & \omega_2 & 0 & 0 & 0 \\ 0 & 0 & 0 & 2\omega_1 & 0 & \sqrt{2}\beta_{12} \\ 0 & 0 & 0 & 0 & 2\omega_2 & \sqrt{2}\beta_{12} \\ 0 & 0 & 0 & \sqrt{2}\beta_{12} & \sqrt{2}\beta_{12} & \omega_1 + \omega_2 \end{pmatrix}, \tag{5.113}$$

which separates into the zero-, the one- and the two-exciton block. So far this is still a harmonic Hamiltonian and therefore the 2D IR signal is zero.

To see a signal, we need to include terms in the Hamiltonian that are higher than second order. Including fourth order terms that account for an anharmonic shift of the energies, similar to a Morse potential, the Hamiltonian reads

$$\hat{H}_q = \hat{H}_h - \frac{\Delta}{2}\hat{a}_1^+\hat{a}_1^+\hat{a}_1\hat{a}_1 - \frac{\Delta}{2}\hat{a}_2^+\hat{a}_2^+\hat{a}_2\hat{a}_2, \tag{5.114}$$

with the local mode anharmonic shift Δ. The corresponding Hamilton matrix is

$$\mathbf{H}_q = \begin{pmatrix} 0 & 0 & 0 & 0 & 0 & 0 \\ 0 & \omega_1 & \beta_{12} & 0 & 0 & 0 \\ 0 & \beta_{12} & \omega_2 & 0 & 0 & 0 \\ 0 & 0 & 0 & 2\omega_1 - \Delta & 0 & \sqrt{2}\beta_{12} \\ 0 & 0 & 0 & 0 & 2\omega_2 - \Delta & \sqrt{2}\beta_{12} \\ 0 & 0 & 0 & \sqrt{2}\beta_{12} & \sqrt{2}\beta_{12} & \omega_1 + \omega_2 \end{pmatrix}, \tag{5.115}$$

where only the two-exciton part of the matrix is modified due to the presence of the quartic terms in the Hamiltonian.

The model presented so far is only good for near resonant vibrational states [44]. We focus on the extreme limiting case of $\omega_1 = \omega_2$. If the Hamiltonian is diagonalized block-wise, we then get (with $\beta = \beta_{12}$)

$$\mathbf{H}_q = \begin{pmatrix} 0 & 0 & 0 & 0 & 0 & 0 \\ 0 & \omega - \beta & 0 & 0 & 0 & 0 \\ 0 & 0 & \omega + \beta & 0 & 0 & 0 \\ 0 & 0 & 0 & 2\omega - \frac{1}{2}\Delta - \eta & 0 & 0 \\ 0 & 0 & 0 & 0 & 2\omega - \frac{1}{2}\Delta + \eta & 0 \\ 0 & 0 & 0 & 0 & 0 & 2\omega_1 - \Delta \end{pmatrix}, \tag{5.116}$$

with $\eta = \frac{1}{2}\sqrt{\Delta^2 + 16\beta^2}$. In Fig. 5.25, the energy levels in the more general case of two slightly different frequencies are depicted.

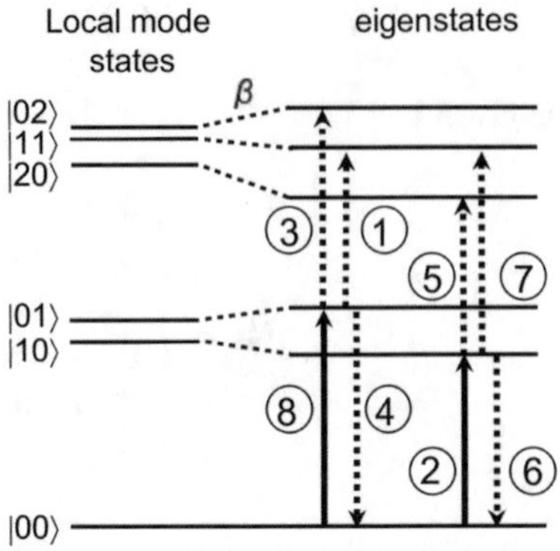

Fig. 5.25 Level scheme of two coupled oscillators emerging from the uncoupled number states with slightly different frequency of the single oscillators [44]

Simplifying the mathematics by looking at the case of equal frequencies again, the frequency of the absorption from the ground state labeled by the number 8 is the same as that of the stimulated emission labeled by the number 4 and is given by $\omega_{\text{probe}} = \omega + \beta$. The absorption from the singly excited state emerging from $|01\rangle$ labeled by the number three is given by $\omega_{\text{probe}} = \omega - \frac{1}{2}\Delta + \frac{1}{2}\sqrt{\Delta^2 + 16\beta^2} - \beta$, which for $\beta \gg \Delta$ is $\omega_{\text{probe}} = \omega + \beta - \frac{1}{2}\Delta$, being smaller than the frequencies of the transitions 4 and 8! This is exactly the same observation that leads to the separation of the diagonal peaks in the case of a single anharmonic molecule.

The other *diagonal* transitions are the ones labeled by 2 and 6 (bleach and stimulated emission) with $\omega_{\text{probe}} = \omega - \beta$ and 5 (excited state absorption, this time from the state emerging from $|10\rangle$), with $\omega_{\text{probe}} = \omega - \beta - \frac{1}{2}\Delta$, They are again separated in probe frequency by $\frac{1}{2}\Delta$.

In addition there are *off-diagonal* peaks appearing now. The peaks labeled by 1 and 2 correspond to a pumping of the excited state level emerging from $|01\rangle$ and are due to the bleach of the absorption to the level emerging from $|10\rangle$ and the excited state absorption from the level emerging from $|01\rangle$. They are separated by Δ. The corresponding 2D IR spectrum is shown in Fig. 5.26.

The distance in pump frequency between the upper and the lower row in the 2D spectrum is given by 2β and is therefore a measure of the interaction between the anharmonic vibrational modes.

5.4.2 *Pump-Probe Photoelectron Spectroscopy of Na_2*

As a first example of a pump-probe experiment involving electron dynamics, we consider the excitation of the $(2)^1\Sigma_u^+$-state of the sodium dimer by a 40 fs laser pulse of the central wavelength 340 nm (pump-pulse) and subsequent ionization by a probe-pulse of wavelength 530 nm, arriving after a variable time delay T_{d}. The energy of the emitted electrons can then be measured as a function of T_{d} [45]. Theoretical

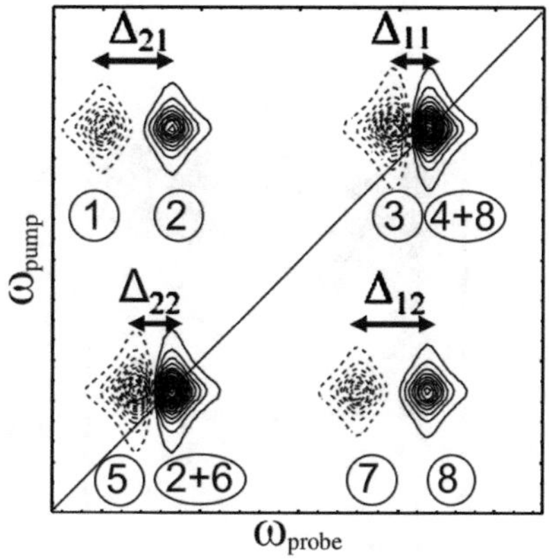

Fig. 5.26 2D IR spectrum of two coupled oscillators with slightly different frequency [44]

investigations of the same system have been performed by Christoph Meier in his PhD thesis [46], which we will follow closely.

Before the action of the pump pulse, the molecule is described by the vibrational ground state wavepacket in the electronic ground state $X^1\Sigma_g^+$, depicted in the left panel of Fig. 5.27. The perturbation of the system by the pump pulse

$$V_{\rm L} = \mu_{10} f(t) \frac{\mathcal{E}_0}{2} (e^{i\omega_{\rm P} t} + e^{-i\omega_{\rm P} t}) \tag{5.117}$$

leads to excitation of the wavefunction onto an excited electronic surface. Using the Condon approximation, the dipole moment is assumed to be independent of position. Furthermore, for the following investigation, perturbation theory is adequate to describe the laser molecule interaction. In a form applicable to a vector-type time-dependent Schrödinger equation as in (5.65) and (5.66) it is reviewed in Appendix 5.B.

In first order in the perturbation and in rotating wave approximation (neglecting the counter-rotating term $\sim e^{i\omega_{\rm P} t}$ of the perturbation)

$$\chi_1(R,t) = \frac{1}{i} \int_0^t dt' e^{-i\hat{H}_1(t-t')} \mu_{10} f(t') \frac{\mathcal{E}_0}{2} e^{-i\omega_{\rm P} t'} e^{-iE_0 t'} \chi_0(R,0) \tag{5.118}$$

for the wavefunction on the excited state surface is found.

For a numerical implementation the integral above has to be discretized according to

$$\chi_1(R,t) = \frac{\Delta t}{i} \sum_{j=0}^{n} e^{-i\hat{H}_1(n-j)\Delta t} \mu_{10} f(j\Delta t) \frac{\mathcal{E}_0}{2} e^{-i(E_0+\omega_{\rm P}) j\Delta t} \chi_0(R,0). \tag{5.119}$$

The vibrational ground state with the energy $E = E_0$ is propagated on the shifted ground state surface until an intermediate time $j\Delta t$ is reached. The resulting state

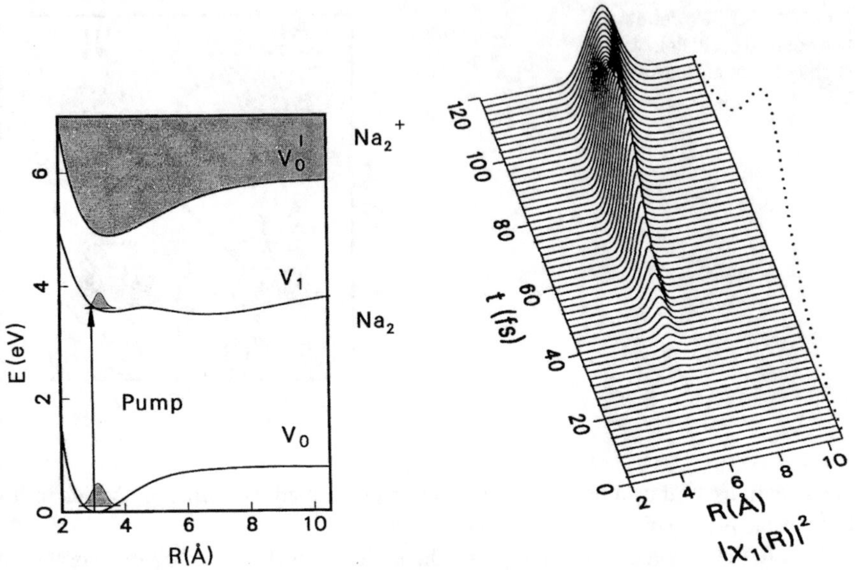

Fig. 5.27 *Left panel*: potential surfaces of the sodium dimer together with the action of the pump pulse onto the initial wavepacket in the electronic ground state; *right panel*: emergence of the wavepacket on the excited state surface, due to the action of the pulse (*dotted line*); adapted from [46]

then is multiplied by the perturbation, is lifted on the excited surface and is propagated with Hamiltonian $\hat{H}_1$ until the final time $n\Delta t$ is reached. It is not known, at what time the photon is being absorbed, however, and therefore all the intermediate times have to be integrated over. This procedure can be also formulated iteratively according to [47]

$$\begin{aligned}\chi_1(R, t_n + \Delta t) &= \mathrm{e}^{-\mathrm{i}\hat{H}_1\Delta t}\chi_1(R, t_n)\\ &\quad + \frac{\Delta t}{\mathrm{i}}\mu_{10} f(t_n + \Delta t)\frac{\mathcal{E}_0}{2}\mathrm{e}^{-\mathrm{i}(E_0+\omega_\mathrm{P})(t_n+\Delta t)}\chi_0(R, 0). \qquad (5.120)\end{aligned}$$

The first term can be calculated with the split-operator method, whereas the second term is given analytically. At each time step, a further part of the wavefunction is lifted on the excited electronic surface. The result of such a calculation is depicted in the right panel of Fig. 5.27, which shows a vibrationally excited wavepacket on the excited electronic surface that moves almost dispersion-less to larger internuclear distances.

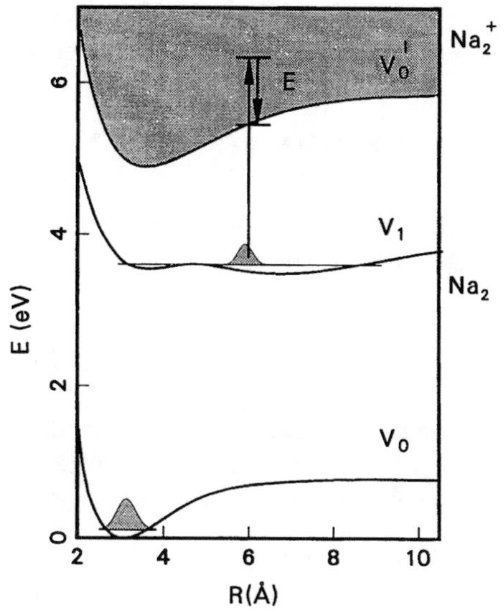

Fig. 5.28 Potential surfaces of the sodium dimer with the action of the probe pulse on the wavepacket in the excited state [46]

5.11. *Rewrite the sum in the discretised representation of first order perturbation theory for* $n = 1, 2, 3$ *and verify the iterative prescription*

$$\begin{aligned}\chi_1(R, t_n + \Delta t) &= \hat{U}_1(\Delta t)\chi_1(R, t_n)\\ &\quad + \frac{\Delta t}{\mathrm{i}}\mu_{10}\mathcal{E}(t_n + \Delta t)U_0(t_n + \Delta t)\chi_0(R, 0)\end{aligned}$$

with $\hat{U}_1(t) = \mathrm{e}^{-\mathrm{i}\hat{H}_1 t}$, $U_0(t) = \mathrm{e}^{-\mathrm{i}(E_0+\omega_\mathrm{P})t}$ *and* $\mathcal{E}(t) = f(t)\mathcal{E}_0/2$ *for the propagation of the component of the wavefunction on the excited surface.*

The probe pulse, delayed by a time T_d and centered around the frequency ω_T, now allows the detection of the nuclear wavepacket motion on the excited potential energy surface via the measurement of the energy of the emitted photo electrons after ionization, as can be seen in Fig. 5.28. The key to the understanding of this measurement is the reflection principle. The use of that principle in the theory of photodissociation is reviewed in Appendix 5.C.

In order to invoke the reflection principle, the wavefunction in the ionization continuum has to be considered. The basis of bound states $\phi_{\mathrm{e},j}(\boldsymbol{r}, R)$ is extended by the continuum states ϕ_{E,V_0^I} (free electron with energy E, ionic core in the ground state with potential V_0^I). In the Born-Oppenheimer approximation the nuclei would then fulfill the uncoupled equations

$$\mathrm{i}\dot{\chi}_j(R, t) = \{\hat{T}_R + V_j(R)\}\chi_j(R, t), \tag{5.121}$$

$$\mathrm{i}\dot{\chi}_E(R, t) = \{\hat{T}_R + V_0^\mathrm{I}(R) + E\}\chi_E(R, t). \tag{5.122}$$

Coupling to a laser field (that starts to interact with the system at time T_d) in length gauge and using again the RWA, we get in first order perturbation theory

$$\chi_E(R,t) = \frac{1}{\mathrm{i}}\int_{T_\mathrm{d}}^{t} \mathrm{d}t' \mathrm{e}^{-\mathrm{i}(\hat{H}_0^\mathrm{I}+E-\omega_\mathrm{T})(t-t')}\mu_{E1} f(t'-T_\mathrm{d})\frac{\epsilon_0}{2}\chi_1(R,t') \quad (5.123)$$

for the wavefunction in the ionized state. From this, the spectrum of the emitted electron can be extracted according to

$$P^\mathrm{I}(E,T_\mathrm{d}) = \lim_{t\to\infty}\int \mathrm{d}R|\chi_E(R,t)|^2 \,. \quad (5.124)$$

Further progress towards an understanding of the process is made by using the short-time approximation, that we have already encountered in Sect. 5.3.3. In this approximation the kinetic energy of the nuclei is neglected, which means for (5.123):

- Replace $\hat{H}_0^\mathrm{I}$ by V_0^I
- Replace $\chi_1(R,t')$ by $\mathrm{e}^{-\mathrm{i}V_1(t'-T_\mathrm{d})}\chi_1(R,T_\mathrm{d})$

Equation (5.123) thus becomes the Fourier transformation of the pulse envelope. Using the definition[11]

$$F(x) = \int \mathrm{d}t\, \mathrm{e}^{\mathrm{i}xt} f(t) \quad (5.125)$$

and the short-time approximation, we get in close analogy to (5.102)

$$P^\mathrm{I}(E,T_\mathrm{d}) \sim \int \mathrm{d}R|\mu_{E1}\chi_1(R,T_\mathrm{d})|^2|F(D(R)+E-\omega_\mathrm{T})|^2 \,. \quad (5.126)$$

Here the definition of the difference potential,

$$D(R) \equiv V_0^\mathrm{I}(R) - V_1(R), \quad (5.127)$$

has been used. The largest contributions to the expression for the electron emission probability come from regions of vanishing argument of the Fourier transform. This is yet another application of the SPA from Sect. 2.2.1. The SPA condition leads to the definition of so-called transient Franck-Condon regions [48]

$$D(R_\mathrm{tr}) \approx \omega_\mathrm{T} - E \,. \quad (5.128)$$

In case of a monotonous function $D(R)$ only a single stationary phase point exists and the remaining integral can be approximated by

---

[11]Note that the integration boundaries can be shifted to infinity due to the envelope, and the terms $\mathrm{e}^{\mathrm{i}V_1T_\mathrm{d}}$ and $\mathrm{e}^{-\mathrm{i}V_0^It}$ drop out by taking the absolute square.

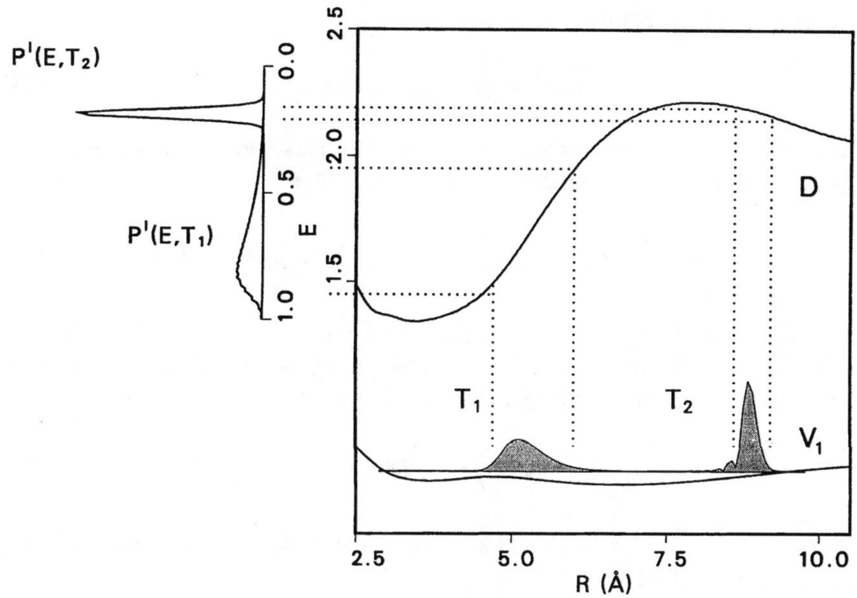

Fig. 5.29 Dynamical reflection principle for two different pulse delays [46]

$$P^{\mathrm{I}}(E, T_{\mathrm{d}}) \sim |\mu_{E1}\chi_1(R_{\mathrm{tr}}(E), T_{\mathrm{d}})|^2 \,. \tag{5.129}$$

This is the mathematical formulation of the dynamical reflection principle, saying that the electron spectrum is proportional to the absolute square of the wavefunction at time T_{d}. The structure of its argument tells us that the squared wavefunction is reflected at the difference potential [46]. If it is steep then P^{I} is broad in energy. If it has a small slope, however, a sharp peak of P^{I} emerges. Both situations are depicted graphically in Fig. 5.29. In this figure two different probe pulse delays, $T_1 = 0.2\,\mathrm{ps}$ and $T_2 = 1\,\mathrm{ps}$ are compared with each other. Due to the motion of the wavepacket on surface V_1, a dramatic change of the electron spectrum occurs. For the longer delay, first it is shifted to smaller energies, and second, it becomes much narrower. By knowing the potential surfaces, one can monitor the motion of the nuclear wavepacket by means of the measured photoelectron spectrum.

The validity of the dynamical reflection principle hinges on the applicability of the short-time approximation. This can be judged by looking at the full time-evolution operators appearing in (5.123)

$$\mathrm{e}^{\mathrm{i}\hat{H}_0^{\mathrm{I}}t'}\mathrm{e}^{-\mathrm{i}\hat{H}_1t'} = \mathrm{e}^{\hat{A}} \,. \tag{5.130}$$

Using the Baker-Campbell-Haussdorff formula from Sect. 2.3.2 in the form

$$\mathrm{e}^{\hat{A}}\mathrm{e}^{\hat{B}} \approx \mathrm{e}^{\hat{A}+\hat{B}+1/2[\hat{A},\hat{B}]} \tag{5.131}$$

one finds with (5.127) that

$$\hat{A} = \mathrm{i}D(R)t' - [\hat{T}_R, D(R)]t'^2 \tag{5.132}$$

holds. Retaining only the first term in this expression leads to the short-time approximation. The term proportional to t'^2 with the prefactor

$$-[\hat{T}_R, D(R)] = \frac{1}{2M_\mathrm{r}}(D''(R) + 2D'(R)\partial_R) \approx \frac{\mathrm{i}}{M_\mathrm{r}}D'p \tag{5.133}$$

should be small. D'' usually is very small and if one interprets $(p/M_\mathrm{r})t'$ classically as the change of the internuclear distance during the pulse, then the condition for the applicability of the short time approximation is

$$D'\frac{p}{M_\mathrm{r}}t' \ll D(R), \tag{5.134}$$

i.e., the difference potential should not change much over the range that the wavepacket crosses during the pulse.

5.4.3 Fluorescence Spectroscopy of ICN

Instead of detecting the motion of the wavepacket on the excited surface via the measurement of the energy of emitted electrons as in the previous case, also the fluorescence after excitation into a second excited state can be monitored. An example that has been studied experimentally as well as theoretically is the ICN molecule. Theoretically it suffices to consider only the dynamics of the C-I stretch coordinate. The corresponding dynamics on the 3 coupled potential surfaces

- Electronic ground state,
- I+CN($X^2\Sigma^+$) excited (dissociative) state,
- I+CN($B^2\Sigma^+$) excited (dissociative) state

two of which can be seen in Fig. 5.30, has been investigated in [49].

The coupled surface time-dependent Schrödinger equation for the laser driven system is given by

$$\mathrm{i}\partial_t \begin{pmatrix} \chi_0 \\ \chi_1 \\ \chi_2 \end{pmatrix} = \begin{pmatrix} \hat{H}_0 & \hat{H}_{01} & 0 \\ \hat{H}_{10} & \hat{H}_1 & \hat{H}_{12} \\ 0 & \hat{H}_{21} & \hat{H}_2 \end{pmatrix} \begin{pmatrix} \chi_0 \\ \chi_1 \\ \chi_2 \end{pmatrix}, \tag{5.135}$$

where in the rotating wave approximation the coupling by the pump, respectively the probe pulse is given by

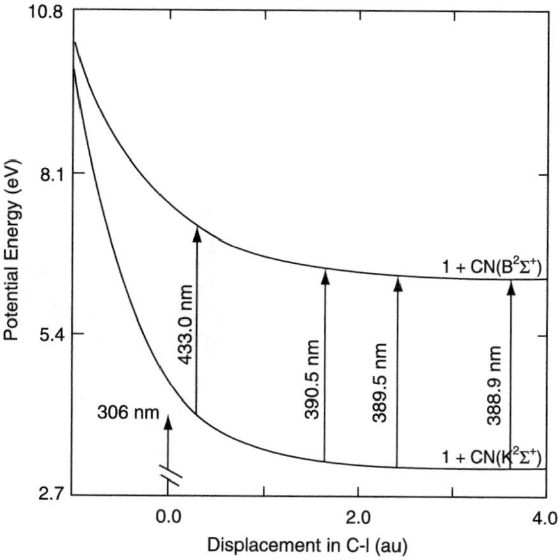

Fig. 5.30 Dissociative potential surfaces and different probe frequencies for ICN; adapted from [49]

$$\hat{H}_{01} = \mu_{01} A_1(t) \mathrm{e}^{-\mathrm{i}\omega_\mathrm{P} t}, \tag{5.136}$$

$$\hat{H}_{12} = \mu_{12} A_2(t - T_\mathrm{d}) \mathrm{e}^{-\mathrm{i}\omega_\mathrm{T} t}. \tag{5.137}$$

The occupation of the second excited state is proportional to the measured fluorescence signal and is therefore to be calculated theoretically. In first order perturbation theory we get for the wavefunctions in the different electronic states (if $E_0 = 0$)

$$|\chi_0(t)\rangle = \mathrm{e}^{-\mathrm{i}E_0 t}|\chi_0(0)\rangle = |\chi_0(0)\rangle, \tag{5.138}$$

$$|\chi_1(t)\rangle \sim \int_{-\infty}^{t} \mathrm{d}t' \mu_{01} A_1(t') \mathrm{e}^{-\mathrm{i}\hat{H}_1'(t-t')}|\chi_0(0)\rangle, \tag{5.139}$$

$$|\chi_2(t)\rangle \sim \int_{-\infty}^{t} \mathrm{d}t' \mu_{12} A_2(t' - T_\mathrm{d}) \mathrm{e}^{-\mathrm{i}\hat{H}_2'(t-t')}|\chi_1(t)\rangle, \tag{5.140}$$

with $\hat{H}_1' = \hat{H}_1 - \omega_\mathrm{P}$ and $\hat{H}_2' = \hat{H}_2 - \omega_\mathrm{T}$, due to the fact that the potentials are shifted in RWA by ω_P, respectively ω_T.[12] The transfer of probability density to an excited surface can only be large if the crossing with the shifted level is at the maximum of the wavepacket. This so-called resonance case was depicted together with the off-resonance case in Fig. 5.19. Only in the case of resonance a total population transfer is possible by a so-called π-pulse (see Sect. 3.2.3).

In the following the results of a simulation [49] of an experiment of the Zewail group [50] will be reviewed. The considered parameters are:

[12]The lower time-integration boundary has been shifted to $-\infty$ to treat pulses that are nonzero at negative times.

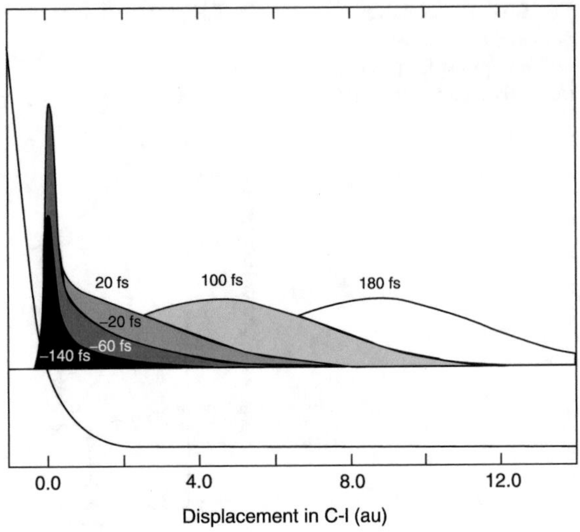

Fig. 5.31 Time evolution on the first excited potential energy surface [49]

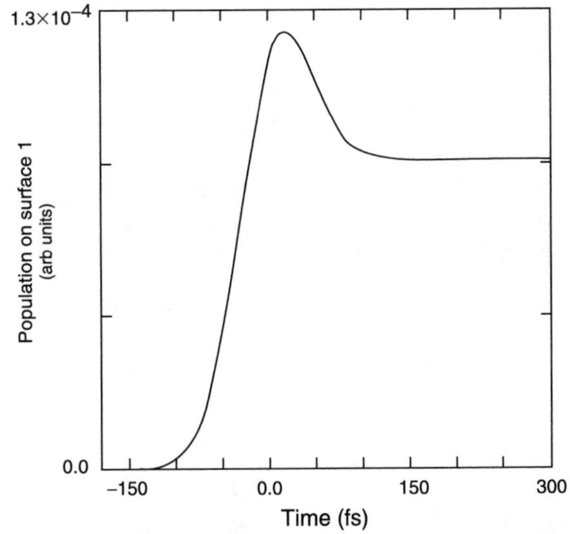

Fig. 5.32 Occupation probability of the first excited surface [49]

- Pump- and probe pulse have a FWHM of 125 fs
- The pump wavelength is fixed at the off-resonant value of 306 nm
- Four different (only three in the experiment) probe wavelengths have been applied

In Fig. 5.31, the time evolution of the wavepacket on the first excited state surface is displayed, with the pump pulse being centered around $t = 0$. The wavepacket dissociates and spreads simultaneously. In Fig. 5.32, the population of the first excited state is depicted. That this population is of the order of a few times 10^{-4} reflects the fact that an off-resonant pump frequency is used in the experiment.

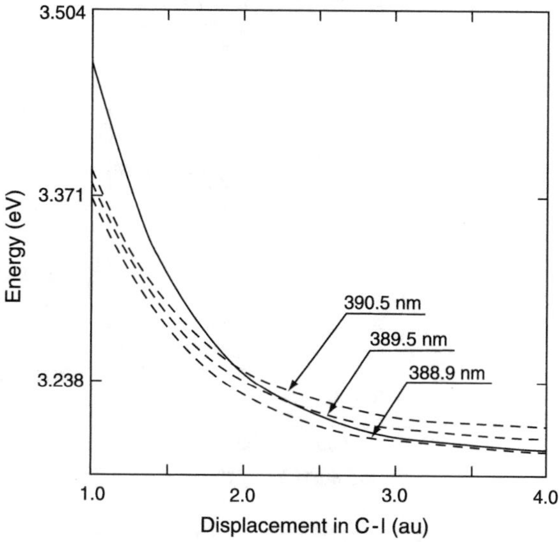

Fig. 5.33 First excited surface and second surface shifted by different amounts, corresponding to different probe frequencies [49]

The probe part of the experiment has been performed with three different probe frequencies, leading to resonance at different internuclear displacements as depicted in Fig. 5.33. The additional probe wavelength of 433 nm is not depicted in this plot. For the long wavelengths the resonance condition is more or less well localized in space and therefore as a function of the probe pulse delay, a peaked structure is to be expected. This is exactly what can be observed in Fig. 5.34! For the additional theoretical wavelength the signal is barely visible but peaked. The peak tends to become a plateau for the shorter wavelengths. In these cases the resonance condition is fulfilled for a long interval of internuclear distances as can be seen in Fig. 5.33. The plateau is perfectly developed in the case of 388.9 nm. The experimental results are very well reproduced by the calculations as can be seen by comparing the two panels in Fig. 5.34.

5.5 Control of Molecular Dynamics

Up to now we have encountered a multitude of partly counter-intuitive phenomena, appearing in atomic or molecular systems exposed to a laser field. Quite naturally one might ask if a suitable field can be found that drives a system into a desired quantum state or steers a chemical reaction into a desired channel.

Let us start to find an answer to that question for a system that exhibits one of the most fundamental quantum phenomena: a symmetric double-well potential allowing for coherent tunneling between its two minima. This system has been studied in detail under the influence of an external periodic laser field. In the following we will then refrain from the restriction to periodic fields and will consider pulsed fields with

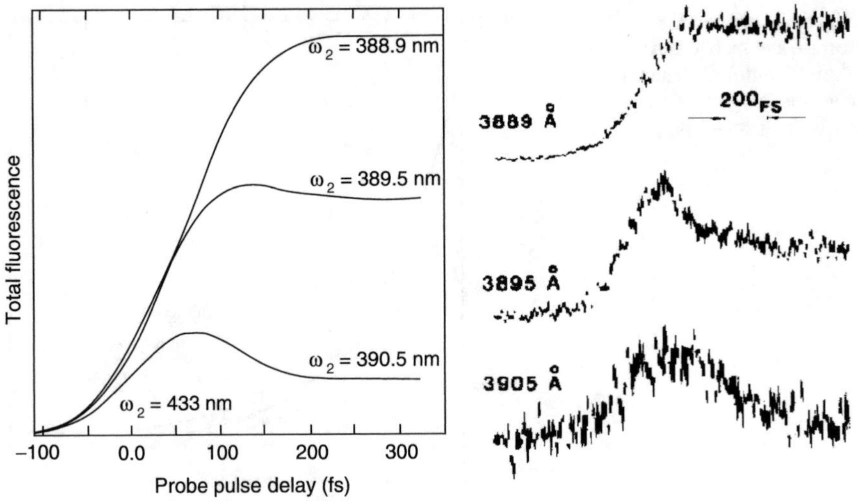

Fig. 5.34 Comparison of calculated (*left panel*) and experimental (*right panel*) fluorescence spectra of ICN for 4 (theory), respectively 3 (experiment) values of the probe wavelength, from [49, 50]

arbitrary pulse shapes that, e.g., allow for the control of chemical reactions or for the selective excitation of vibrational modes.

5.5.1 Control of Tunneling

Tunneling in a stationary double well is a phenomenon, which is discussed in almost every textbook on quantum mechanics. In the heyday of quantum theory it has been used to explain the vibrational spectrum of pyramidal molecules, like NH_3, by F. Hund [51]. The influence of a periodic external force (mediated, e.g., by a cw-laser) on coherent tunneling has been investigated in my PhD thesis and the results have been published in [52]. The basis for the understanding of those results is Floquet theory as will be seen in the following.

5.5.1.1 The Model System

A model potential for a particle of mass M_r moving in a symmetric quartic double well is given by

$$V_{\mathrm{DW}}(R) \equiv -\frac{M_\mathrm{r}\omega_\mathrm{e}^2}{4}R^2 + \frac{M_\mathrm{r}^2\omega_\mathrm{e}^4}{64E_\mathrm{B}}R^4. \tag{5.141}$$

The frequency of small oscillations around the minima

$$R_{r,1} = \pm\sqrt{\frac{8E_B}{M_r\omega_e^2}} \tag{5.142}$$

of this potential is ω_e, and E_B denotes the height of the barrier between the two wells. The dynamics of a wavepacket that can be written as a superposition of the ground and the first excited state of that single surface, is the well-known coherent tunneling dynamics, reviewed in Appendix 5.D.

A realization of the potential of (5.141) is given by the pyramidal NH_3 molecule. The relevant coordinate R refers to the umbrella mode (see Fig. 1.4 of Chap. 1) and measures the distance between the nitrogen atom and the hydrogen plane. The reduced mass is $M_r = M_N M_{3H}/(M_N + M_{3H})$ (see footnote on p. 566 of [2]). In this system an external periodic force can be generated by a monochromatic laser field of amplitude $\mathcal{E}_0$, coupling to the dipole moment

$$\mu(R) = \mu' R \tag{5.143}$$

of the molecule with the dipole gradient μ'. The amplitude of the external force is then given by

$$F_0 = \mu'\mathcal{E}_0. \tag{5.144}$$

In this subsection, we measure energies in units of $\hbar\omega_e$, such that $D = E_B/\hbar\omega_e$. Time is measured in units of $1/\omega_e$, $x = \sqrt{(M_r\omega_e)/\hbar}R$, and the dimensionless amplitude is given by $S = F_0/\sqrt{\hbar M_r\omega_e^3}$. The dimensionless Hamiltonian is then given by

$$\hat{H}(x,t) = -\frac{1}{2}\partial_x^2 - \frac{1}{4}x^2 + \frac{1}{64D}x^4 + xS\sin(wt), \tag{5.145}$$

where $w = \omega/\omega_e$ is the dimensionless ratio of driving frequency and harmonic well frequency.

5.5.1.2 Coherent Destruction of Tunneling

One of the most counter-intuitive effects that external forcing can have is the suppression of the tunneling dynamics in a double well. For frequencies in the middle of the interval

$$\frac{\Delta}{2} \le w \le w_{\rm res}, \tag{5.146}$$

with $\Delta = E_2 - E_1$ and $w_{\rm res} = E_3 - E_1$ (E_n being the unperturbed eigenvalues of the double well) this suppression is found along a one-dimensional curve in the (w, S) parameter space [53].

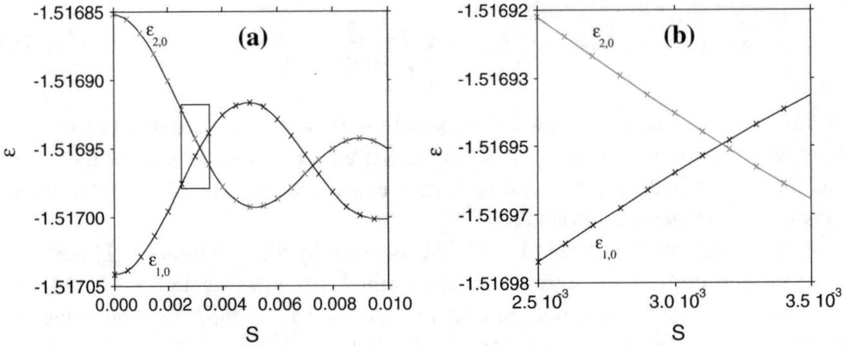

Fig. 5.35 Exact crossing of two quasi-energies of the driven double well ($D = 2, w = 0.01$) **a** $0 < S < 0.01$; **b** zoomed region around $S \approx 3 \times 10^{-3}$ [53]

The reason for this behavior is an exact crossing of the quasi-energies emerging from the lowest two unperturbed energies as a function of the amplitude. The quasi-energies can be determined by diagonalizing the Floquet matrix of Sect. 2.2.8. For $D = 2$ (which is close to the NH_3 value of 2.18) and an external frequency of $w = 0.01$, near the geometric mean of the unperturbed tunneling frequency $\Delta = 1.895 \times 10^{-4}$ and the first resonance frequency $w_{res} = 0.876$, the behavior shown in Fig. 5.35, is found. In panel (a) of this figure, the quasi-energies cross at two different values of the external force. These crossings are exact (see panel (b)), due to the fact that the quasi-eigenfunctions have different symmetry under the generalized parity transformation defined in (3.80). The non-crossing rule does therefore not hold, and as a function of a parameter the quasi-energies may approach each other arbitrarily closely.

Furthermore, the time-evolution of a Gaussian wavepacket, $\chi_l^{GW}(x)$, initially centered in the left well,[13] has been investigated for the parameters at the first exact crossing. Quantities of interest are the probability to stay (survival probability) in the initial state, i.e., the absolute square of the auto-correlation function of that wavefunction

$$P(t) := |\langle \chi_l^{GW}(t)|\chi_l^{GW}(0)\rangle|^2 \tag{5.147}$$

and the probability to be to the left of the barrier

$$\rho_l(t) := \int_{-\infty}^{0} \mathrm{d}x |\chi_l^{GW}(x,t)|^2. \tag{5.148}$$

These quantities are displayed in Fig. 5.36.

The unperturbed tunneling period for the parameters chosen here ($D = 2, w = 0.01$) is at around 50 periods, T, of the external field. In Fig. 5.36, panels (a–b), we can see, however, that in the presence of driving the particle is almost completely

[13]Note that this initial state is slightly different from the one studied in Appendix 5.D.

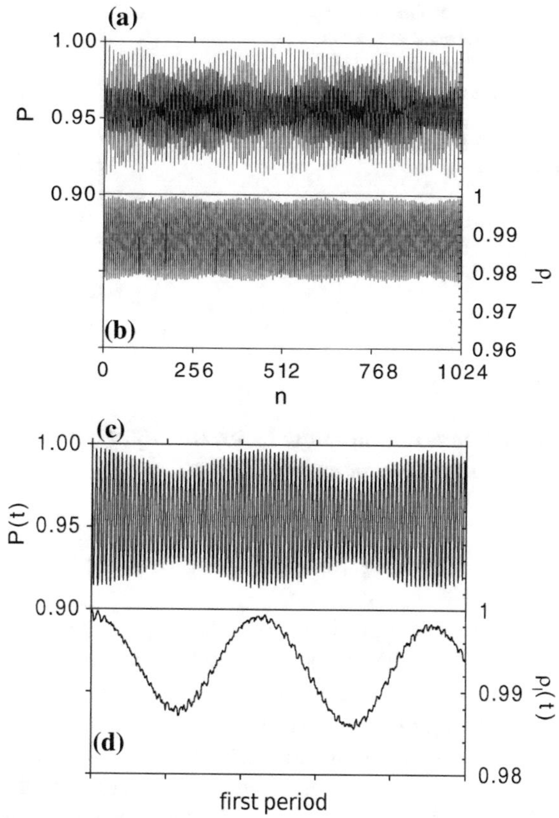

Fig. 5.36 Staying probability and probability to be to the left of the barrier ($D = 2$) for the exact crossing parameters $S = 3.171 \times 10^{-3}$, $w = 0.01$: **(a–b)** Stroboscopic time evolution over 2^{10} periods $\frac{2\pi}{w}$; **(c–d)** time evolution inside the first period, adapted from [53]

localized at the initial position even after $2^{10}T$. This is due to the fact that the initial wavefunction consists mainly of two Floquet functions, whose quasi-energies cross exactly. Small deviations of the staying probability from unity are due to the finite overlap of the initial state with other Floquet functions, whose energies are not crossing exactly. The dynamics has been considered only stroboscopically, so far. Time-evolution *during a period of the external field* is shown in panels (c–d) of Fig. 5.36. It can be observed that the periodic time-dependence of the quasi-eigenfunctions does not destroy the tunneling suppression for the present parameters.

In order to illustrate the localization effect in position space, in Fig. 5.37, the absolute square of the initial state $\chi_1^{\mathrm{GW}}(x, 0)$ and the time-evolved state with the lowest overlap (occurring at $t = 458\ T$) during the first 1,024 periods is depicted. Apart from a small shift of the center of the wavepacket to the right, there is almost no dynamics observable. This picture also explains, why the probability to stay left to the barrier deviates less from unity than the staying probability. In calculating ρ_1, one has to integrate over the whole range $-\infty < x \leq 0$, and motion of the wavefunction in the left well will not show up directly in the dynamics of ρ_1. The probability to

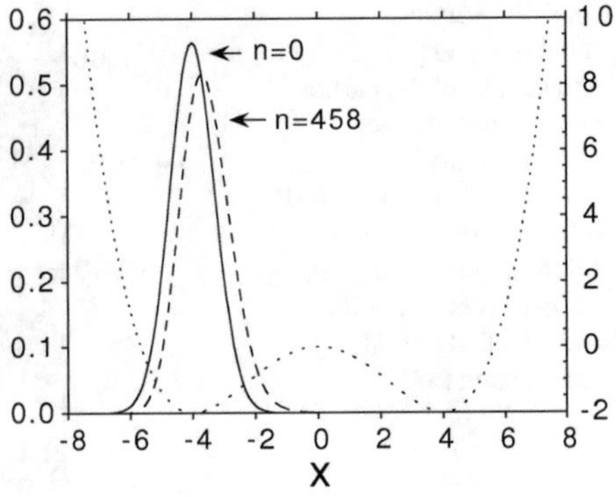

Fig. 5.37 Absolute square of wavefunctions at $t = 0$ (*solid line*) and at $t = 458T$ (*dashed line*), and unperturbed potential (*dotted line*) with $D = 2$, adapted from [53]

stay left thus deviates only by maximally 2% from unity, whereas $P(n)$ maximally looses around 8% of its initial value.

5.5.1.3 Crossing Manifold and Two-Level System

As already mentioned, the localization phenomenon occurs along a 1D manifold in the (w, S) parameter space, along which two relevant quasi-energies, having different parity, cross. This manifold has been determined in [53] and is shown in Fig. 5.38. The crossing of the two quasi-energies is a necessary but not a sufficient condition for the localization phenomenon, however. This is studied in great detail in [53], where it is shown how the old unperturbed tunneling behavior is recovered for small driving frequencies $w \approx \Delta$ and what happens at resonance $w \approx w_{\mathrm{res}}$, where the third level comes into play.

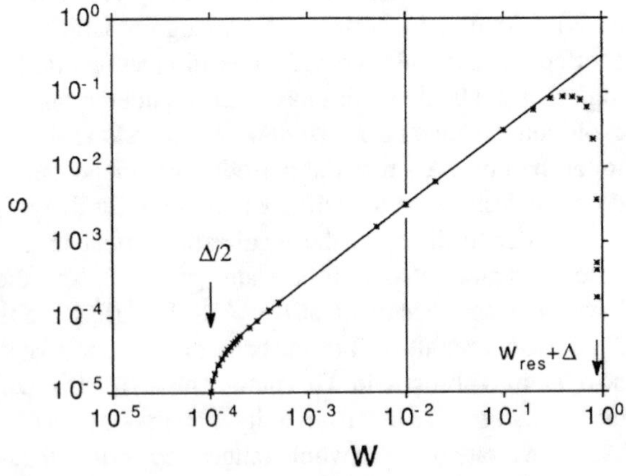

Fig. 5.38 Double logarithmic plot of the one dimensional manifold in (w, S) parameter space, along which the relevant quasi-energies cross for the first time. Driven double well ($D = 2$): *crosses*; driven two-level system: *solid line*. The *vertical line* crosses the manifold at $w = 0.01$, $S = 3.171 \times 10^{-3}$ [53]

A deeper understanding of the linear part of the manifold, along which real localization is observed, can be gained by studying the two-level system describing just the lowest two levels of the double-well problem [54]. The corresponding time-dependent Schrödinger equation is given by

$$\mathrm{i}\dot{c}_1(t) = E_1 c_1(t) + \langle\chi_1|x|\chi_2\rangle S\sin(wt)c_2(t), \tag{5.149}$$

$$\mathrm{i}\dot{c}_2(t) = E_2 c_2(t) + \langle\chi_1|x|\chi_2\rangle S\sin(wt)c_1(t), \tag{5.150}$$

with $c_{1,2}(t) \equiv \langle\chi_{1,2}|\chi(t)\rangle$. The unperturbed Hamiltonian is now a 2×2 matrix and the quasi-energies can again be determined according to the scheme reviewed in Sect. 2.2.8. The location in parameter space of the exact crossing of the two quasi-energies, emerging out of the two lowest unperturbed states, is plotted in Fig. 5.38 as a solid line. For frequencies $w \ll w_{\rm res}$ it is very close to the manifold of the full problem and it stays a perfect straight line also for frequencies $w > w_{\rm res}$ due to the non-existence of a third unperturbed level.

The slope of the manifold in the linear range can be determined analytically for a two-level system. In the case of $w \gg \Delta$ (defining the linear region), it has been shown by Shirley that the first crossing is approximately given by the first zero of the Bessel function $J_0\left(\frac{2b}{w}\right)$ [55]. Here b denotes the field strength multiplied by the dipole matrix element

$$b \equiv \langle\chi_1|x|\chi_2\rangle S \overset{D=2}{\approx} 3.791\, S\,. \tag{5.151}$$

Using the simple form of the argument of the Bessel function and its first zero [56] the straight line

$$S = \frac{2.40482\ldots}{2\langle\chi_1|x|\chi_2\rangle} w \overset{D=2}{\approx} 0.3172\, w \tag{5.152}$$

in (w, S) parameter space is found. Higher zeroes of the Bessel function give straight lines along which the quasi-energies cross each other exactly again (see Fig. 5.35a). Tuning the parameters to an exact crossing, localization is also found in the time-dependent two-level Schrödinger equation (5.149) and (5.150) with the initial conditions $c_1(0) = -c_2(0) = 1/\sqrt{2}$.

Furthermore, it is worthwhile to note that in the strong field limit, the suppression phenomenon can also be understood in the transfer matrix formalism [57].

5.5.1.4 The Asymmetric Double-Well Potential

What is the effect of a finite asymmetry on the localization phenomenon? In order to study this question, a potential of the form

$$V_\sigma(x,t) = -\frac{1}{4}x^2 + \frac{1}{64D}x^4 + \sigma x + xS\sin(wt), \tag{5.153}$$

Fig. 5.39 Quasi-energies as a function of S for $D = 2$ and $w = 0.01$ in the asymmetric ($\sigma = 10^{-6}$) double well [53]

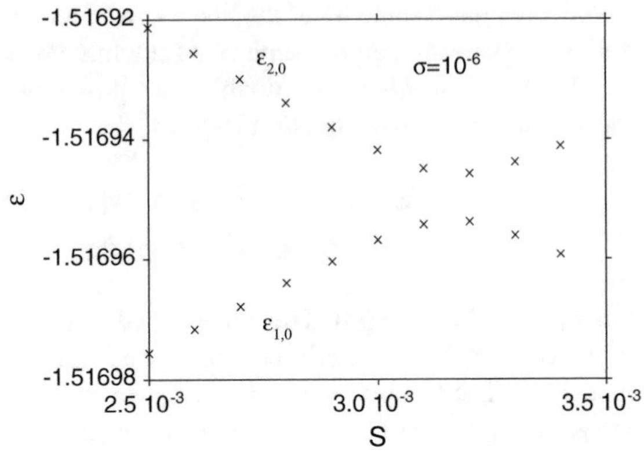

with a static asymmetry, $\sigma > 0$, can be used. In this case no symmetry under the generalized parity transformation (3.80) does exist any more. This leads to the fact that all Floquet energies, due to the non-crossing rule, do not cross exactly (except maybe at singular points). Therefore no 1D manifold along which two quasi-energies cross does exist.

What happens to the allowed crossings that we have observed in the symmetric case? In order to answer this question a relatively small asymmetry with parameter σ can be applied. In Fig. 5.39, the same field and potential parameters have been used as in Fig. 5.35b, except for the asymmetry. We can see that the allowed exact crossing becomes an avoided crossing in the presence of asymmetry. Localization therefore goes away gradually. The splitting of the levels is rather small and the wavepacket would still be localized for relatively long times. It would not be localized forever any more as in the symmetric case, however. For strong asymmetry two effects have to be considered. First the avoided crossing becomes broader but secondly a partial localization does result from the fact that the lowest eigenstate becomes similar to the coherent state in the lower well.

5.5.1.5 More Driven Double-Well Systems

The realization of driven double-well systems is possible in many different branches of physics. Recently, driven double wells have, e.g., been realized in optical fiber systems. In such a system a light beam propagating through a periodically curved waveguide is coupled to a parallel fiber. In this setup the first experimental realization of the effect of coherent destruction of tunneling has been performed [58]!

Another physical system whose dynamics can be described with the help of a multistable potential is the rf-SQUID. There the macroscopic flux through the ring is the tunneling degree of freedom. An external perturbation may be given by a magnetic field. This is an example from the realm of solid state physics, however, and shall not be dealt with here. Very recently, the direct observation of suppression of single particle tunneling of atoms in light shift double-well potentials has been reported [59].

Let us finally come back to molecular physics. Apart from the NH_3-molecule where the nitrogen atom experiences a double-well potential, also the electron in the H_2^+- (resp. D_2^+-) molecule sees a double-well potential, due to the electron-nuclear interaction. Recently it has been shown both theoretically as well as experimentally that the dissociation of the electron can be steered by the carrier envelope phase such that the electron is localized preferentially at a specific proton (deuteron) [16, 60].

5.5.2 *Control of Population Transfer*

The transfer of population into a desired state is one of the central challenges of control theory. Before we discuss a direct approach to that field using optimal control theory, a counter-intuitive method to control population transfer shall be reviewed.

In molecular systems, this is the stimulated Raman adiabatic passage or short STIRAP method. In this scheme a three level system, displayed in Fig. 5.40, is coupled via two different laser pulses. A direct coupling of level $|1\rangle$, which might by a rotational level in the vibrational ground state and the highly excited vibrational state $|3\rangle$ shall be dipole forbidden. The methodology is used experimentally to selectively excite vibrational states [61].

The pump-pulse couples levels 1 and 2, while the Stokes pulse couples levels 2 and 3. The total Hamilton matrix $\mathbf{H} = \mathbf{H_0} + \mathbf{W}$ is given by

$$\mathbf{H} = \begin{pmatrix} E_1 & -\mu_{12} E_{\mathrm{P}} \cos(\omega_{\mathrm{P}} t) & 0 \\ -\mu_{21} E_{\mathrm{P}} \cos(\omega_{\mathrm{P}} t) & E_2 & -\mu_{23} E_{\mathrm{S}} \cos(\omega_{\mathrm{S}} t) \\ 0 & -\mu_{32} E_{\mathrm{S}} \cos(\omega_{\mathrm{S}} t) & E_3 \end{pmatrix}. \quad (5.154)$$

Transformation into the interaction picture (see Sect. 2.2.4) with the help of the unperturbed Hamiltonian

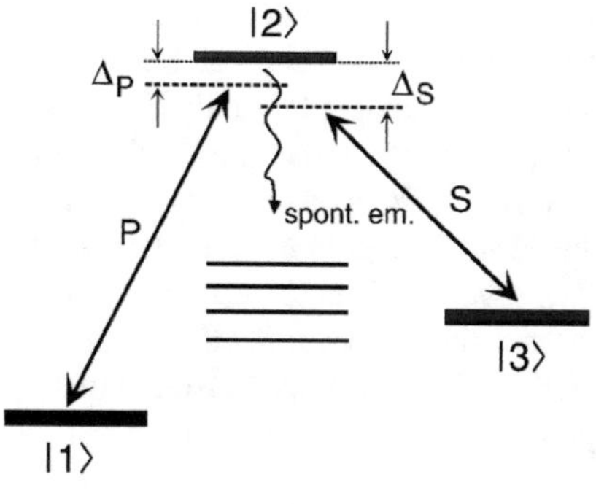

Fig. 5.40 A three level system (so-called Λ-system) coupled via pump and Stokes pulse with the respective detunings Δ_{P} and Δ_{S} [61]

$$\mathbf{H_0} = \begin{pmatrix} E_1 & 0 & 0 \\ 0 & E_2 & 0 \\ 0 & 0 & E_3 \end{pmatrix}, \tag{5.155}$$

and invoking the rotating wave approximation leads to the STIRAP matrix

$$\mathbf{W_I} = \mathbf{U_0^\dagger W U_0} = -\frac{1}{2}\begin{pmatrix} 0 & \Omega_P e^{-i\Delta_P t} & 0 \\ \Omega_P e^{i\Delta_P t} & 0 & \Omega_S e^{i\Delta_S t} \\ 0 & \Omega_S e^{-i\Delta_S t} & 0 \end{pmatrix} \tag{5.156}$$

for the Hamiltonian in the interaction representation. Here the abbreviations $\Omega_P(t) = \mu_{21}E_P(t)$, $\Omega_S(t) = \mu_{32}E_S(t)$ for the Rabi frequencies without detuning and $\Delta_P = (\omega_2 - \omega_1) - \omega_P$, $\Delta_S = (\omega_2 - \omega_3) - \omega_S$ for the detunings that are plotted in Fig. 5.40 have been used.

In the case of vanishing detunings, the time-dependent eigenvalues and eigenstates (dressed states) of the STIRAP matrix are given by

$$\omega_{0,\pm}(t) = 0, \pm\frac{\Omega(t)}{2}, \tag{5.157}$$

$$|g_0\rangle(t) = \cos[\Theta(t)]|1\rangle - \sin[\Theta(t)]|3\rangle, \tag{5.158}$$

$$|g_\pm\rangle(t) = \frac{1}{\sqrt{2}}\left(\sin[\Theta(t)]|1\rangle + \cos[\Theta(t)]|3\rangle \mp |2\rangle\right), \tag{5.159}$$

with $\Omega(t) = \sqrt{\Omega_P(t)^2 + \Omega_S(t)^2}$ and the definition of the mixing angle

$$\Theta(t) \equiv \arctan\left(\frac{\Omega_P(t)}{\Omega_S(t)}\right). \tag{5.160}$$

Cosines and sines of this angle can be resolved by using the relations given in Footnote 6 of Chap. 3.

5.12. *Calculate the eigenvalues and eigenvectors of the STIRAP matrix in the case of vanishing detunings.*

With the help of the dressed states and of the quantum mechanical adiabatic theorem of Appendix 5.E, the pulse sequence can be understood. Starting from state $|1\rangle$, only the dressed state $|g_0\rangle$ is occupied initially if $\Omega_S \gg \Omega_P$. This amounts to the counter-intuitive pulse sequence depicted in panel (a) of Fig. 5.41, where the Stokes pulse precedes the pump pulse! If the field changes adiabatically,[14] then according to the adiabatic theorem the system stays in the dressed state $|g_0\rangle$ of the instantaneous Hamiltonian. For large positive times $\Omega_P \gg \Omega_S$ holds, however, and thus the system finally is in state $|3\rangle$, without having occupied the "dark state" $|2\rangle$ in the meantime. This dynamics is depicted in Fig. 5.41, where also the mixing angle and the dressed eigenvalues are displayed.

[14] Exercise 15.11 in [24] sheds more light on what "adiabatically" means in this context.

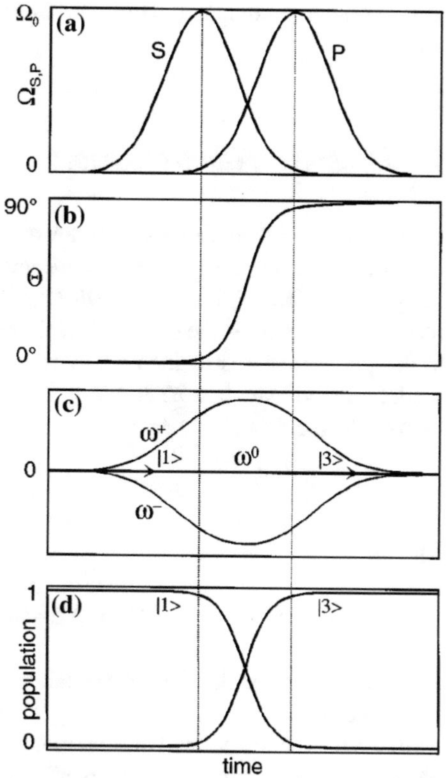

Fig. 5.41 STIRAP dynamics: **a** Pump- and Stokes-pulse, **b** angle Θ, **c** dressed eigenvalues, **d** occupation probabilities [61]

The ordering of the pulses is counter-intuitive. They have to have a non-vanishing overlap, however, for population transfer to be achieved (see also Fig. 5.41). This can be seen by doing Exercise 15.8 in [24], the reference we are following closely throughout this subsection. Furthermore, an alternative perspective on STIRAP can be gained by demanding constant probability to be in the second (dark) state. For this, the time derivative

$$\frac{\mathrm{d}|a_2|^2}{\mathrm{d}t} = 2\mathrm{Re}[a_2^*\dot{a}_2] = -[\Omega_\mathrm{P}(t)\mathrm{Im}(a_2^*a_1) + \Omega_\mathrm{S}(t)\mathrm{Im}(a_2^*a_3)] \tag{5.161}$$

must vanish and again, we have assumed resonance $\Delta_\mathrm{S} = \Delta_\mathrm{P} = 0$. This requirement leads to the conditions

$$\Omega_\mathrm{P} = -\Omega_0(t)\mathrm{Im}[a_3^*(t)a_2(t)], \tag{5.162}$$

$$\Omega_\mathrm{S} = \Omega_0(t)\mathrm{Im}[a_1^*(t)a_2(t)]. \tag{5.163}$$

The two terms on the RHS of (5.161) then cancel each other. The counter-intuitive ordering of the pulses follows from the fact that a_1 is initially large and therefore also Ω_S is large compared to Ω_P. When the system is in state $|3\rangle$, the pump pulse takes over.

5.5.3 *Optimal Control Theory*

Optimal control theory deals with the search for external fields that steer a system into a desired state. This can be a certain vibrational excitation, which was also the goal of STIRAP, just discussed. One of the most demanding goals that can be reached with lasers is the control of a chemical reaction, however. One can, e.g., try to design laser pulses in such a way that chemically bound species dissociate in a predetermined way. In a triatomic system several different reaction channels exist. Two of them are

$$\mathrm{ABC} \to \mathrm{A} + \mathrm{BC} \qquad \text{channel 1}$$

and

$$\mathrm{ABC} \to \mathrm{AB} + \mathrm{C} \qquad \text{channel 2}$$

and a laser field that discriminates channel 1 in favor of channel 2 might for example be looked for.

In the following, we will discuss two important scenarios in the field of chemical reactions, starting with the "precursor" of the optimal control schemes, the so-called "pump-dump"-scheme and then reviewing in detail the Krotov method, which gives a mathematical prescription to find the optimal field. Finally, we will come back to the question of steering a system into a desired quantum state.

5.5.3.1 Pump-Dump Control

The so-called pump-dump method is a very intuitive way to approach the field of optimal control [62]. One tries to steer the breaking of a specific bond by first lifting the system onto an electronically excited state and then using the motion of the nuclei in that state in such a way that the system is deexcited exactly at a time when the subsequent motion in the electronic ground state leads to dissociation in the desired channel.

In order to understand the physics behind the pump-dump method we first look at a typical potential landscape of a collinear ABC system, shown in Fig. 5.42. The potentials are drawn as functions of two degrees of freedom corresponding to the two interatomic distances. The lower surface has a local minimum and two channels which are separated from the minimum via saddle points. The upper surface is almost harmonic. The basic idea of how to steer the reaction into a desired channel becomes clear, if we consider the classical Lissajous motion on the electronically excited

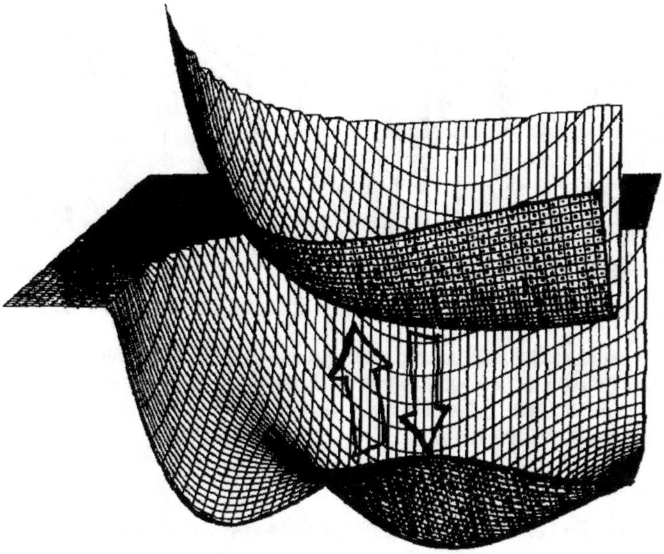

Fig. 5.42 Electronic potential curves of a collinear ABC system with pictorial representation of the pump and the dump pulse [63]

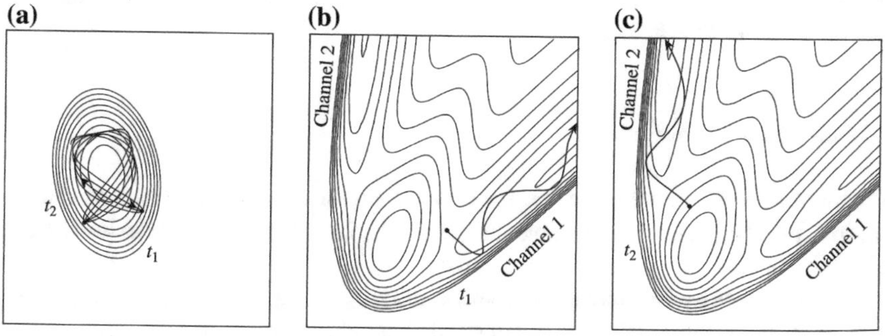

Fig. 5.43 Classical mechanical understanding of the pump-dump scenario in the ABC system: **a** Lissajous motion on upper surface, **b** classical trajectory exiting in channel 1, **c** classical trajectory exiting in channel 2 [24]

surface after excitation with the pump pulse. This motion is depicted in the leftmost panel in Fig. 5.43. Now the dump-pulse arrives with a specific time delay. The Husimi transform of a typical pump-dump pulse sequence has been displayed in Fig. 1.10 of Chap. 1 already. Choosing the time delay accordingly, the Lissajous motion can be intercepted at any desired point. If it is intercepted at t_1, such that the motion on the electronic ground state continues in channel 1, then the dissociation has been steered to proceed in this channel, as depicted in the middle panel of Fig. 5.43.

Quantum mechanically, the dynamics of wavefunctions and not of a single classical trajectory has to be considered. It turns out, however, that due to the harmonic

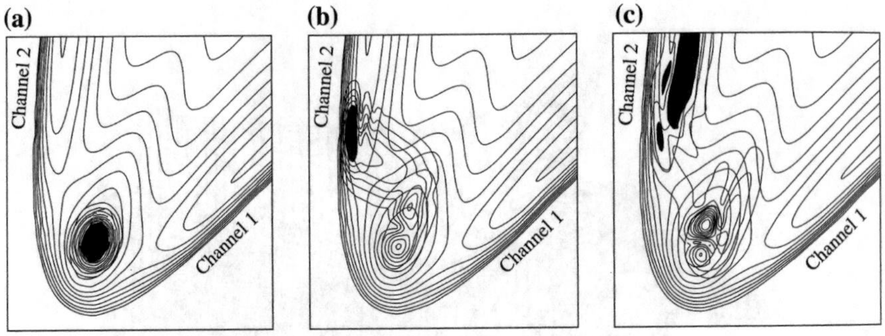

Fig. 5.44 Quantum mechanical pump-dump scenario in the HHD system for a time delay of 810 a.u.: **a** wave function in the electronic ground state at the initial time $t = 0$ a.u., **b** wave function in the electronic ground state at the time $t = 1,000$ a.u., **c** wave function in the electronic ground state at the time $t = 1,200$ a.u. [24]

nature of the excited electronic state, the physical picture of the pump-dump method stays intact [64]. The wavepacket evolves almost dispersionless on the upper surface and the description in terms of classical trajectories is sufficient. After action of a dump pulse with a time delay of 810 a.u., the wavepacket exits in channel 2 on the electronic ground state, as depicted in Fig. 5.44. This case corresponds to the rightmost panel of Fig. 5.43.

5.5.3.2 Krotov Method

The pump-dump method that we have just discussed is the precursor of modern control methods that try to achieve higher yields, i.e., to achieve the desired goal to a higher degree.

The goal can by formulated mathematically by using a projection operator $\hat{P}_\alpha$, projecting the wavefunction on the desired channel and trying to maximize

$$J_P = \langle \chi(T_t)|\hat{P}_\alpha|\chi(T_t)\rangle. \tag{5.164}$$

Here T_t is the total time allowed for the control process.

In order that the energy content of the field does not grow indefinitely, the functional above is usually augmented by a term

$$J_\mathcal{E} = \lambda \int_0^{T_t} \mathrm{d}t |\mathcal{E}(t)|^2, \tag{5.165}$$

proportional to a Lagrange multiplier λ. Furthermore, the time-dependent Schrödinger equation is introduced again via a Lagrange multiplier $\langle\xi(t)|$ into the functional by the real term

$$J_H = 2\mathrm{Re}\int_0^{T_t} \mathrm{d}t\, \langle\xi(t)| \left(-\partial_t + \frac{\hat{H}}{\mathrm{i}}\right) |\chi(t)\rangle, \tag{5.166}$$

which deconstrains $\mathcal{E}$ and χ to first order in the field [24]. The functional to be extremized is finally given by

$$\tilde{J} \equiv J_P + J_H - J_{\mathcal{E}}. \tag{5.167}$$

To be specific, the case of a two component wavefunction $\boldsymbol{\chi} = (\chi_{\mathrm{g}}, \chi_{\mathrm{e}})$ and a corresponding 2×2 Hamilton matrix operator (see also Appendix 5.B)

$$\hat{\mathbf{H}} = \begin{pmatrix} \hat{H}_{\mathrm{g}} & \mu\mathcal{E}^*(t) \\ \mu\mathcal{E}(t) & \hat{H}_{\mathrm{e}} \end{pmatrix} \tag{5.168}$$

is considered in the following. After integration by parts of the J_H term,

$$\tilde{J} = J_P - 2\mathrm{Re}\langle\boldsymbol{\xi}|\boldsymbol{\chi}\rangle|_0^{T_t} + 2\mathrm{Re}\int_0^{T_t} \mathrm{d}t \left\{ \langle\boldsymbol{\xi}(t)|\frac{\hat{\mathbf{H}}}{\mathrm{i}}|\boldsymbol{\chi}(t)\rangle + \langle\dot{\boldsymbol{\xi}}|\boldsymbol{\chi}\rangle \right\} - J_\epsilon \tag{5.169}$$

is found. The variation of this expression can now be performed according to the rules that are gathered in Appendix 2.B. Extremalizing with respect to $\boldsymbol{\chi}$, i.e., the condition

$$\frac{\delta\tilde{J}}{\delta|\boldsymbol{\chi}(t)\rangle} = 0 \tag{5.170}$$

leads to the equation

$$-\mathrm{i}\langle\dot{\boldsymbol{\xi}}| = \langle\boldsymbol{\xi}|\hat{\mathbf{H}}, \tag{5.171}$$

which is a backward Schrödinger equation for the Lagrange parameter. Its final condition is found by doing the variation

$$\frac{\delta\tilde{J}}{\delta|\boldsymbol{\chi}(T_t)\rangle} = 0, \tag{5.172}$$

leading to

$$\langle\boldsymbol{\xi}(T_t)| = \langle\boldsymbol{\chi}(T_t)|\hat{P}_\alpha. \tag{5.173}$$

In addition to this equation, also the initial value equation

$$\mathrm{i}|\dot{\boldsymbol{\chi}}\rangle = \hat{\mathbf{H}}|\boldsymbol{\chi}\rangle, \tag{5.174}$$

$$|\boldsymbol{\chi}(0)\rangle = |\boldsymbol{\chi}_0\rangle \tag{5.175}$$

has to hold.

Extremalizing with respect to $\mathcal{E}^*$, i.e., the condition

$$\frac{\delta \tilde{J}}{\delta \mathcal{E}^*(t)} = 0 \tag{5.176}$$

leads to

$$\mathcal{E}(t) = \frac{-\mathrm{i}}{\lambda}[\langle \xi_\mathrm{g}|\mu|\chi_\mathrm{e}\rangle - \langle \chi_\mathrm{g}|\mu|\xi_\mathrm{e}\rangle] \tag{5.177}$$

for the field [63].

5.13. *Show that setting the variation of $J_H - J_\mathcal{E}$ with respect to $\mathcal{E}^*$ equal to zero leads to the expression (5.177) for the field.*

The five equations. (5.171), (5.173), (5.174), (5.175) and (5.177) contain a double sided boundary value problem. The easiest solution procedure is given by the following steps:

1. Propagate $\boldsymbol{\chi}(t)$ from $t = 0$ to $t = T_\mathrm{t}$ forward in time
2. Apply $\hat{P}_\alpha$ to $\boldsymbol{\chi}(T_\mathrm{t})$ yielding $\boldsymbol{\xi}(T_\mathrm{t})$
3. Propagate $\boldsymbol{\xi}$ from $t = T_\mathrm{t}$ to $t = 0$ backward in time

The field has to be guessed, however, and does not necessarily fulfill the equation coming out of the variation procedure! Therefore the scheme above has to be augmented by an iterative procedure, due to Krotov [65]:

1. Choose an initial field $\mathcal{E}^0(t)$
2. Propagate $\boldsymbol{\chi}(t)$ under $\mathcal{E}^0(t)$ forward in time
3. Projection of $\boldsymbol{\chi}(T_\mathrm{t})$ gives $\boldsymbol{\xi}(T_\mathrm{t})$
4. Propagate $\boldsymbol{\xi}$ backward in time
5. Commonly propagate $\boldsymbol{\xi}(t)$ (with the old field) and $\boldsymbol{\chi}(t)$ with the new instantaneously calculated field

$$\mathcal{E}^1(t) = \frac{-\mathrm{i}}{\lambda}[\langle \xi_\mathrm{g}^0|\mu|\chi_\mathrm{e}^1\rangle - \langle \chi_\mathrm{g}^1|\mu|\xi_\mathrm{e}^0\rangle] \tag{5.178}$$

   forward in time
6. Project $\boldsymbol{\chi}(T_\mathrm{t})$ and continue the procedure until convergence is achieved.

The propagation of $\boldsymbol{\xi}(t)$ forward in time seems to be superfluous, because the result is already known. Keeping the wavefunction in computer memory would be barely possible for most cases of interest, however, and therefore it is cheaper to calculate $\boldsymbol{\xi}(t)$ once more. In general the propagated wavefunction has amplitude in both the desired and the undesired channel, see, e.g., Fig. 2f in [63]. The above procedure iterates the field in such way that the undesired portion of the wavefunction is minimized. Whether this minimum is an absolute or a local one is a question that goes far beyond the scope of this book. The method just laid out goes back to Krotov.

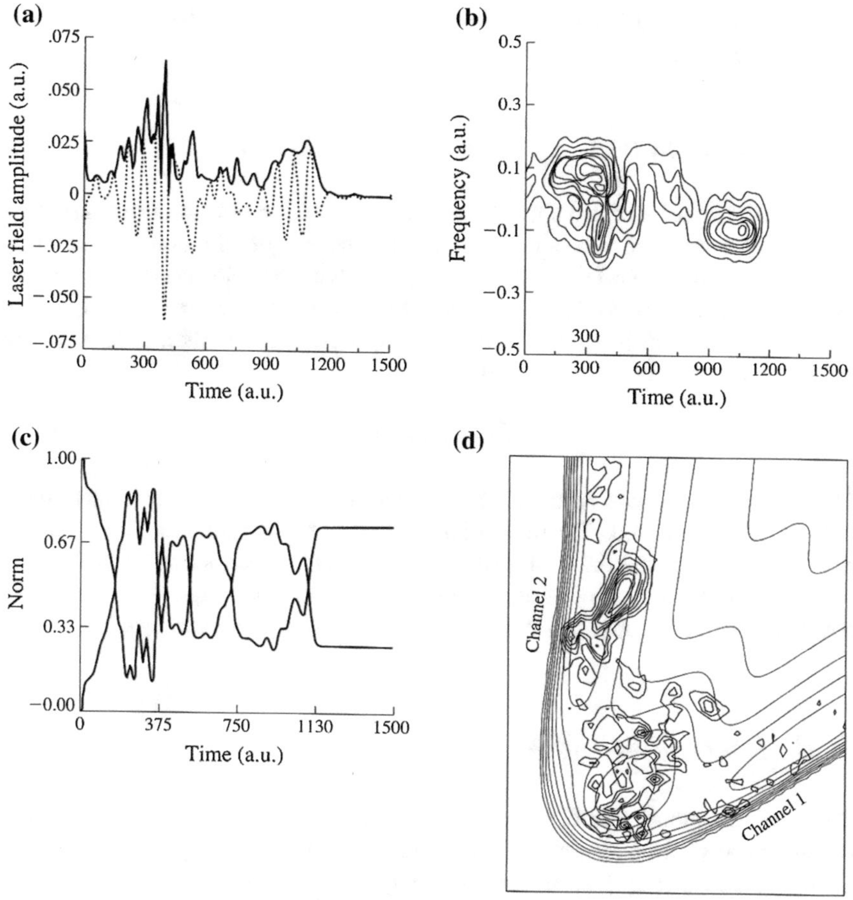

Fig. 5.45 Steering the breakup of the ABC system into the second channel: **a** optimal field, **b** Husimi transform of the field, **c** norm of the wavefunction in the ground and excited state, **d** final wavefunction, from [24]

Other methods to solve for the optimal field have been devised, however, see [24, 66] and the references therein. Whereas in the simple pump-dump scheme, low yields of 10^{-2} and selectivity ratios of 3:2 have been reported, in the optimally controlled case, typical yields increase to more than 10% and selectivity ratios can be as high as 13:3 [63].

As an example, let us review results that were obtained for an ABC system. As the initial guess for the electric field in the iterative process a pump-dump pulse as depicted in Fig. 1.10 has been used. One can try to steer the reaction either into channel 1 or into channel 2. In the first case the resulting optimal field is still rather similar to the original pump-dump pulse [24], whereas in the second case the field displayed in Fig. 5.45 is resulting. Around 50 iterations are typically necessary to converge the results.

5.5.3.3 Optimally Controlled Excitation of Quantum States

Optimal control schemes do not only work for the breakup reaction just considered. They have been shown to be applicable also for the case of vibrational excitation.

In [33] optimal control theory has been applied with the objective to steer a Morse oscillator, representing a CH-stretch, with Morse parameters $D_e = 0.199$, $R_e = 1.5$, and $\alpha = 0.9386$, all in atomic units, into a specific excited target state $|n_T\rangle$. The projection operator therefore is given by

$$\hat{P}_T = |n_T\rangle\langle n_T|. \tag{5.179}$$

The dipole moment was assumed to be of Mecke form with the parameters $\mu_0 = 1.76$ a.u. and $R^* = 1$ a.u.. The result of the optimization starting from the vibrational ground state and fixing the final time to be $T_t = 0.1$ ps are shown in Fig. 5.46.

Optimal control theory has been applied in a lot of other physical systems. One out of many other examples is the control of cis-trans isomerization [67].

5.5.4 *Genetic Algorithms*

The theory of optimal control of the last section rests on the availability of analytically (or numerically) given potential energy surfaces and on the validity of the underlying Born-Oppenheimer approximation. Both requirements may be violated, however, and even the Hamiltonian might not be known. Therefore, alternative control schemes are sought for.

A recent development in the field of control therefore is the application of genetic algorithms. Their application is based on an experimental "analog computer". The system to be controlled is exposed to a laser whose temporal shape can be varied. By a feedback mechanism, a digital computer using a genetic (evolutionary) algorithm can vary the field iteratively in such a way that the desired goal is reached. The principal setup is displayed in Fig. 5.47.

The first theoretical study that showed the feasibility of such an approach is due to Judson and Rabitz [69]. These authors have shown that it is based on the following three paradigms:

- "Survival of the fittest"
- Crossover
- Mutation

As an example, the transition from the $n = j = 0$ ground vibrational state to the rotationally excited vibrational ground state $n = 0, j' = 3$ of the KCl molecule

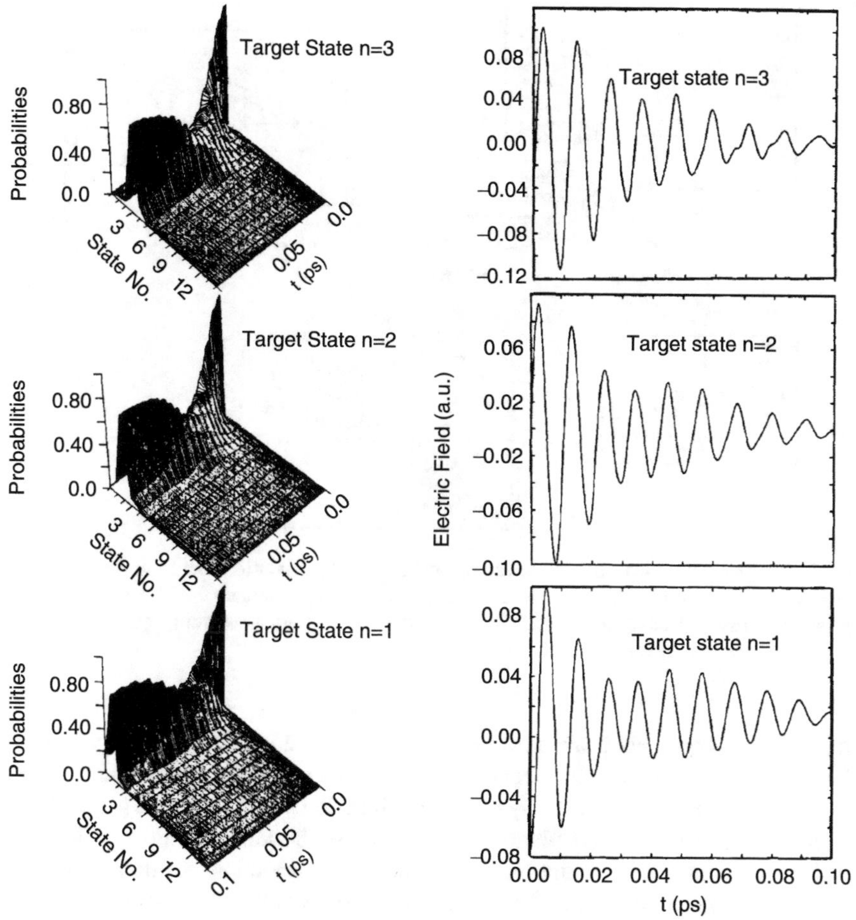

Fig. 5.46 Steering a Morse oscillator into selected excited states via optimal control theory, *left panels* show probabilities as a function of time and quantum number, *right panels* show the corresponding electric fields [33]

was investigated. An "individuum" of the genetic algorithm is a specific laser pulse sequence. Its initial gene consists of $N_{gene} = 128$ entries of random numbers, uniformly distributed between zero and one, and was then scaled to a maximum field strength of $5\,kV\,cm^{-1}$. The total number of individuals was chosen as $N_{pop} = 50$. The first paradigm can be tested by introducing the "cost-function"

$$\sum_j (\delta_{jj'} - \rho_j)^2,$$

where ρ_j is the occupation probability of state j. It is zero if the desired state with label j' is fully populated. Individuals can then be ranked according to their fitness. The

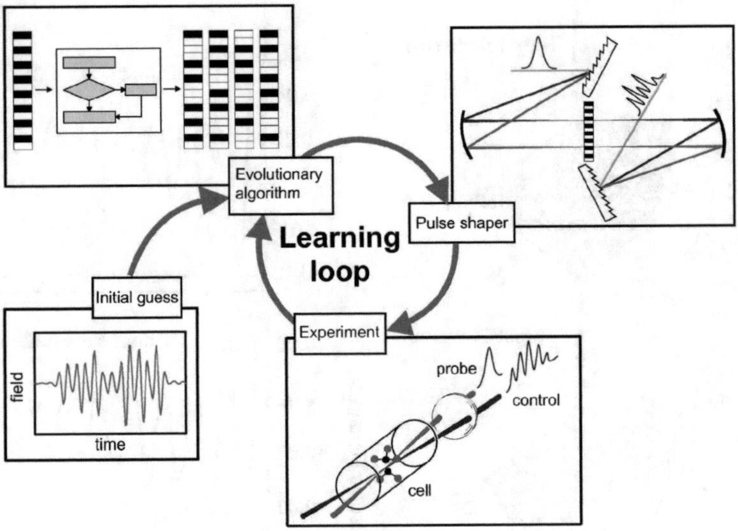

Fig. 5.47 Principal setup of an “analog computer” for feedback control. The control pulse excites the system and the probe pulse measures the outcome (the measurement could, e.g., also be performed by a mass spectrometer), which is fed into the evolutionary algorithm; a good initial guess helps to achieve convergence quickly [68]

highest ranked ones were taken over to the next generation without change, whereas the other ones had to undergo crossover and a small probability of mutation. In the upper panel of Fig. 5.48 the decay of the cost function as a function of generation is displayed for the average population, as well as for the best individuum. Furthermore, in the lower panel of that figure, the spectrum corresponding to the best gene is shown. It displays maxima at the resonant transitions between the rotational states $j = 0, 1, 2, 3$. It is important to stress that the spectral information was not input into the generation of the optimal field but was found by the learning loop.

An experimental realization of control based on evolutionary algorithms was performed by the Gerber group [70]. The goal of the experiment was to steer the photo-fragmentation of $CpFe(CO)_2Cl$ into a desired channel. A pulse shaper that allows to split the laser light into 128 spectral components and vary them separately has been used. Selectivity ratios of about 5 have been achieved.

5.5.5 *Towards Quantum Computing with Molecules*

A recent new development in the field of laser-molecule interaction is the realization of quantum logic operations with the help of molecular vibrational states. We will not deal with that exciting new field in much detail but will discuss the realization of the basic ingredient of every setup used for computing: the flipping of a bit. As we

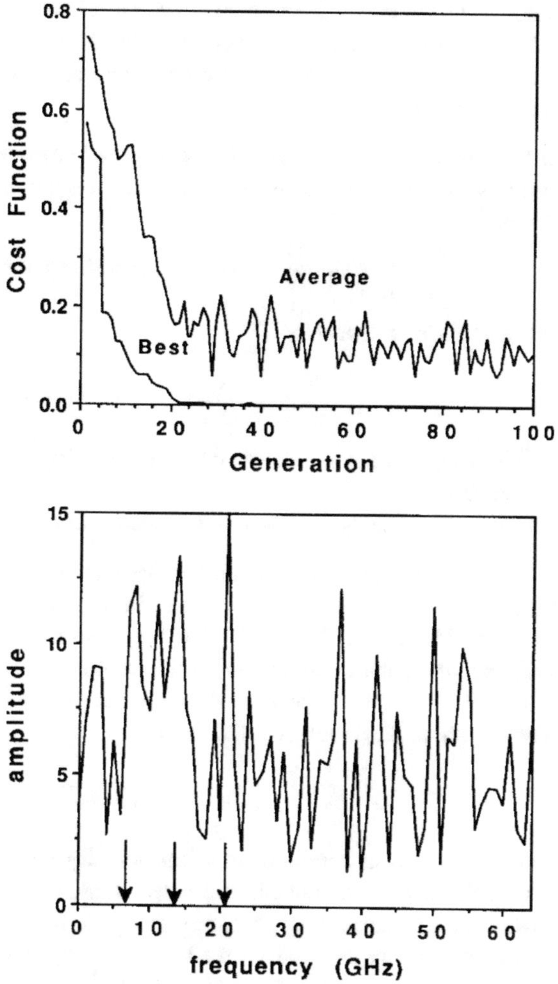

Fig. 5.48 *Upper panel*: Cost function of the average population and the best gene; *lower panel*: Spectrum of the optimal pulse (*arrows* indicate positions of resonant transitions between rotational sublevels of the vibrational ground state of the KCl molecule), from [69]

will see, anharmonic vibrational modes have to be used to this end. Using the OH diatomic, this has been shown by Babikov [71] and by Cheng and Brown [72].

The flipping of a bit is based on the realization of the NOT-Operation. In a two-level system this corresponds to the complete transfer of population from level 0 to level 1 or vice versa

$$\mathrm{NOT}|0\rangle = |1\rangle, \tag{5.180}$$

$$\mathrm{NOT}|1\rangle = |0\rangle. \tag{5.181}$$

Each deviation from the complete population transfer reduces the so-called fidelity, defined as the occupation probability of the initially unpopulated level. Unfortunately, a harmonic oscillator cannot be controlled to switch completely to a desired state, because of the equidistance of its levels. One has to choose two levels of an anharmonic system, as e.g., the Morse potential of Sect. 5.1.2 in order to realize the NOT operation.

5.14. *Derive the maximal probability to populate the n-th excited state of a harmonic oscillator by using an external field in length gauge and starting from the (vibrational) ground state* χ_0.

(a) *Calculate the time-dependent wavefunction* $\chi_0(x,t)$ *under the influence of the external field.*
(b) *Determine the overlap*

$$a_{n0}(t) = \langle \chi_n | \chi_0(t) \rangle,$$

by using $\int \mathrm{d}x \exp[-(x-y)^2] H_n(x) = \pi^{1/2}(2y)^n$ *with the Hermite polynomial* $H_n(x)$.
(c) *Show that the maximum of the absolute value* $|a_{n0}(t)|^2$ is given by $n^n \mathrm{e}^{-n}/n!$

As a specific example, the driven OH stretch with the Morse parameters $D_\mathrm{e} = 0.1994$, $R_\mathrm{e} = 1.821$, and $\alpha = 1.189$ in atomic units and the Mecke parameters $\mu_0 = 1.634$ a.u. and $R^* = 1.134$ a.u. (see also caption of Fig. 5.16) has been used. Although the anharmonicity constant of that molecule is fixed in nature at approximately 90 cm^{-1}, it can be viewed as a parameter in theoretical considerations [72]. These authors have looked at the fidelity

$$P_{10}(T_\mathrm{t}) = |\langle 1 | 0(T_\mathrm{t}) \rangle|^2, \tag{5.182}$$

with $T_\mathrm{t} = 750$ fs as a function of anharmonicity, and found the results reproduced in Fig. 5.49. Two different results are shown there. Firstly, the system has been exposed to an optimal control pulse, in close analogy to the work of Shi and Rabitz [33], and secondly to a simple π-pulse, we know already from Sect. 3.2.3. For large anharmonicity it can be seen that the π-pulse is superior to the "optimal" pulse! Furthermore, the statement that the harmonic oscillator cannot be controlled to 100% can be read off from the results at small anharmonicity.

The anharmonic properties of suitable candidates for molecular quantum computing and the realization of additional gates are discussed in [73].

5.6 Notes and Further Reading

Hydrogen molecular ion

The authoritative reference on the hydrogen molecular ion is the classic book by Slater [1]. More general material on molecular spectra and molecular structure can be found in the textbook by Bransden and Joachain [2]. In our presentation of the

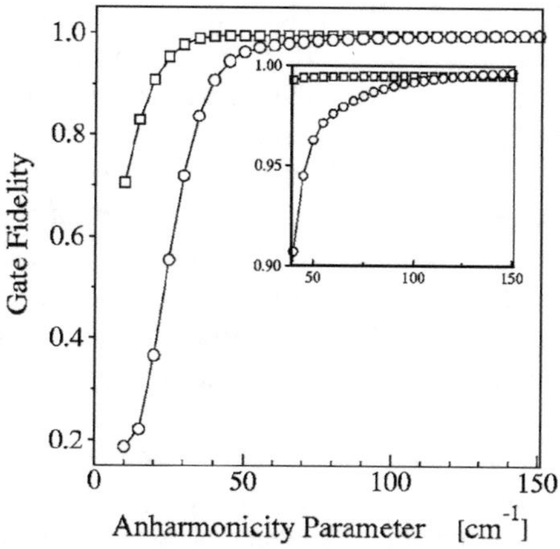

Fig. 5.49 Fidelity of the NOT gate as a function of the anharmonicity $\omega_e x_e$ for an optimal pulse (*squares*) and for the π-pulse (*circles*); adapted from [72]

LCAO solution of the electronic Schrödinger equation for H_2^+, we closely followed the book by Haken and Wolf [22]. Spectroscopic constants that can be used for the generation of analytic Morse potentials for diatomic molecules are contained in [8]. Further information on diatomic molecules is available in the references given in [74]. A modern exposition of quantum chemistry, which is the main theoretical tool for the calculation of electronic potential energy surfaces, is given in the book by Szabo and Ostlund [3].

Reviews covering both theoretical as well as experimental facts on the dynamics of H_2^+ in intense laser fields are given in [10, 75]. More information on molecules in laser fields can also be found in the book edited by Bandrauk [76]. The review by Posthumus [75] contains a very insightful discussion of field dressed states (for strong fields these are Floquet states) of H_2^+ and their use to explain phenomena like molecular stabilization (bond hardening) and bond softening.

Computational aspects of the Morse potential and its eigenfunctions and eigenvalues are discussed in Chap. 13 of [77].

Born-Oppenheimer approximation

The original work of Born and Oppenheimer [78] is very hard to read, involving an expansion of the energy in terms of the fourth root of the mass ratio. A later version of the theory along the lines of Appendix VIII of [79] is much simpler to understand. Additional historical remarks can be found at the end of Chap. 12 in [24] and in [80]. The Born-Huang approximation (or adiabatic approximation) [79] goes beyond the Born-Oppenheimer approximation by taking into account diagonal electronic matrix elements of the nuclear Laplace operator.[15] Frequently, also the Born-Oppenheimer approximation of the main text is termed adiabatic. While the Born-Oppenheimer approximation of the text gives a lower bound of the ground state energy (see Exercise 5.5), the Born-Huang approximation leads to an upper bound,

[15]The diagonal matrix elements of the nabla operator vanish exactly, see, e.g., the discussion in Sect. 12.2.1 of [24].

however. Explicit calculations of the diagonal corrections to the Born-Oppenheimer potential energy surfaces for H_3 are given in [81].

Furthermore, in the derivation in the main text, we have suppressed the discussion of the mathematical subtleties that arise due to the existence of continuous parts in the spectrum of the electronic Hamiltonian [82–84]. For an earlier review of these issues see also [80]. The superiority of atomic over nuclear masses in the study of nuclear dynamics is hinted at in [80] and discussed in [85]. A mathematical treatise on the time-dependent Born-Oppenheimer approximation is given in [86].

The transformation from adiabatic to diabatic states and the non-uniqueness of that transformation are discussed in Sect. 15.2 of [23]. In this reference, the specific example of a triatomic collinear molecule is discussed explicitly. Additional material on the adiabtic/diabatic transformation can be found in Sect. 12.2 of [24]. Furthermore, a strictly diabatic treatment of polyatomic molecules is shown to be impossible in [87]. Nonadiabatic molecular dynamics can be tackled in many different ways [88]. Further information on that topic with additional references can be found for the case without an external laser in [89] and with an external laser in [90]. The semiclassical initial value method applied to the problem of coupled surfaces is reviewed in [91].

An interesting twist is the treatment of the excited states of two electron atoms by Feagin and Briggs [92]. These authors have studied atoms like H^- in a manner similar in spirit to the Born-Oppenheimer approximation for H_2^+. More recently, an old idea of Hunter [93], who pioneered the exact factorization of the molecular wavefunction in terms of a single product Ansatz, has come in the focus of intense investigations [94, 95].

Pump-Probe spectroscopy

For the presentation of 2D IR spectroscopy in the frequency domain, we closely followed the book by Hamm and Zanni [44]. A time-domain approach to 2D IR spectroscopy is also possible. It consists of the application of two pump pulses and after a waiting time, a probe pulse hits the molecule. This is explained in greater detail in [96] as well as in Chap. 4 of [44]. The authoritative reference for the general field of nonlinear spectroscopy is the book by Mukamel [97].

Coherent Control

Driven quantum tunneling is reviewed in depth in [98] and STIRAP is discussed in greater detail than here in the books by Rice and Zhao [66] and by Tannor [24]. Some experimental aspects of STIRAP can be found in the overview article by Bergmann et al. [61]. The formulation of most of the section on optimal control is based on Chap. 16 of [24]. Tannor's book as well as [66, 99] contain a wealth of additional material on the coherent control of quantum dynamics. Also in Tannor's book (Chap. 15.6 of [24]) an intuitive local control scheme for the heating of an electronic ground state wavepacket without substantial excitation of higher electronic states is reviewed.

Miscellaneous Topics

Reviews of early experimental and theoretical approaches to femtosecond chemistry are collected in [100]. More recently, a similar collection of articles has appeared in [101].

A natural extension of the scope of this chapter would be to study clusters in intense laser fields. A recent review of that field is given in [102]. Furthermore, the phenomenon of HHG in molecules (and solids) has become a hot topic which is not covered here. A tutorial review of that field, including the concept of molecular orbital tomography is contained in [103].

5.A Relative and Center of Mass Coordinates for H_2^+

In order to derive the Hamiltonian for H_2^+ in a laser field, we follow Hiskes's general treatment of diatomic molecules [104]. Specializing to the single electron case, the Hamiltonian in length gauge and atomic units is first expressed by using the coordinates of the nuclei $\boldsymbol{R}_a$, $\boldsymbol{R}_b$ and of the electron, $\boldsymbol{r}_\mathrm{e}$, according to

$$\hat{H}_\mathrm{mol} = -\frac{1}{2}\left\{\Delta_a/M_\mathrm{p} + \Delta_b/M_\mathrm{p} + \Delta_\mathrm{e}\right\} + V_1 - \mathcal{E}(t)[Z_a + Z_b - z_\mathrm{e}], \quad (5.183)$$

with the proton mass M_p, where $V_1 = V_\mathrm{CC} + 1/R$, with V_CC from (5.25), contains all Coulomb interaction terms and the laser is polarized in z-direction.

Now center of mass and relative coordinates

$$\boldsymbol{R}_\mathrm{S} = \frac{M_\mathrm{p}\boldsymbol{R}_a + M_\mathrm{p}\boldsymbol{R}_b + \boldsymbol{r}_\mathrm{e}}{M_\mathrm{S}}, \quad (5.184)$$

$$\boldsymbol{R} = \boldsymbol{R}_a - \boldsymbol{R}_b, \quad (5.185)$$

$$\boldsymbol{r}_i = \boldsymbol{r}_\mathrm{e} - \frac{\boldsymbol{R}_a + \boldsymbol{R}_b}{2} \quad (5.186)$$

are introduced. $M_\mathrm{S} = 2M_\mathrm{p} + 1$ is the total mass of the system (in a.u.) and the coordinate of the electron is measured relative to the center of mass of the nuclei.

In matrix form the old and the new coordinates are related by

$$\begin{pmatrix} \boldsymbol{R}_\mathrm{S} \\ \boldsymbol{R} \\ \boldsymbol{r}_i \end{pmatrix} = \begin{pmatrix} \frac{M_\mathrm{p}}{M_\mathrm{S}} & \frac{M_\mathrm{p}}{M_\mathrm{S}} & \frac{1}{M_\mathrm{S}} \\ 1 & -1 & 0 \\ -1/2 & -1/2 & 1 \end{pmatrix} \begin{pmatrix} \boldsymbol{R}_a \\ \boldsymbol{R}_b \\ \boldsymbol{r}_\mathrm{e} \end{pmatrix}. \quad (5.187)$$

With the help of the inverse matrix, the back transformation can be derived, which amounts to

$$\begin{pmatrix} \boldsymbol{R}_a \\ \boldsymbol{R}_b \\ \boldsymbol{r}_\mathrm{e} \end{pmatrix} = \begin{pmatrix} 1 & 1/2 & -\frac{1}{M_\mathrm{S}} \\ 1 & -1/2 & -\frac{1}{M_\mathrm{S}} \\ 1 & 0 & \frac{2M_\mathrm{p}}{M_\mathrm{S}} \end{pmatrix} \begin{pmatrix} \boldsymbol{R}_\mathrm{S} \\ \boldsymbol{R} \\ \boldsymbol{r}_i \end{pmatrix}. \quad (5.188)$$

Finally, not only the old coordinates, but also their time derivatives in the classical form of the Hamiltonian are expressed in terms of the new coordinates. It turns out that the center of mass with charge e moves "freely" in the electrical field [104]. The relative motion, however, is governed by the Hamiltonian

$$\hat{H}_{\rm rel} = -\frac{1}{2}\{\Delta_R/M_{\rm r} + \Delta_i/m_i\} + V_1 + \mathcal{E}(t)\,[1 + 1/M_{\rm S}]\,z_i, \tag{5.189}$$

where the reduced masses $M_{\rm r} = M_{\rm p}/2$ and $m_i = \frac{2M_{\rm p}}{M_{\rm S}}$ have been introduced and the Coulomb interaction is expressed in terms of the relative coordinates.

In the case of H_2^+, no kinetic couplings between the different degrees of freedom are present, nor does the field couple directly to the relative coordinate of the nuclei, which is not true in the general case [104].

5.B Perturbation Theory for Two Coupled Surfaces

In the case of a laser driven two level system with a 2×2 (matrix-) Hamilton operator

$$\hat{\mathbf{H}} = \hat{\mathbf{H}}_0 + \hat{\mathbf{W}}(t), \tag{5.190}$$

perturbation theory is best performed in the interaction picture of Sect. 2.2.4. In first order and after back transformation to the Schrödinger picture, analogous to (2.104),

$$|\boldsymbol{\chi}(t)\rangle = \mathrm{e}^{-\mathrm{i}\hat{\mathbf{H}}_0 t}|\boldsymbol{\chi}(0)\rangle + \frac{1}{\mathrm{i}}\int_0^t \mathrm{d}t' \mathrm{e}^{-\mathrm{i}\hat{\mathbf{H}}_0(t-t')}\hat{\mathbf{W}}(t')\mathrm{e}^{-\mathrm{i}\hat{\mathbf{H}}_0 t'}|\boldsymbol{\chi}(0)\rangle \tag{5.191}$$

can be written for the vector valued wavefunction in atomic units.

Invoking the Born-Oppenheimer approximation, we assume that the unperturbed Hamiltonian $\hat{\mathbf{H}}_0$ has only diagonal elements $\hat{H}_{\rm g}$, $\hat{H}_{\rm e}$. Furthermore, the initial wavefunction shall be restricted to the electronic ground state

$$|\boldsymbol{\chi}(0)\rangle = \begin{pmatrix} |\chi_{\rm g}(0)\rangle \\ 0 \end{pmatrix}. \tag{5.192}$$

Under the influence of the perturbation (having only off diagonal elements), the component of the wavefunction

$$|\boldsymbol{\chi}(t)\rangle = \begin{pmatrix} |\chi_{\rm g}(t)\rangle \\ |\chi_{\rm e}(t)\rangle \end{pmatrix} \tag{5.193}$$

in the excited electronic state as a function of time is the desired quantity. Under the assumptions mentioned above, for this quantity the golden rule expression in position representation

$$\chi_e(\boldsymbol{R}, t) = \frac{1}{i}\int_0^t dt' e^{-i\hat{H}_e(t-t')}\hat{W}_{e.g.}(\boldsymbol{R}, t')e^{-i\hat{H}_g t'}\chi_g(\boldsymbol{R}, 0) \tag{5.194}$$

is found. Here the definitions

$$\hat{W}_{e.g.}(\boldsymbol{R}, t) = \langle e(\boldsymbol{R})|\hat{W}(t)|g(\boldsymbol{R})\rangle \qquad \hat{H}_j = \langle j(\boldsymbol{R})|\hat{H}_0|j(\boldsymbol{R})\rangle \tag{5.195}$$

of the matrix elements of the (electronic plus nuclear) Hamiltonian[16] have been introduced, and j can either be e(xited) or g(round).

The physical interpretation of this final result is straightforward. The wavefunction propagates for a time t' on the lower surface is then multiplied by the perturbation and propagates on the upper surface until the final time t. All the possibilities to split the time interval $[0, t]$ have to be integrated over.

5.C Reflection Principle of Photodissociation

The dynamical reflection principle plays a major role for the interpretation of the photoelectron spectrum in a pump-probe experiment. It has an analog in the field of photodissociation. In [23] it is shown that the absorption spectrum of photodissociation is given by the Fourier transform of the auto-correlation function of the ground state wavepacket,

$$\chi_g(R) = \left(\frac{2\alpha_R}{\pi}\right)^{1/4} e^{-\alpha_R(R-R_e)^2}, \tag{5.196}$$

approximated by a Gaussian centered around R_e and with inverse width parameter α_R, that is instantaneously lifted to the excited state, where it is evolving in time.

In order to perform analytic calculations, this antibinding surface is approximated by a straight line

$$V(R) \approx V_e - V_R(R - R_e). \tag{5.197}$$

Using the short-time approximation (i.e., neglecting the kinetic energy) the wavefunction on the antibinding surface is given by

$$\chi_e(R, t) \sim e^{-i[V_e - V_R(R-R_e)]t} e^{-\alpha_R(R-R_e)^2}. \tag{5.198}$$

The Fourier transformation of the auto-correlation $c(t) = \langle\chi_e(0)|\chi_e(t)\rangle$ can be done analytically, yielding the absorption spectrum

[16]Note that the matrix elements are still operators (as indicated by the hat), due to the fact that the integrations in (5.195) are only over electronic coordinates.

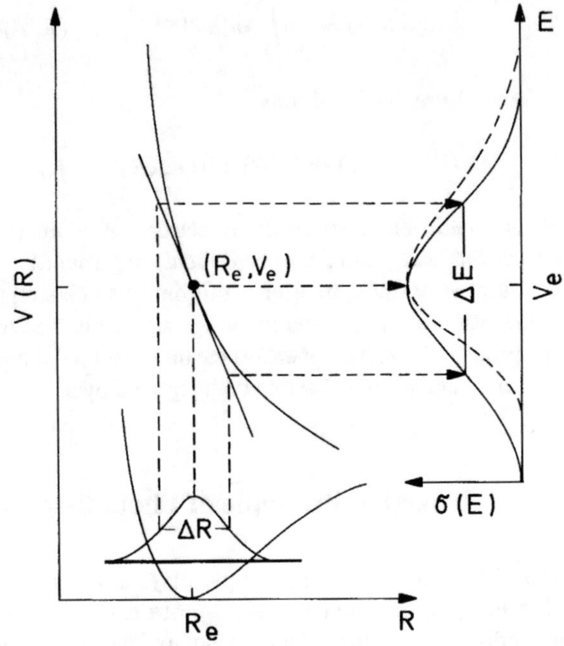

Fig. 5.50 Reflection principle of photodissociation [23]. ΔE is the FWHM of the absorption spectrum, whereas ΔR is the FWHM of the absolute square of the initial wavepacket

$$\sigma(E) \sim \frac{e^{-2\beta(E-V_e)^2}}{V_R} \tag{5.199}$$

with $\beta = (V_R^2/\alpha_R)^{-1}$. The same result can also be obtained by a purely classical calculation [23]. The maximum of the spectrum is at $E = V_e$ and its FWHM $\Delta E = V_R \Delta R$ is proportional to the negative slope of the antibinding surface and the FWHM of the squared initial wavepacket. In Fig. 5.50 it is shown that the reflection of the squared initial wavepacket at the antibinding surface yields the spectrum.

5.D The Undriven Double-Well Problem

Figure 5.51 shows the unperturbed double-well potential,

$$V_{DW}(x) \equiv -\frac{1}{4}x^2 + \frac{1}{64D}x^4, \tag{5.200}$$

in the units introduced in Sect. 5.5.1 for $D = 2$, including the five energy eigenvalues which lay below the barrier.

The coherent tunneling of a particle in the double-well potential emerges by considering an initial state that is a superposition of the two lowest (real-valued)

Table 5.4 Numerical and semiclassical (index s) results for the tunneling splitting as a function of the barrier height [53]

D	Δ	Δ_s	$\frac{\Delta_s-\Delta}{\Delta}$ (%)
1	2.392×10^{-2}	3.082×10^{-2}	28.8
1.5	2.262×10^{-3}	2.623×10^{-3}	15.9
2	1.895×10^{-4}	2.104×10^{-4}	11.0
2.5	1.507×10^{-5}	1.635×10^{-5}	8.5
3	1.164×10^{-6}	1.244×10^{-6}	6.8

eigenfunctions, $\chi_1(x)$, $\chi_2(x)$, depicted in Fig. 5.52, with the energies E_1, E_2.[17] At time $t = 0$ this leads to a state that is localized in the left well

$$\chi_l(x, 0) = \frac{1}{\sqrt{2}}\left[\chi_1(x) - \chi_2(x)\right]. \tag{5.201}$$

In the case $D \to \infty$ it is identical to the ground state of the harmonic approximation to the left well. Its absolute value has the time evolution

$$|\chi_l(x,t)|^2 = \frac{1}{2}\left\{|\chi_1(x)|^2 + |\chi_2(x)|^2 - 2\chi_1(x)\chi_2(x)\cos[(E_2 - E_1)t]\right\}. \tag{5.202}$$

Defining the tunneling splitting as

$$\Delta = E_2 - E_1, \tag{5.203}$$

the corresponding tunneling time

$$T_{\rm tu} = \frac{2\pi}{\Delta} \tag{5.204}$$

follows. The eigenvalues of the time-independent Schrödinger equation with a quartic potential and thus also the tunneling splitting are not available exactly analytically. Using a semiclassical approximation, Δ can be determined, however. The result of such a calculation is [105]

$$\Delta_{\rm s} = 8\sqrt{\frac{2D}{\pi}}\exp\left(-\frac{16D}{3}\right), \tag{5.205}$$

depending exponentially on the dimensionless barrier height. In Table 5.4 some values of Δ for different barrier heights can be found.

[17] Please note that in this appendix $\chi_j(x)$ denotes the j-th eigenfunction in the same electronic state (the double well).

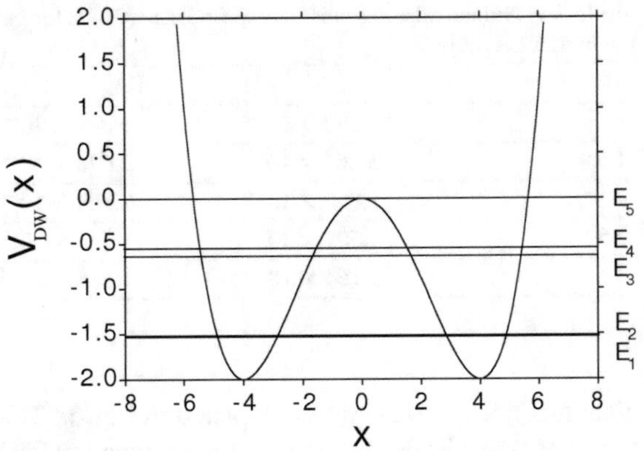

Fig. 5.51 Symmetric double-well potential with five energy eigenvalues for $D = 2$

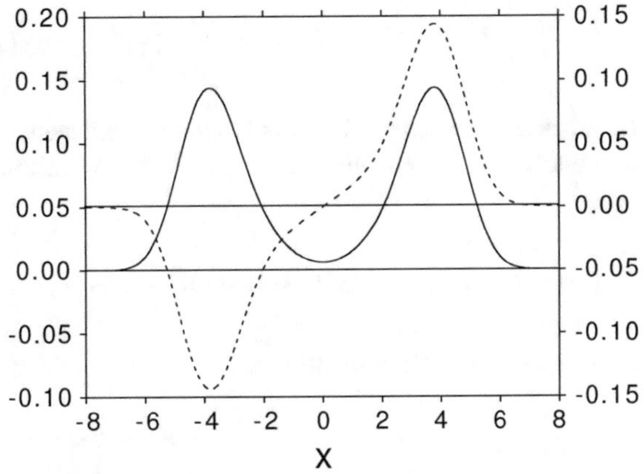

Fig. 5.52 Eigenfunctions of the lowest two eigenvalues for $D = 2$. *Solid line*: $\chi_1(x)$, *dashed line*: $\chi_2(x)$

5.E The Quantum Mechanical Adiabatic Theorem

To derive the adiabatic theorem in quantum theory, let us consider a system with discrete levels, whose state vector is given by

$$|\boldsymbol{\Psi}(t)\rangle = \begin{pmatrix} |\psi_1(t)\rangle \\ |\psi_2(t)\rangle \\ \vdots \end{pmatrix} . \tag{5.206}$$

In case of a time-dependent perturbation, the Hamilton matrix is given by

$$\mathbf{H}(t) = \begin{pmatrix} E_1 & V_{12}(t) & \cdots \\ V_{21}(t) & E_2 & \cdots \\ \vdots & \vdots & \ddots \end{pmatrix} . \tag{5.207}$$

Clearly, an eigenstate stays an eigenstate without the external perturbation. Even in the presence of a slowly changing perturbation an analogous statement holds, however.

To show this, one first defines a unitary transformation, diagonalizing the instantaneous Hamiltonian

$$\mathbf{U}^{-1}(t)\mathbf{H}(t)\mathbf{U}(t) = \mathbf{D}(t)\,. \tag{5.208}$$

The transformed state vector is given by

$$|\boldsymbol{\Psi}'(t)\rangle = \mathbf{U}^{-1}|\boldsymbol{\Psi}(t)\rangle\,. \tag{5.209}$$

It fulfills the time-dependent Schrödinger equation

$$\mathrm{i}|\dot{\boldsymbol{\Psi}}'(t)\rangle = \mathbf{D}|\boldsymbol{\Psi}'(t)\rangle - \mathrm{i}\mathbf{U}^{-1}\dot{\mathbf{U}}|\boldsymbol{\Psi}'(t)\rangle\,. \tag{5.210}$$

If the Hamilton matrix $\mathbf{H}$ is slowly time-dependent, then also $\mathbf{U}$ depends only weakly on time and the second term on the right hand side of the equation above can be neglected. An eigenfunction of the original Hamiltonian thus stays an eigenfunction of the instantaneous Hamiltonian. In [106] it has been shown that for periodically driven systems the adiabatic theorem has to be modified.

Finally, it is worthwhile to mention that by choosing the perturbation in such a way that the Hamiltonian switches from a simple to a complex one, the eigenstates of the complex Hamiltonian can be gained numerically [107]. In addition, an application to two-level systems has been given in [108].

References

1. J.C. Slater, *Quantum Theory of Molecules and Solids*, vol. 1 (McGraw-Hill, New York, 1963)
2. B.H. Bransden, C.J. Joachain, *Physics of Atoms and Molecules*, 2nd edn. (Pearson Education, Harlow, 2003)
3. A. Szabo, N.S. Ostlund, *Modern Quantum Chemistry* (Dover, Mineola, 1996)
4. B.N. Finkelstein, G.E. Horowitz, Z. Phys. **48**, 118 (1928)
5. H. Wind, J. Chem. Phys. **42**, 2371 (1965)
6. P.M. Morse, Phys. Rev. **34**, 57 (1929)
7. D. ter Haar, Phys. Rev. **70**, 222 (1946)
8. G. Herzberg, *Molecular Spectra and Molecular Structure: I. Spectra of Diatomic Molecules*, 2nd edn. (Krieger, Malabar, 1989)
9. H. Wind, J. Chem. Phys. **43**, 2956 (1965)
10. A. Giusti-Suzor, F.H. Mies, L.F. DiMauro, E. Charron, B. Yang, J. Phys. B At. Mol. Opt. Phys. **28**, 309 (1995)
11. S. Chelkowski, T. Zuo, A.D. Bandrauk, Phys. Rev. A **46**, R5342 (1992)
12. S. Chelkowski, T. Zuo, O. Atabek, A.D. Bandrauk, Phys. Rev. A **52**, 2977 (1995)
13. M. Uhlmann, T. Kunert, R. Schmidt, Phys. Rev. A **72**, 045402 (2005)
14. J.D. Jackson, *Klassische Elektrodynamik*, 2nd edn. (Walter de Gruyter, Berlin, 1983)
15. T. Zuo, A.D. Bandrauk, Phys. Rev. A **52**, R2511 (1995)

16. V. Roudnev, B.D. Esry, I. Ben-Itzhak, Phys. Rev. Lett. **93**, 163601 (2004)
17. A.D. Bandrauk, in *Molecules in Laser Fields*, ed. by A.D. Bandrauk (Dekker, New York, 1994), chap. 1, pp. 1–69
18. A.D. Bandrauk, E.E. Aubanel, J.M. Gauthier, *Molecules in Laser Fields*, ed. by A.D. Bandrauk (Dekker, New York, 1994), chap. 3, pp. 109–179
19. B. Feuerstein, U. Thumm, Phys. Rev. A **67**, 043405 (2003)
20. A. Kästner, F. Grossmann, S. Krause, R. Schmidt, A. Kenfack, J.M. Rost, New J. Phys. **11**, 083014 (2009)
21. A. Palacios, H. Bachau, F. Martin, Phys. Rev. Lett. **96**, 143001 (2006)
22. H. Haken, H.C. Wolf, *Molekülphysik und Quantenchemie* (Springer, Heidelberg, 1992)
23. R. Schinke, *Photodissociation Dynamics* (Cambridge University Press, Cambridge, 1993)
24. D.J. Tannor, *Introduction to Quantum Mechanics: A Time-Dependent Perspective* (University Science Books, Sausalito, 2007)
25. S.T. Epstein, J. Chem. Phys. **44**, 836 (1966)
26. R. Requist, E.K.U. Gross, Phys. Rev. Lett. **117**, 193001 (2016)
27. R.B. Walker, R.K. Preston, J. Chem. Phys. **67**, 2017 (1977)
28. R. Mecke, Z. Elektrochemie **54**, 38 (1950)
29. M.J. Davis, R.E. Wyatt, Chem. Phys. Lett. **86**, 235 (1982)
30. S. Chelkowski, A.D. Bandrauk, P.B. Corkum, Phys. Rev. Lett. **65**, 2355 (1990)
31. W. Jakubetz, J. Manz, V. Mohan, J. Chem. Phys **90**, 3683 (1989)
32. S. Krempl, T. Eisenhammer, A. Hübler, G. Mayer-Kress, P.W. Milonni, Phys. Rev. Lett. **69**, 430 (1992)
33. S. Shi, H. Rabitz, J. Chem. Phys. **92**, 364 (1990)
34. B.M. Garraway, K.A. Suominen, Rep. Prog. Phys. **58**, 365 (1995)
35. J.C. Tully, J. Chem. Phys. **93**, 1061 (1990)
36. H.D. Meyer, W.H. Miller, J. Chem. Phys. **72**, 2272 (1980)
37. G. Stock, M. Thoss, Phys. Rev. Lett. **78**, 578 (1997)
38. J.J. Sakurai, *Modern Quantum Mechanics* (Addison-Wesley, Reading, 1994)
39. K.A. Suominen, B.M. Garraway, S. Stenholm, Phys. Rev. A **45**, 3060 (1992)
40. J. Franck, Trans. Faraday Soc. **21**, 536 (1926)
41. E.U. Condon, Phys. Rev. **28**, 1182 (1926)
42. I.S. Gradshteyn, I.M. Ryzhik, *Table of Integrals Series and Products*, 5th edn. (Academic Press, New York, 1994)
43. F. Grossmann, Phys. Rev. A **60**, 1791 (1999)
44. P. Hamm, M. Zanni, *Concepts and Methods of 2D IR Spectroscopy* (Cambridge University Press, Cambridge, 2011)
45. A. Assion, M. Geisler, J. Helbing, V. Seyfried, T. Baumert, Phys. Rev. A **54**, R4605 (1996)
46. C. Meier, *Theoretische Untersuchungen zur Photoelektronenspektroskopie kleiner Moleküle mit kurzen und intensiven Laserpulsen*. Ph.D. thesis, Universität Freiburg (1995)
47. V. Engel, Comp. Phys. Comm. **63**, 228 (1991)
48. V. Engel, H. Metiu, J. Chem. Phys. **100**, 5448 (1994)
49. S.O. Williams, D.G. Imre, J. Phys. Chem. **92**, 6648 (1988)
50. M. Dantus, M.J. Rosker, A.H. Zewail, J. Chem. Phys. **87**, 2395 (1987)
51. F. Hund, Z. Phys. **43**, 805 (1927)
52. F. Grossmann, T. Dittrich, P. Jung, P. Hänggi, Phys. Rev. Lett. **67**, 516 (1991)
53. F. Grossmann, *Der Tunneleffekt in periodisch getriebenen Quantensystemen*. Ph.D. thesis, Universität Augsburg (1992)
54. F. Grossmann, P. Hänggi, Europhys. Lett. **18**, 1 (1992)
55. J.H. Shirley, Phys. Rev. **138**, B979 (1965)
56. M. Abramowitz, I.A. Stegun, *Handbook of Mathematical Functions* (Dover Publications, New York, 1965)
57. Y. Kayanuma, Phys. Rev. A **50**, 843 (1994)
58. G.D. Valle, M. Ornigotti, E. Cianci, V. Foglietti, P. Laporta, S. Longhi, Phys. Rev. Lett. **98**, 263601 (2007)

59. E. Kierig, U. Schnorrberger, A. Schietinger, J. Tomkovic, M.K. Oberthaler, Phys. Rev. Lett. **100**, 190405 (2008)
60. M.F. Kling, C. Siedschlag, A.J. Verhoef, J.I. Khan, M. Schultze, T. Uphues, Y. Ni, M. Uiberacker, M. Drescher, F. Krausz, M.J.J. Vrakking, Science **312**, 246 (2006)
61. K. Bergmann, H. Theuer, B.W. Shore, Rev. Mod. Phys. **70**, 1003 (1998)
62. D.J. Tannor, S.A. Rice, Adv. Chem. Phys. **70**, 441 (1988)
63. R. Kosloff, S.A. Rice, P. Gaspard, S. Tersigni, D.J. Tannor, Chem. Phys. **139**, 201 (1989)
64. D.J. Tannor, R. Kosloff, S.A. Rice, J. Chem. Phys. **85**, 5805 (1986)
65. V.F. Krotov, I.N. Feldman, Engrg. Cybernetics **21**, 123 (1984)
66. S.A. Rice, M. Zhao, *Optical Control of Molecular Dynamics* (Wiley, New York, 2000)
67. F. Grossmann, L. Feng, G. Schmidt, T. Kunert, R. Schmidt, Europhys. Lett. **60**, 201 (2002)
68. D. Zeidler, S. Frey, K.L. Kompa, M. Motzkus, Phys. Rev. A **64**, 023420 (2001)
69. R.S. Judson, H. Rabitz, Phys. Rev. Lett. **68**, 1500 (1992)
70. A. Assion, T. Baumert, M. Bergt, T. Brixner, B. Kiefer, V. Seyfried, M. Strehle, G. Gerber, Science **282**, 919 (1998)
71. D. Babikov, J. Chem. Phys. **121**, 7577 (2004)
72. T. Cheng, A. Brown, J. Chem. Phys. **124**, 034111 (2006)
73. C. Gollub, R. de Vivie-Riedle, J. Chem. Phys. **126**, 204102 (2008)
74. P. Passarinho, M.L. da Silva, J. Mol. Spec. **236**, 148 (2006)
75. J.H. Posthumus, Rep. Prog. Phys. **67**, 623 (2004)
76. A.D. Bandrauk (ed.), *Molecules in Laser Fields* (Dekker, New York, 1994)
77. J.M. Feagin, *Quantum Methods with Mathematica* (Springer, New York, 1994)
78. M. Born, R. Oppenheimer, Ann. Phys. (Leipzig) **84**, 457 (1927)
79. M. Born, K. Huang, *Dynamical Theory of Crystal Lattices* (Clarendon Press, Oxford, 1954)
80. W. Kutzelnigg, Mol. Phys. **90**, 909 (1997)
81. S.L. Mielke, D.W. Schwenke, G.C. Schatz, B.C. Garrett, K.A. Peterson, J. Phys. Chem. **113**, 4479 (2009)
82. B.T. Sutcliffe, R.G. Woolley, J. Chem. Phys. **137**, 22A544 (2012)
83. B.T. Sutcliffe, R.G. Woolley, J. Chem. Phys. **140**, 037101 (2014)
84. T. Jecko, J. Math. Phys. **55**, 053504 (2014)
85. N.C. Handy, A.M. Lee, Chem. Phys. Lett. **252**, 909 (1996)
86. G. Panati, H. Spohn, S. Teufel, ESAIM: Mathematical Modelling and Numerical Analysis **41**, 297 (2007)
87. B.K. Kendrick, C. Alden Mead, D.G. Truhlar, Chem. Phys. Lett **330**, 629 (2000)
88. M. Persico, G. Granucci, Theor. Chem. Acc. **133**, 1526 (2014)
89. U. Saalmann, R. Schmidt, Z. Phys. D **38**, 153 (1996)
90. T. Kunert, R. Schmidt, Eur. Phys. J. D **25**, 15 (2003)
91. M. Thoss, H. Wang, Ann. Rev. Phys. Chem. **55**, 299 (2004)
92. J.M. Feagin, J.S. Briggs, Phys. Rev. Lett. **57**, 984 (1986)
93. G. Hunter, Int. J. Quantum Chem. **9**, 237 (1975)
94. A. Abedi, N. Maitra, E.K.U. Gross, Phys. Rev. Lett. **105**, 123002 (2010)
95. L. Cederbaum, J. Chem. Phys. **138**, 224110 (2013)
96. Y. Tanimura, J. Phys. Soc. Jpn. **75**, 082001 (2006)
97. S. Mukamel, *Principles of Nonlinear Optical Spectroscopy* (Oxford University Press, New York, 1995)
98. M. Grifoni, P. Hänggi, Phys. Rep. **304**, 229 (1998)
99. P. Brumer, M. Shapiro, *Principles of the Quantum Control of Molecular Processes* (Wiley-VCH, Berlin, 2003)
100. J. Manz, L.Wöste (eds.), *Femtosecond Chemistry*, vol. 1 and 2 (VCH, Weinheim, 1995)
101. R. de Nalda, L. Banares (eds.), *Ultrafast Phenomena in Molecular Sciences: Femtosecond Physics and Chemistry*, Springer Series in Chemical Physics, vol. 107 (Springer International Publishing, 2014)
102. T. Fennel, K.H. Meiwes-Broer, J. Tiggesbäumker, P.G. Reinhard, P.M. Dinh, E. Suraud, Rev. Mod. Phys. **82**, 1793 (2010)

103. S. Haessler, J. Caillat, P. Salières, J. Phys. B At. Mol. Opt. Phys. **44**, 203001 (2011)
104. J.R. Hiskes, Phys. Rev. **122**, 1207 (1961)
105. U. Weiss, W. Häffner, Phys. Rev. D **27**, 2916 (1983)
106. D.W. Hone, R. Ketzmerick, W. Kohn, Phys. Rev. A **56**, 4045 (1997)
107. D. Kohen, D.J. Tannor, J. Chem. Phys. **98**, 3168 (1993)
108. A. Emmanouilidou, X.G. Zhao, P. Ao, Q. Niu, Phys. Rev. Lett. **85**, 1626 (2000)

Part III
Supplements

Chapter 6
Solutions to Problems

In this chapter, detailed solutions to the numbered problems given in Chaps. 1 through 5 are gathered. Where appropriate, references to related, additional material in the literature are given.

6.1 Solutions to Problems in Chap. 1

1.1. The rate equations in the stationary case for three levels are

$$0 = -\Gamma N_1 + \gamma_{12}N_2 + \gamma_{13}N_3, \tag{6.1}$$

$$0 = -\gamma_{12}N_2 + \gamma_{23}N_3, \tag{6.2}$$

$$0 = \Gamma N_1 - (\gamma_{13} + \gamma_{23})N_3. \tag{6.3}$$

Eliminating N_2 from (6.1) using (6.2) and from (6.2) using $N = N_1 + N_2 + N_3$ leads to

$$0 = -\Gamma N_1 + (\gamma_{13} + \gamma_{23})N_3, \tag{6.4}$$

$$0 = -\gamma_{12}N + \gamma_{12}N_1 + (\gamma_{12} + \gamma_{23})N_3. \tag{6.5}$$

Resolving the last equation for N_3 and using it in (6.4) resolved for N_1, leads to the intermediate result

$$N_1 = \frac{(\gamma_{13} + \gamma_{23})\gamma_{12}}{(\gamma_{12} + \gamma_{23})\Gamma}(N - N_1), \tag{6.6}$$

which finally gives

F. Grossmann, *Theoretical Femtosecond Physics*, Graduate Texts in Physics,
https://doi.org/10.1007/978-3-319-74542-8_6

$$N_1 = \frac{(\gamma_{13} + \gamma_{23})\gamma_{12}}{(\gamma_{13} + \gamma_{23})\gamma_{12} + (\gamma_{12} + \gamma_{23})\Gamma} N. \tag{6.7}$$

Analogously

$$N_2 = \frac{\gamma_{23}\Gamma}{(\gamma_{13} + \gamma_{23})\gamma_{12} + (\gamma_{12} + \gamma_{23})\Gamma} N \tag{6.8}$$

is found. For N_2 to be larger than N_1, the condition

$$\Gamma > \gamma_{12}\left(1 + \frac{\gamma_{13}}{\gamma_{23}}\right) \tag{6.9}$$

therefore has to hold! To achieve population inversion with moderate pumping, γ_{12} has to be small and γ_{23} has to be large compared to γ_{13} [1].

1.2. The electric field is given by

$$\mathcal{E}(t) = \sum_{n=-p}^{p} \mathcal{E}_n \cos[(\omega + n\Omega)t], \tag{6.10}$$

with $\Omega = 2\pi\delta\nu$. For $\mathcal{E}_n = \mathcal{E}_0$ and using the exponential function and the geometric series

$$\begin{aligned}
\mathcal{E}(t) &= \mathrm{Re}[\mathcal{E}_0 \mathrm{e}^{\mathrm{i}\omega t} \sum_{n=-p}^{p} \mathrm{e}^{\mathrm{i}n\Omega t}] \\
&\overset{n'=n+p}{=} \mathrm{Re}[\mathcal{E}_0 \mathrm{e}^{\mathrm{i}\omega t - \mathrm{i}p\Omega t} \sum_{n'=0}^{2p} \mathrm{e}^{\mathrm{i}n'\Omega t}] \\
&= \mathrm{Re}\left[\mathcal{E}_0 \mathrm{e}^{\mathrm{i}\omega t - \mathrm{i}p\Omega t} \frac{1 - \mathrm{e}^{\mathrm{i}(2p+1)\Omega t}}{1 - \mathrm{e}^{\mathrm{i}\Omega t}}\right] \\
&= \mathrm{Re}\left[\mathcal{E}_0 \mathrm{e}^{\mathrm{i}\omega t} \frac{\sin[(2p+1)\Omega t/2]}{\sin(\Omega t/2)}\right] \\
&= \mathcal{E}_0 \cos(\omega t) \frac{\sin[(2p+1)\Omega t/2]}{\sin(\Omega t/2)}
\end{aligned} \tag{6.11}$$

follows and from this the corresponding intensity (1.25).

1.3. For the spectrogram, we calculate the integral

$$\begin{aligned}
I(\tau, \Omega) = \int \mathrm{d}t \exp\Big\{ &- \frac{(t-\tau)^2}{2\sigma^2} - \mathrm{i}\Omega(t-\tau) - \frac{(t-t_0)^2}{2\sigma^2} \\
&+ \mathrm{i}\omega_0(t-t_0) \pm \mathrm{i}\frac{\lambda}{2}(t-t_0)^2 \Big\},
\end{aligned} \tag{6.12}$$

with an exponential of the form $\exp\{-a(t-t_0)^2 + b(t-t_0) + c\}$, where

$$a = \frac{1}{\sigma^2} \mp \mathrm{i}\frac{\lambda}{2}, \tag{6.13}$$

$$b = \mathrm{i}(\omega_0 - \Omega) - \frac{1}{\sigma^2}(t_0 - \tau), \tag{6.14}$$

$$c = -\frac{1}{2\sigma^2}(t_0 - \tau)^2. \tag{6.15}$$

Using the Gaussian integral (1.32), we calculate $\frac{b^2}{4a} + c$ and get

$$F(\tau, \Omega) \sim |I(\tau, \Omega)|^2 \sim \mathrm{e}^{-[A(\Omega-\omega_0)^2 + B(\Omega-\omega_0)(\tau-t_0) + C(\tau-t_0)^2]}, \tag{6.16}$$

with

$$A = \frac{8}{\sigma^2 \kappa}, \tag{6.17}$$

$$B = \mp\frac{8\lambda}{\sigma^2 \kappa}, \tag{6.18}$$

$$C = \frac{8}{\sigma^6 \kappa} + \frac{4\lambda^2}{\sigma^2 \kappa}, \tag{6.19}$$

where $\kappa = (4/\sigma^2)^2 + 4\lambda^2$. Spectrograms for an unchirped, an up-chirped and a down-chirped pulse are shown in Fig. 6.1.

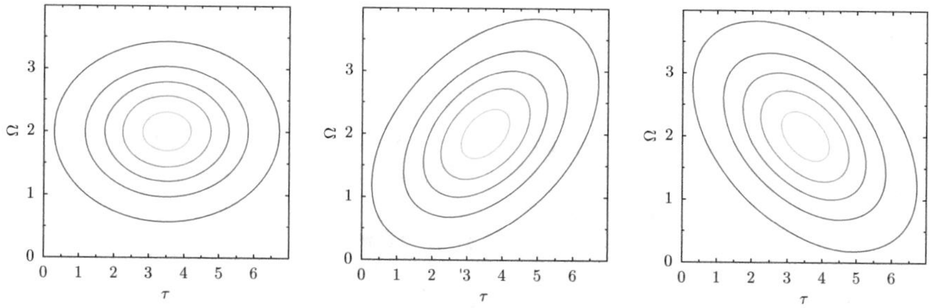

Fig. 6.1 Spectrograms for an unchirped pulse $\lambda = 0$ (*left panel*), an up-chirped pulse (plus sign) with $\lambda = 0.5$ (*middle panel*) and a down-chirped pulse (minus sign) with $\lambda = 0.5$ (*right panel*); in all cases $\sigma = 1.5$

6.2 Solutions to Problems in Chap. 2

2.1. If $\boldsymbol{j}$ falls to zero faster than $1/r^2$, then Gauss's theorem applied to the equation of continuity integrated over all space gives

$$\frac{\partial}{\partial t}\int \mathrm{d}^3 r\,\rho + \oint \mathrm{d}\boldsymbol{f}\cdot\boldsymbol{j} = \frac{\partial}{\partial t}\int \mathrm{d}^3 r\,\rho = 0\,, \tag{6.20}$$

showing that if the wavefunction was normalized at $t = 0$ it remains normalized at all times.

2.2. The time-evolution operator

$$\hat{U}(t,t_0) = \left[\hat{1} - \frac{\mathrm{i}}{\hbar}\hat{H}(t_{N-1})\Delta t\right]\left[\hat{1} - \frac{\mathrm{i}}{\hbar}\hat{H}(t_{N-2})\Delta t\right]\ldots\left[\hat{1} - \frac{\mathrm{i}}{\hbar}\hat{H}(t_0)\Delta t\right] \tag{6.21}$$

- up to zeroth order in Δt is

$$\hat{U}(t,t_0) = \hat{1}\,, \tag{6.22}$$

- up to first order in Δt is

$$\hat{U}(t,t_0) = \hat{1} - \frac{\mathrm{i}}{\hbar}\sum_{\nu=0}^{N-1}\hat{H}(t_\nu)\Delta t \overset{\substack{N\to\infty\ \Delta t\to 0\\ N\Delta t = t-t_0}}{\longrightarrow} \hat{1} - \frac{\mathrm{i}}{\hbar}\int_{t_0}^{t}\mathrm{d}t'\,\hat{H}(t') \tag{6.23}$$

- and up to second order in Δt is

$$\begin{aligned}\hat{U}(t,t_0) &= \hat{1} - \frac{\mathrm{i}}{\hbar}\sum_{\nu=0}^{N-1}\hat{H}(t_\nu)\Delta t \\ &+ \left(-\frac{\mathrm{i}}{\hbar}\right)^2\sum_{\nu=1}^{N-1}\sum_{\nu'=0}^{\nu-1}\hat{H}(t_\nu)\hat{H}(t_{\nu'})\Delta t^2 \\ &\overset{\substack{N\to\infty\ \Delta t\to 0\\ N\Delta t = t-t_0}}{\longrightarrow} \hat{1} - \frac{\mathrm{i}}{\hbar}\int_{t_0}^{t}\mathrm{d}t'\,\hat{H}(t') \\ &+ \left(-\frac{\mathrm{i}}{\hbar}\right)^2\int_{t_0}^{t}\mathrm{d}t''\int_{t_0}^{t''}\mathrm{d}t'\,\hat{H}(t'')\hat{H}(t')\,.\end{aligned} \tag{6.24}$$

For the Hermitian conjugate of the time-evolution operator, in case of a Hermitian Hamiltonian, we get

$$\hat{U}^\dagger(t,t_0) = \left(\left[\hat{1} - \frac{i}{\hbar}\hat{H}(t_{N-1})\Delta t\right]\left[\hat{1} - \frac{i}{\hbar}\hat{H}(t_{N-2})\Delta t\right]\dots\left[\hat{1} - \frac{i}{\hbar}\hat{H}(t_0)\Delta t\right]\right)^\dagger$$
$$= \left[\hat{1} + \frac{i}{\hbar}\hat{H}(t_0)\Delta t\right]\left[\hat{1} + \frac{i}{\hbar}\hat{H}(t_1)\Delta t\right]\dots\left[\hat{1} + \frac{i}{\hbar}\hat{H}(t_{N-1})\Delta t\right]. \quad (6.25)$$

For $\Delta t \to 0$ we consider only terms up to order Δt, finding that

$$\hat{U}^\dagger(t,t_0)\hat{U}(t,t_0) = \left[\hat{1} + \frac{i}{\hbar}\hat{H}(t_0)\Delta t\right]\left[\hat{1} + \frac{i}{\hbar}\hat{H}(t_1)\Delta t\right]\dots\left[\hat{1} + \frac{i}{\hbar}\hat{H}(t_{N-1})\Delta t\right]$$
$$\left[\hat{1} - \frac{i}{\hbar}\hat{H}(t_{N-1})\Delta t\right]\left[\hat{1} - \frac{i}{\hbar}\hat{H}(t_{N-2})\Delta t\right]\dots\left[\hat{1} - \frac{i}{\hbar}\hat{H}(t_0)\Delta t\right]$$
$$= \hat{1} + \frac{i}{\hbar}\sum_{\nu=0}^{N-1}[\hat{H}(t_\nu) - \hat{H}(t_\nu)]\Delta t$$
$$= \hat{1} \quad (6.26)$$

holds and unitarity is shown.

2.3. We verify the closed form expression of the time-evolution operator (as well as of its time derivative) using the time-ordering operator.

(a) Taylor expansion of the exponential function allows to write

$$\frac{d}{dt}\exp\{\hat{B}(t)\} = \sum_{n=0}^{\infty}\frac{1}{n!}\frac{d}{dt}\hat{B}^n(t). \quad (6.27)$$

The derivative of the n-th power is the sum of n terms (product rule!)

$$\frac{d}{dt}\hat{B}^n(t) = \sum_{k=0}^{n-1}\hat{B}^k\frac{d}{dt}\hat{B}(t)\hat{B}^{n-k-1}, \quad (6.28)$$

which cannot be simplified in general, due to the non-commutativity of the operator $\hat{B}$ and its derivative [2].

(b) For $\hat{B}(t) \equiv -\frac{i}{\hbar}\hat{H}_0 t$, we have

$$\frac{d}{dt}\hat{B}(t) = -\frac{i}{\hbar}\hat{H}_0, \quad (6.29)$$

which is commuting with $\hat{B}(t)$ and therefore

$$\frac{d}{dt}\hat{B}^n(t) = n\frac{d}{dt}\hat{B}(t)\hat{B}^{n-1}(t), \quad (6.30)$$

leading to the TDSE for the time-evolution operator

$$\frac{\mathrm{d}}{\mathrm{d}t}\hat{U}(t) = -\frac{\mathrm{i}}{\hbar}\hat{H}_0\hat{U}(t)\,. \tag{6.31}$$

(c) For $\hat{B}(t) \equiv -\frac{\mathrm{i}}{\hbar}\int_0^t \mathrm{d}t'\hat{H}(t')$, however,

$$\frac{\mathrm{d}}{\mathrm{d}t}\hat{B}(t) = -\frac{\mathrm{i}}{\hbar}\hat{H}(t) \tag{6.32}$$

does not necessarily commute with $\hat{B}(t)$ and we do not find a TDSE of the form (6.31).

(d) In order to prove the given relation, we first differentiate the operator

$$\begin{aligned}\hat{\mathcal{H}}_n(t) &\equiv \hat{T}\left(\int_0^t \mathrm{d}t'\hat{H}(t')\right)^n \\ &= \hat{T}\left[\int_0^t \mathrm{d}t_n\hat{H}(t_n)\int_0^t \mathrm{d}t_{n-1}\hat{H}(t_{n-1})\ldots\int_0^t \mathrm{d}t_1\hat{H}(t_1)\right],\end{aligned} \tag{6.33}$$

leading to

$$\frac{\mathrm{d}}{\mathrm{d}t}\hat{\mathcal{H}}_n(t) = n\hat{H}(t)\hat{T}\left(\int_0^t \mathrm{d}t'\hat{H}(t')\right)^{n-1} = n\hat{H}(t)\hat{\mathcal{H}}_{n-1}(t), \tag{6.34}$$

due to the fact that the time-ordering operator moves all occurrences of $\hat{H}(t)$ all the way to the left. This is a recurrence-differential equation and the expression

$$\hat{\mathcal{H}}_n(t) = n!\left[\int_0^t \mathrm{d}t_n\hat{H}(t_n)\int_0^{t_n} \mathrm{d}t_{n-1}\hat{H}(t_{n-1})\ldots\int_0^{t_2} \mathrm{d}t_1\hat{H}(t_1)\right] \tag{6.35}$$

fulfills the same recurrence-differential equation and has the same initial conditions $\hat{\mathcal{H}}_1(t)$ and $\hat{\mathcal{H}}_n(0)$. The given relation is thus proven.

By comparison with (2.31), we see that the construction $\hat{U}(t) = \hat{T}\exp[\hat{B}(t)]$ therefore is a closed form expression for the time-evolution operator, and furthermore it fulfills the TDSE

$$\frac{\mathrm{d}}{\mathrm{d}t}\hat{U}(t) = -\frac{\mathrm{i}}{\hbar}\hat{H}\hat{U}(t). \tag{6.36}$$

We have closely followed the presentation of Appendix B in [2].

2.4. Taking the trace of the time-evolution operator of the harmonic oscillator, we get

$$
\begin{aligned}
\mathrm{Tr}\hat{U}(t,0) &= \int \mathrm{d}x\, K(x,t;x,0) \\
&= \sqrt{\frac{m\omega_\mathrm{e}}{2\pi\mathrm{i}\hbar\sin(\omega_\mathrm{e}t)}}\int \mathrm{d}x \exp\left\{\frac{\mathrm{i}m\omega_\mathrm{e}}{\hbar\sin(\omega_\mathrm{e}t)}x^2\left[\cos(\omega_\mathrm{e}t)-1\right]\right\} \\
&\overset{(1.29)}{=} \frac{1}{\sqrt{2\left[\cos(\omega_\mathrm{e}t)-1\right]}} \\
&= \frac{1}{\mathrm{e}^{\mathrm{i}\omega_\mathrm{e}t/2}-\mathrm{e}^{-\mathrm{i}\omega_\mathrm{e}t/2}} \\
&= \sum_{n=0}^{\infty}\mathrm{e}^{-\mathrm{i}(n+1/2)\omega_\mathrm{e}t}, \qquad (6.37)
\end{aligned}
$$

where the geometric series $\sum_{n=0}^{\infty}x^n = \frac{1}{1-x}$ with $x = \mathrm{e}^{-\mathrm{i}\omega_\mathrm{e}t}$ was used (on the border of the radius of convergence). Therefore the eigenvalue spectrum is given by $E_n = \left(n+\frac{1}{2}\right)\hbar\omega_\mathrm{e}$, $n = 0, 1, 2, \ldots$.

2.5. The GWD allows for an exact analytical solution of the TDSE for maximally quadratic potentials.

(a) The probability density and probability flux density for the Gaussian wavepacket (2.48) are given by

$$\rho = \Psi^*\Psi = \left(\frac{2\alpha_0}{\pi}\right)^{1/2}\exp\left\{-2\mathrm{Re}\alpha_t(x-q_t)^2 - \frac{2\mathrm{Im}\delta_t}{\hbar}\right\}, \qquad (6.38)$$

$$j = \frac{\hbar}{m}\mathrm{Im}[\Psi^*\partial_x\Psi] = \frac{\hbar}{m}\left\{\frac{p_t}{\hbar} - 2\mathrm{Im}\alpha_t(x-q_t)\right\}\rho. \qquad (6.39)$$

From (2.60) and (2.61) we get

$$\mathrm{Im}\dot{\delta}_t = -\frac{\hbar^2}{m}\mathrm{Im}\alpha_t, \qquad (6.40)$$

$$\mathrm{Re}\dot{\alpha}_t = \frac{4\hbar}{m}\mathrm{Re}\alpha_t\mathrm{Im}\alpha_t. \qquad (6.41)$$

Using this together with Hamilton's equations, the continuity equation in 1D

$$\dot{\rho} = -\partial_x j \qquad (6.42)$$

can be shown to hold by comparing coefficients in front of equal powers of $(x - q_t)$.

(b) For the free particle, $V(x) = 0$, the solutions of Hamilton's equations are

$$q_t = q_0 + \frac{p_0}{m}t, \qquad (6.43)$$

$$p_t = p_0. \qquad (6.44)$$

The Riccati equation (2.61) can be integrated by separation of variables

$$\int_{\alpha_0}^{\alpha_t} \frac{d\alpha}{\alpha^2} = -\frac{2i\hbar}{m} \int_0^t dt' , \tag{6.45}$$

leading to

$$\alpha_t = \frac{\alpha_0}{1 + \frac{2i\hbar}{m}\alpha_0 t} \tag{6.46}$$

and

$$\delta_t = \frac{p_0^2}{2m}t + \frac{i\hbar}{2}\ln\left\{1 + \frac{2i\hbar}{m}\alpha_0 t\right\} . \tag{6.47}$$

(c) For the harmonic oscillator, $V(x) = \frac{1}{2}m\omega_e^2 x^2$, the solutions of Hamilton's equations are

$$q_t = q_0 \cos(\omega_e t) + \frac{p_0}{m\omega_e}\sin(\omega_e t), \tag{6.48}$$

$$p_t = p_0 \cos(\omega_e t) - m\omega_e q_0 \sin(\omega_e t). \tag{6.49}$$

The Riccati equation (2.61) is

$$\dot{\alpha}_t = -\frac{2i\hbar}{m}\alpha_t^2 + \frac{i}{2\hbar}m\omega_e^2 . \tag{6.50}$$

It can again be integrated using separation of variables, leading to (see also [3])

$$\alpha_t = a\frac{\alpha_0 \cos(\omega_e t) + ia\sin(\omega_e t)}{a\cos(\omega_e t) + i\alpha_0 \sin(\omega_e t)}, \tag{6.51}$$

with $a = m\omega_e/(2\hbar)$. In the special case $\alpha_0 = a$, the inverse width parameter stays constant and the wavepacket is a coherent state of the harmonic oscillator (see also below).
The phase factor is given by

$$\delta_t = \frac{p_t q_t - p_0 q_0}{2} + \frac{i\hbar}{2}\ln\left\{\frac{i\alpha_0 \sin(\omega_e t) + a\cos(\omega_e t)}{a}\right\} , \tag{6.52}$$

which in the special case $\alpha_0 = a$ leads to

$$\delta_t = \frac{p_t q_t - p_0 q_0}{2} - \frac{\hbar\omega_e t}{2}. \tag{6.53}$$

(d) Using the Gaussian integral (1.29), from normalization we get

$$\begin{aligned}\langle\hat{1}\rangle &= \int \mathrm{d}x\, \Psi^*(x,t)\Psi(x,t) \\ &= \left(\frac{2\alpha_0}{\pi}\right)^{1/2} \int \mathrm{d}x \exp\{-2\mathrm{Re}\alpha_t(x-q_t)^2 - 2\mathrm{Im}\delta_t/\hbar\} \\ &= \left(\frac{\alpha_0}{\mathrm{Re}\alpha_t}\right)^{1/2} \exp\{-2\mathrm{Im}\delta_t/\hbar\} \equiv 1\,. \end{aligned} \tag{6.54}$$

Note that norm conservation does not have to be postulated here. Normalization can also be shown to hold using (6.40) and (6.41).

The expectation value and the (square root of the) variance for position can be calculated using the identity $x = (x - q_t) + q_t$ together with the Gaussian integrals (1.29) and (1.30, 1.31)

$$\begin{aligned}\langle\hat{x}\rangle &= \int \mathrm{d}x\, \Psi^*(x,t)x\Psi(x,t) \\ &= \left(\frac{2\alpha_0}{\pi}\right)^{1/2} \int \mathrm{d}x\, x \exp\{-2\mathrm{Re}\alpha_t(x-q_t)^2 - 2\mathrm{Im}\delta_t/\hbar\} \\ &= \left(\frac{\alpha_0}{\mathrm{Re}\alpha_t}\right)^{1/2} \exp\{-2\mathrm{Im}\delta_t/\hbar\}q_t \\ &\overset{(6.54)}{=} q_t, \end{aligned} \tag{6.55}$$

$$\begin{aligned}\langle\hat{x}^2\rangle &= \int \mathrm{d}x\, \Psi^*(x,t)x^2\Psi(x,t) \\ &= \left(\frac{2\alpha_0}{\pi}\right)^{1/2} \int \mathrm{d}x\, x^2 \exp\{-2\mathrm{Re}\alpha_t(x-q_t)^2 - 2\mathrm{Im}\delta_t/\hbar\} \\ &= q_t^2 + \frac{1}{4\mathrm{Re}\alpha_t}, \end{aligned} \tag{6.56}$$

$$\Delta x(t) = \sqrt{\langle\hat{x}^2\rangle - \langle\hat{x}\rangle^2} = \sqrt{\frac{1}{4\mathrm{Re}\alpha_t}}. \tag{6.57}$$

Similarly, for the momentum we get

$$\Delta p(t) = \sqrt{\langle\hat{p}^2\rangle - \langle\hat{p}\rangle^2} = \frac{\hbar|\alpha_t|}{\sqrt{\mathrm{Re}\alpha_t}}\,, \tag{6.58}$$

leading to the uncertainty product

$$\Delta x(t)\Delta p(t) = \frac{\hbar}{2}\frac{|\alpha_t|}{\mathrm{Re}\alpha_t} \geq \frac{\hbar}{2} . \tag{6.59}$$

In the free particle case

$$\mathrm{Re}\alpha_t = \frac{\alpha_0}{1 + 4\hbar^2\alpha_0^2 t^2/m^2} \tag{6.60}$$

and the square root of the position variance increases (asymptotically linearly) in time as depicted in Fig. 6.2.
Furthermore, the momentum variance, due to

$$|\alpha_t| = \frac{\alpha_0}{(1 + 4\hbar^2\alpha_0^2 t^2/m^2)^{1/2}} , \tag{6.61}$$

stays constant in time. The uncertainty product

$$\Delta x(t)\Delta p(t) = \frac{\hbar}{2}\sqrt{1 + 4\hbar^2\alpha_0^2 t^2/m^2} \tag{6.62}$$

increases with time.
In the harmonic oscillator case we find

$$\mathrm{Re}\alpha_t = a\frac{a\alpha_0}{a\cos^2(\omega_\mathrm{e}t) + \alpha_0^2\sin^2(\omega_\mathrm{e}t)} \tag{6.63}$$

and thus

$$\Delta x^2(t) = \frac{1}{4\mathrm{Re}\alpha_t} = \frac{\hbar}{2m\omega_\mathrm{e}}\left[\left(\frac{a}{\alpha_0} - \frac{\alpha_0}{a}\right)\cos^2(\omega_\mathrm{e}t) + \frac{\alpha_0}{a}\right]. \tag{6.64}$$

This result corroborates that for $\alpha_0 = a$, in accord with (6.51), also $\alpha_t = \alpha_0 = a$. In this special case both variances as well as the uncertainty product are independent of time (coherent state). In the more general case of a squeezed state the variances as well as the uncertainty product are functions of time, see Fig. 6.3. A pictorial representation in phase space of the dynamics of a squeezed state is given in the appendix of [4].

2.6. We use the eigenstates (2.62) and energies (2.63) of the box potential to find the probabilities and the time-evolution of a special initial state.

(a) The overlap is given by

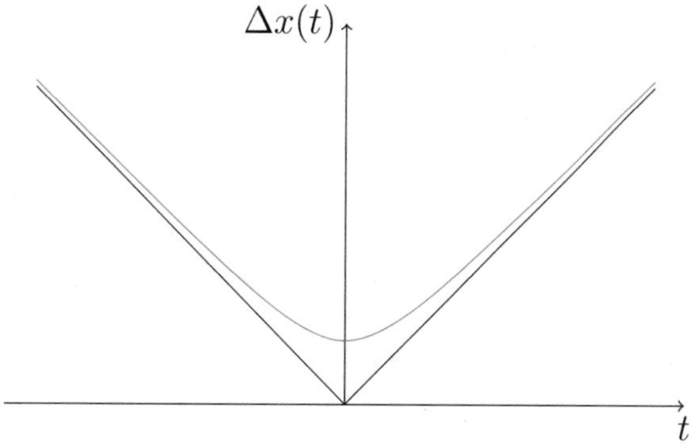

Fig. 6.2 Sketch of $\Delta x(t)$ (*in red*) for the free particle, together with its straight line asymptotes

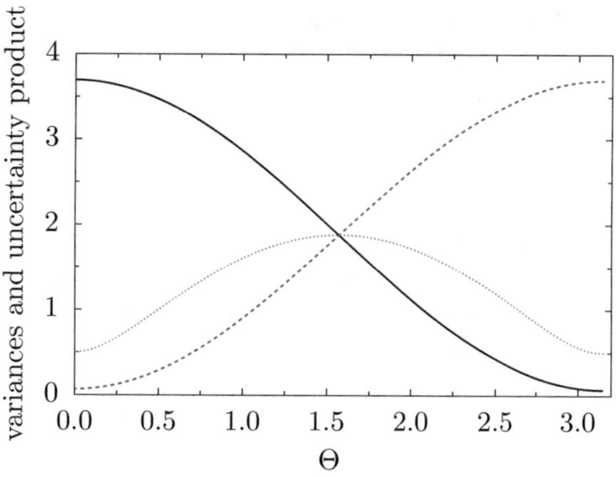

Fig. 6.3 Variances of position (*solid black line*) momentum (*dashed blue line*) and uncertainty product (*dotted red line*) for a harmonic oscillator with initial inverse width parameter $\alpha_0 = a\mathrm{e}^{-2}$ as a function of $\Theta = 2\omega_\mathrm{e} t$ (we have set $m = \hbar = \omega_\mathrm{e} = 1$)

$$
\begin{aligned}
\langle \psi'_m | \psi_n \rangle &= \frac{\sqrt{2}}{L} \int_0^L \mathrm{d}x \sin\left(\frac{m\pi x}{2L}\right) \sin\left(\frac{n\pi x}{L}\right) \\
&= \frac{\sqrt{2}}{\pi} \int_0^\pi \mathrm{d}z \sin\left(\frac{m}{2}z\right) \sin(nz) \\
&= \frac{1}{\sqrt{2}\pi} \left\{ \frac{\sin\left(\frac{m}{2} - n\right)\pi}{\frac{m}{2} - n} - \frac{\sin\left(\frac{m}{2} + n\right)\pi}{\frac{m}{2} + n} \right\} .
\end{aligned}
\tag{6.65}
$$

The sought for probability is given by $P_m = |\langle \psi'_m | \psi_n \rangle|^2$. We can distinguish two cases:

(i) even $m \neq 2n$ leads to $P_m = 0$, for $m = 2n$, see (b)
(ii) for odd $m = 2k + 1$, we find that

$$
\sin\left(\frac{m}{2} - n\right)\pi = \sin\left(\frac{m}{2} + n\right)\pi = \pm 1 \tag{6.66}
$$

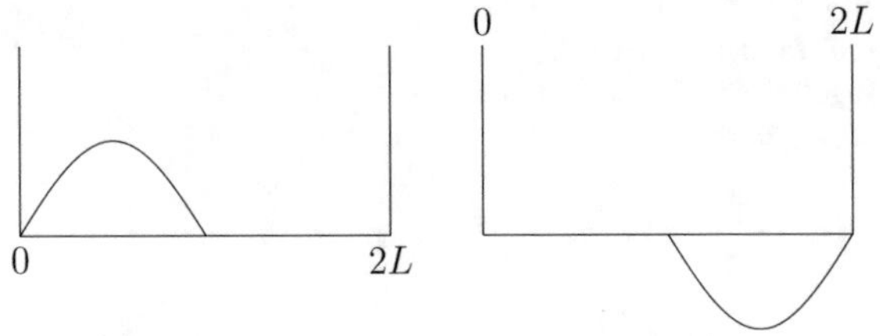

Fig. 6.4 Ground state of the small box potential (*left panel*); time-evolved state at $t_{\min}/2$ (*right panel*)

and $P_m = \frac{2n^2}{\pi^2\left[\frac{m}{2}^2 - n^2\right]^2}$

(b) From the second line of (6.65), we find directly that $P_m = \left|\frac{\sqrt{2}}{\pi}\frac{\pi}{2}\right| = \frac{1}{2}$.

(c) The $n = 1$ eigenfunction in the small well is a superposition of eigenfunctions of the large well. For the large well, the fundamental frequency, defined in (2.65) is

$$\omega' = \frac{\omega}{4} \tag{6.67}$$

and the corresponding time $t_{\min} = \frac{2\pi}{\omega'} = \frac{8\pi}{\omega}$. At this time all phases are integer multiples of 2π and the initial wavepacket has completely revived.

(d) Except for $m = 2$ only odd states of the large well contribute to the $n = 1$ state of the original well according to

$$\Psi(x,t) = \sum_m a_m \sin\left(\frac{m\pi x}{2L}\right) \mathrm{e}^{-\mathrm{i}m^2\omega' t} \,. \tag{6.68}$$

At $t_{\min}/2$ all the odd functions have acquired a minus sign, whereas the $m = 2$ function remains unchanged. The cancellation of the wavefunction at $t = 0$ in the right half of the well has thus turned into a cancellation in the left half as can be seen in Fig. 6.4 (see also [5]).

2.7. We set up the short-time propagator and use it to derive the time-dependent Schrödinger equation.

(a) Using (2.32) for the time-evolution operator, the infinitesimal short-time propagator is

$$\begin{aligned}
\langle x_{i+1}|\hat{U}(\Delta t)|x_i\rangle &= \langle x_{i+1}|\mathrm{e}^{-\frac{\mathrm{i}}{\hbar}\hat{H}\Delta t}|x_i\rangle \\
&\approx \langle x_{i+1}|x_i\rangle - \frac{\mathrm{i}}{\hbar}\Delta t\langle x_{i+1}|\hat{H}|x_i\rangle \\
&= \delta(x_{i+1}-x_i) - \frac{\mathrm{i}}{\hbar}\Delta t\int \frac{\mathrm{d}p}{2\pi\hbar}\mathrm{e}^{\frac{\mathrm{i}}{\hbar}p(x_{i+1}-x_i)}\left[\frac{p^2}{2m}+V(x_i)\right] \\
&\approx \int \frac{\mathrm{d}p}{2\pi\hbar}\mathrm{e}^{\frac{\mathrm{i}}{\hbar}[p(x_{i+1}-x_i)-\Delta t H(p,x_i)]}\,,
\end{aligned} \tag{6.69}$$

where in the second line, Taylor expansion of the exponential function has been used and in the third line, unity has been inserted in terms of momentum eigenstates. In the last line, the Fourier representation of the δ-function and re-exponentiation have been used. Alternatively, one could have used the Trotter product splitting of the time-evolution operator, which is also exact up to first order in Δt.
The p-integral is a Gaussian integral and can be done by using (1.32), leading to

$$\langle x_{i+1}|\hat{U}(\Delta t)|x_i\rangle = \sqrt{\frac{m}{2\pi \mathrm{i}\hbar\Delta t}}\mathrm{e}^{\frac{\mathrm{i}\Delta t}{\hbar}\left[\frac{m}{2}\left(\frac{x_{i+1}-x_i}{\Delta t}\right)^2-V(x_i)\right]} \tag{6.70}$$

and we have also derived the prefactor.
Applying this result in a concatenation of short-time propagators, one can derive the Riemannian form (2.75) of the path integral.

(b) Using the infinitesimal short-time propagator to propagate a wavefunction leads to

$$\Psi(x,t+\Delta t) = \int \mathrm{d}y A^{-1}\mathrm{e}^{\frac{\mathrm{i}\Delta t}{\hbar}\left[\frac{m(x-y)^2}{2\Delta t^2}-V(y)\right]}\Psi(y,t)\,, \tag{6.71}$$

where we have left the prefactor unspecified. Due to the smallness of Δt only y-values close to x contribute significantly to the integral (this is yet another stationary phase approximation type argument) and we substitute $y = x + \eta$ leading to

$$\Psi(x,t+\Delta t) = \int \mathrm{d}\eta A^{-1}\mathrm{e}^{\frac{im\eta^2}{2\hbar\Delta t}-\frac{\mathrm{i}}{\hbar}\Delta t V(x+\eta)}\Psi(x+\eta,t)\,. \tag{6.72}$$

Expanding to first order in Δt and to second order in η leads to

$$\Psi(x,t) + \Delta t\frac{\partial\Psi}{\partial t} =$$
$$\int \mathrm{d}\eta A^{-1}\mathrm{e}^{\frac{im\eta^2}{2\hbar\Delta t}}\left[1-\frac{\mathrm{i}\Delta t}{\hbar}V(x)\right]\left[\Psi(x,t)+\eta\frac{\partial\Psi}{\partial x}+\frac{\eta^2}{2}\frac{\partial^2\Psi}{\partial x^2}\right]. \tag{6.73}$$

Using the Gaussian integrals (1.29), (1.30), (1.31) and comparing coefficients in front of equal powers of Δt leads again to the prefactor

$$A^{-1} = \sqrt{\frac{m}{2\pi \mathrm{i}\hbar\Delta t}}\,. \tag{6.74}$$

Furthermore, from the coefficients of the first order terms

$$\frac{\partial\Psi(x,t)}{\partial t} = -\frac{\mathrm{i}}{\hbar}V(x)\Psi(x,t) - \frac{\hbar}{2im}\frac{\partial^2\Psi(x,t)}{\partial x^2} \tag{6.75}$$

follows, i.e., the TDSE is shown to hold. We closely followed the lines of Feynman and Hibbs [6]. Furthermore, we note that this procedure to derive the TDSE can easily be extended to the case of time-dependent potentials!

2.8. Taking the time derivative of $\hat{U}_I = \hat{U}_0^\dagger(t)\hat{U}(t)$ with the help of (6.31) and (6.36), as well as of $(\hat{A}\hat{B})^\dagger = \hat{B}^\dagger\hat{A}^\dagger$ leads to

$$\begin{aligned} i\hbar\dot{\hat{U}}_I(t) &= i\hbar\dot{\hat{U}}_0^\dagger(t)\hat{U}(t) + i\hbar\hat{U}_0^\dagger(t)\dot{\hat{U}}(t) \\ &= -(\hat{H}_0\hat{U}_0)^\dagger(t)\hat{U}(t) + \hat{U}_0^\dagger(t)\hat{H}\hat{U}(t) \\ &= -\hat{U}_0^\dagger(t)\hat{H}_0\hat{U}(t) + \hat{U}_0^\dagger(t)\hat{H}_0\hat{U}(t) + \hat{U}_0^\dagger(t)\hat{W}\hat{U}(t) \\ &= \hat{U}_0^\dagger(t)\hat{W}\hat{U}(t) = \hat{U}_0^\dagger(t)\hat{W}\hat{U}_0(t)\hat{U}_0^\dagger(t)\hat{U}(t) \\ &= \hat{W}_I\hat{U}_I(t), \end{aligned} \tag{6.76}$$

which is the expected TDSE of the time-evolution operator in the interaction picture.

2.9. From the time-evolution operator in the interaction picture (2.102), we can identify $\lambda = -\frac{i}{\hbar}$ and

$$A_1 \to \int_0^t dt_1\, \hat{W}_I(t_1), \tag{6.77}$$

$$\begin{aligned} A_2 &\to \int_0^t dt_2 \int_0^{t_2} dt_1\, \hat{W}_I(t_2)\hat{W}_I(t_1) \\ &= \int_0^t dt_2 \int_0^t dt_1 \Theta(t_2 - t_1)\hat{W}_I(t_2)\hat{W}_I(t_1). \end{aligned} \tag{6.78}$$

Therefore

$$\begin{aligned} A_2 - \frac{1}{2}A_1^2 = -\frac{1}{2}\int_0^t dt_2 \int_0^t dt_1 \Big[&\hat{W}_I(t_2)\hat{W}_I(t_1) \\ &- 2\Theta(t_2 - t_1)\hat{W}_I(t_2)\hat{W}_I(t_1)\Big]. \end{aligned} \tag{6.79}$$

If $t_2 \geq t_1$, the integrand is

$$\hat{W}_I(t_2)\hat{W}_I(t_1) - 2\hat{W}_I(t_2)\hat{W}_I(t_1) = -\hat{W}_I(t_2)\hat{W}_I(t_1). \tag{6.80}$$

For $t_2 \leq t_1$ the integral can be rewritten as

$$\begin{aligned} \int_0^t dt_2 \int_{t_2}^t dt_1\, \hat{W}_I(t_2)\hat{W}_I(t_1) &= \int_0^t dt_1 \int_0^{t_1} dt_2\, \hat{W}_I(t_2)\hat{W}_I(t_1) \\ &= \int_0^t dt_2 \int_0^{t_2} dt_1\, \hat{W}_I(t_1)\hat{W}_I(t_2), \end{aligned} \tag{6.81}$$

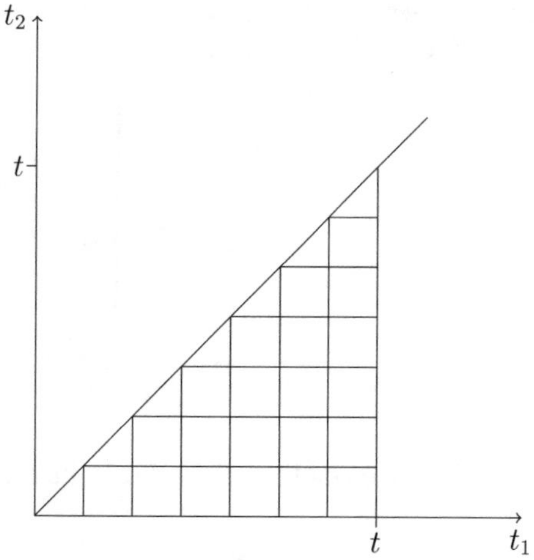

Fig. 6.5 Two-dimensional integration area: whether the integrals are performed in the order $\int_0^t dt_1 \int_0^{t_1} dt_2 \ldots$ (indicated by the *vertical lines*) or in the order $\int_0^t dt_2 \int_{t_2}^t dt_1 \ldots$ (*horizontal lines*) does not matter

where we have renamed time variables in the last step such that again $t_2 \geq t_1$. The correctness of the first step can be seen by looking at Fig. 6.5. This procedure then leads to

$$\hat{\mathrm{H}}_2 = 2[A_2 - \frac{1}{2}A_1^2] = \int_0^t dt_2 \int_0^{t_2} dt_1 [\hat{W}_\mathrm{I}(t_2), \hat{W}_\mathrm{I}(t_1)] \tag{6.82}$$

for the second order Magnus term. The same result could have been obtained by comparing terms proportional to $(\mathrm{i}/\hbar)^2$ in (2.102) and (2.109).

2.10. In this exercise, we study the Ehrenfest method.

(a) Inserting the expansion (2.133) into the TDSE for the light particle yields

$$\mathrm{i}\hbar \sum_k \left[\dot{c}_k(t)\phi_k + c_k(t)\dot{\phi}_k(x|X(t))\right] = \sum_k c_k(t)\hat{H}^0(x|X(t))\phi_k\,. \tag{6.83}$$

Using the eigenvalue equation of the light particle (2.125) and multiplying with $\int dx \phi_j(x|X(t))$ from the left leads to

$$\mathrm{i}\hbar\dot{c}_j(t) = \epsilon_j^0 c_j - \mathrm{i}\hbar \sum_k c_k \int dx \phi_j(x|X(t))\dot{\phi}_k(x|X(t))\,. \tag{6.84}$$

Using the chain rule $\dot{\phi} = \frac{\partial X}{\partial t}\frac{\partial \phi}{\partial X}$, we find

$$\mathrm{i}\hbar\dot{c}_j(t) = \epsilon_j^0 c_j - \mathrm{i}\hbar\dot{X} \sum_k c_k \int dx \phi_j(x|X(t))\frac{\partial \phi_k}{\partial X}\,. \tag{6.85}$$

(b) First we note that $d_{jk} = \int \mathrm{d}x \phi_j \frac{\partial \phi_k}{\partial X} = -d_{kj}$ follows from partial integration. The effective force is given by

$$
\begin{aligned}
F_{\text{eff}} &= -\frac{\partial}{\partial X}\left\{V + \int \mathrm{d}x \Phi^* \hat{H}^0 \Phi\right\} \\
&= -\frac{\partial}{\partial X}\left\{V + \sum_j |c_j|^2 \epsilon_j^0(X)\right\} \\
&= -\frac{\partial V}{\partial X} - \sum_j |c_j|^2 \frac{\partial \epsilon_j^0}{\partial X} - \frac{1}{\dot{X}} \sum_j \epsilon_j^0 \frac{\partial |c_j|^2}{\partial t} .
\end{aligned}
\tag{6.86}
$$

Using the result from (a), the last term in the equation above can be written as

$$
\begin{aligned}
-\frac{1}{\dot{X}} \sum_j \epsilon_j^0 \frac{\partial |c_j|^2}{\partial t} &= \sum_{jk} [c_j^* c_k + c_k^* c_j] \epsilon_j^0 d_{jk} \\
&= \sum_{j,k<j} [c_j^* c_k + c_k^* c_j][\epsilon_j^0 - \epsilon_k^0] d_{jk} ,
\end{aligned}
\tag{6.87}
$$

where for the last line $d_{jk} = -d_{kj}$ has been used.

2.11. Using the extended Hamiltonian, the Floquet Schrödinger equation and its adjunct read

$$
\hat{\mathcal{H}}|\psi_\alpha\rangle = \epsilon_\alpha |\psi_\alpha\rangle, \tag{6.88}
$$

$$
\langle\psi_\beta|\hat{\mathcal{H}}^\dagger = \epsilon_\beta^* \langle\psi_\beta|. \tag{6.89}
$$

Multiplying the first equation from the left with $\langle\psi_\beta|$, integrating over one period $\frac{1}{T}\int \mathrm{d}t \ldots$, and using an analogous procedure for the second equation leads to

$$
\langle\langle\psi_\beta \hat{\mathcal{H}}|\psi_\alpha\rangle\rangle = \epsilon_\alpha \langle\langle\psi_\beta|\psi_\alpha\rangle\rangle, \tag{6.90}
$$

$$
\langle\langle\psi_\beta|\hat{\mathcal{H}}^\dagger|\psi_\alpha\rangle\rangle = \epsilon_\beta^* \langle\langle\psi_\beta|\psi_\alpha\rangle\rangle. \tag{6.91}
$$

The claim is that if $\hat{H}$ is Hermitian then also $\hat{\mathcal{H}}$ is Hermitian. For the time-differentiation part of $\hat{\mathcal{H}}$ hermiticity can be shown by an integration by parts and using the periodicity of the Floquet functions, leading to the cancellation of the boundary term.

Subtracting the second from the first equation above then yields

$$
0 = (\epsilon_\alpha - \epsilon_\beta^*) \langle\langle\psi_\beta|\psi_\alpha\rangle\rangle , \tag{6.92}
$$

which for $\alpha = \beta$ leads to $\epsilon_\alpha = \epsilon_\alpha^*$.

2.12. With $\hat{y} \equiv -\mathrm{i}\hat{V}\Delta t/\hbar$ and $\hat{x} \equiv -\mathrm{i}\hat{T}_{\mathrm{k}}\Delta t/\hbar$, "the symmetric" Strang splitting can be written as

$$\begin{aligned}
\mathrm{e}^{\hat{y}/2+\hat{x}+\hat{y}/2} &= \mathrm{e}^{\hat{y}/2}\mathrm{e}^{\hat{x}+\hat{y}/2}\mathrm{e}^{-\frac{1}{2}[\hat{y}/2,\hat{x}]}\mathrm{e}^{O(\Delta t^3)} \\
&\approx \mathrm{e}^{\hat{y}/2}\mathrm{e}^{\hat{x}}\mathrm{e}^{\hat{y}/2}\mathrm{e}^{-\frac{1}{2}[\hat{x},\hat{y}/2]}\mathrm{e}^{O(\Delta t^3)}\mathrm{e}^{-\frac{1}{2}[\hat{y}/2,\hat{x}]} \\
&\approx \mathrm{e}^{\hat{y}/2}\mathrm{e}^{\hat{x}}\mathrm{e}^{\hat{y}/2}\mathrm{e}^{-\frac{1}{2}([\hat{x},\hat{y}/2]+[\hat{y}/2,\hat{x}])+O(\Delta t^4)} \\
&\approx \mathrm{e}^{\hat{y}/2}\mathrm{e}^{\hat{x}}\mathrm{e}^{\hat{y}/2}\,,
\end{aligned} \tag{6.93}$$

where the Zassenhaus formula has been used in the first two lines and the Baker-Campbell-Haussdorff formula has been used in the third line. In the second and third line, terms of order Δt^3 and in the fourth line terms of order Δt^4 have been neglected.

The Strang splitting is therefore second order accurate in Δt. This result could have also been proven by using Taylor expansion of the exponential functions (strictly keeping track of the operator ordering), see also the solution to Exercise 2.14.

2.13. The SOD propagation and its adjunct can be written as

$$|\Psi(t+\Delta t)\rangle = |\Psi(t-\Delta t)\rangle - \frac{2\mathrm{i}\Delta t}{\hbar}\hat{H}|\Psi(t)\rangle, \tag{6.94}$$

$$\langle\Psi(t+\Delta t)| = \langle\Psi(t-\Delta t)| + \frac{2\mathrm{i}\Delta t}{\hbar}\langle\Psi(t)|\hat{H}^{\dagger}. \tag{6.95}$$

(a) Using hermiticity of the Hamiltonian and multiplying the first equation above by $\langle\Psi(t)|$ from the left and the second one by $|\Psi(t)\rangle$ from the right gives

$$\langle\Psi(t)|\Psi(t+\Delta t)\rangle = \langle\Psi(t)|\Psi(t-\Delta t)\rangle - \frac{2\mathrm{i}\Delta t}{\hbar}\langle\Psi(t)|\hat{H}|\Psi(t)\rangle, \tag{6.96}$$

$$\langle\Psi(t+\Delta t)|\Psi(t)\rangle = \langle\Psi(t-\Delta t)|\Psi(t)\rangle + \frac{2\mathrm{i}\Delta t}{\hbar}\langle\Psi(t)|\hat{H}|\Psi(t)\rangle. \tag{6.97}$$

Adding both equations above leads to $\mathrm{Re}\langle\Psi(t-\Delta t)|\Psi(t)\rangle = \mathrm{Re}\langle\Psi(t)|\Psi(t+\Delta t)\rangle = \text{const.}$

(b) Analogously, multiplying the first equation above by $\langle\Psi(t)|\hat{H}$ from the left and the second one by $\hat{H}|\Psi(t)\rangle$ from the right leads to $\mathrm{Re}\langle\Psi(t-\Delta t)|\hat{H}|\Psi(t)\rangle = \mathrm{Re}\langle\Psi(t)|\hat{H}|\Psi(t+\Delta t)\rangle = \text{const.}$

(c) In the limit of small $\Delta t \to 0$, cum grano salis, we write the result from (a) as $\langle\Psi(t-\Delta t/2)|\Psi(t+\Delta t/2)\rangle = \langle\Psi(t+\Delta t/2)|\Psi(t+\Delta t/2)\rangle$, which is norm conservation, whereas (b) is energy conservation (if $\hat{H}$ is time-independent), although for finite Δt energy and norm conservation are not exact [7].

(d) An eigenstate evolves according to

$$|\Psi(t+\Delta t)\rangle = \mathrm{e}^{-\mathrm{i}E_{\mathrm{app}}\Delta t/\hbar}|\Psi(t)\rangle, \tag{6.98}$$

with some approximate eigenvalue E_{app}, due to the fact that (6.94) is invariant under discrete time translations $t \to t \pm n\Delta t$. Also from (6.94), we have

$$|\Psi(t+\Delta t)\rangle = |\Psi(t)\rangle e^{iE_{\rm app}\Delta t/\hbar} - 2iE\Delta t/\hbar|\Psi(t)\rangle \,. \tag{6.99}$$

Equating the LHSs of (6.98), (6.99) and eliminating $|\Psi(t)\rangle$ leads to

$$\sin(E_{\rm app}\Delta t/\hbar) = E\Delta t/\hbar \,, \tag{6.100}$$

from which it is clear that $|E\Delta t| < \hbar$ has to hold for stability, in accord with the statement in Chap. 2.

2.14. The expansion of the LHS of $e^{-\Delta t\hat{H}} = e^{-\Delta t\hat{T}_k}e^{-\Delta t\hat{V}}$ gives

$$e^{-\Delta t\hat{H}} = 1 - \Delta t\hat{H} + \frac{\Delta t^2}{2}\hat{H}^2 + \ldots \tag{6.101}$$

whereas on the RHS we have

$$e^{-\Delta t\hat{T}_k} = 1 - \Delta t\hat{T}_k + \frac{\Delta t^2}{2}\hat{T}_k^2 + \ldots, \tag{6.102}$$

$$e^{-\Delta t\hat{V}} = 1 - \Delta t\hat{V} + \frac{\Delta t^2}{2}\hat{V}^2 + \ldots. \tag{6.103}$$

Multiplying out the RHS gives up to second order

$$e^{-\Delta t\hat{T}_k}e^{-\Delta t\hat{V}} \approx 1 - \Delta t(\hat{T}_k + \hat{V}) + \frac{\Delta t^2}{2}(\hat{T}_k^2 + \hat{V}^2 + 2\hat{T}_k\hat{V}), \tag{6.104}$$

whereas the LHS is

$$e^{-\Delta t\hat{H}} \approx 1 - \Delta t(\hat{T}_k + \hat{V}) + \frac{\Delta t^2}{2}(\hat{T}_k + \hat{V})^2. \tag{6.105}$$

Now

$$(\hat{T}_k + \hat{V})^2 = \hat{T}_k^2 + \hat{V}^2 + \hat{T}_k\hat{V} + \hat{V}\hat{T}_k \,. \tag{6.106}$$

For the last term we employ the Jacobi identity $\{A,\{B,C\}\} + \{B,\{C,A\}\} + \{C,\{A,B\}\} = 0$ and find (for a single DOF)

$$\begin{aligned} \hat{V}\hat{T}_k\boldsymbol{\eta} = \{V,\{T_k,\boldsymbol{\eta}\}\} &= -\{\boldsymbol{\eta},\{V,T_k\}\} - \{T_k,\{\boldsymbol{\eta},V\}\} \\ &= \{\{V,T_k\},\boldsymbol{\eta}\} + \{T_k,\{V,\boldsymbol{\eta}\}\} \\ &= \left\{\frac{\partial V}{\partial q}\frac{p}{m},\boldsymbol{\eta}\right\} + \hat{T}_k\hat{V}\boldsymbol{\eta} \end{aligned} \tag{6.107}$$

and thus the difference between the LHS and the RHS in order Δt^2 is

$$\frac{\Delta t^2}{2}\left(\widehat{\frac{\partial V}{\partial q}\frac{p}{m}}\right), \tag{6.108}$$

where the hat again denotes the Poisson bracket operation.

2.15. Starting from (2.222) for $N = 1$ and using

$$\left(\hat{a}^\dagger\right)^n|0\rangle = \sqrt{n!}|n\rangle \tag{6.109}$$

leads to

$$|z\rangle = \mathrm{e}^{-1/2|z|^2}\sum_{n=0}^{\infty}\frac{z^n}{\sqrt{n!}}|n\rangle\ . \tag{6.110}$$

The completeness integral (2.226) is

$$\int\frac{\mathrm{d}^2z}{\pi}|z\rangle\langle z| = \sum_{n=0}^{\infty}\sum_{m=0}^{\infty}\frac{|n\rangle\langle m|}{\pi\sqrt{n!m!}}\int\mathrm{d}^2z\,\mathrm{e}^{-|z|^2}(z^*)^m z^n\ . \tag{6.111}$$

Using the polar representation $z = r\mathrm{e}^{\mathrm{i}\varphi}$ leads to [8]

$$\begin{aligned}\int\frac{\mathrm{d}^2z}{\pi}|z\rangle\langle z| &= \sum_{n=0}^{\infty}\sum_{m=0}^{\infty}\frac{|n\rangle\langle m|}{\pi\sqrt{n!m!}}\int_0^\infty\mathrm{d}r\,r\mathrm{e}^{-r^2}r^{n+m}\int_0^{2\pi}\mathrm{d}\varphi\,\mathrm{e}^{\mathrm{i}(n-m)\varphi}\\ &= \sum_{n=0}^{\infty}\frac{|n\rangle\langle n|}{\pi n!}\int_0^\infty\mathrm{d}\xi\,\mathrm{e}^{-\xi}\xi^n\\ &= \sum_{n=0}^{\infty}|n\rangle\langle n| = \hat{1}\ ,\end{aligned} \tag{6.112}$$

where we have used

$$\int_0^{2\pi}\mathrm{d}\varphi\mathrm{e}^{\mathrm{i}(n-m)\varphi} = 2\pi\,\delta_{nm} \tag{6.113}$$

and the substitution $\xi = r^2$ and

$$\int_0^\infty\mathrm{d}\xi\mathrm{e}^{-\xi}\xi^n = n!, \tag{6.114}$$

as well the completeness of the harmonic oscillator eigenfunctions in the very last step.

2.16. Here we show that the stability matrix elements can be combined in such way that the result fulfills a Riccati equation.

(a) The time-dependent inverse width parameter

$$\gamma_t = \gamma \frac{m_{11} + \frac{1}{\mathrm{i}\gamma\hbar} m_{12}}{m_{22} + \mathrm{i}\gamma\hbar m_{21}} \tag{6.115}$$

of the TGWD can be differentiated leading to

$$\begin{aligned}\dot{\gamma}_t = \gamma\Big[&\big(\dot{m}_{11} + \frac{1}{\mathrm{i}\gamma\hbar}\dot{m}_{12}\big)(m_{22} + \mathrm{i}\gamma\hbar m_{21}) \\ &-\big(m_{11} + \frac{1}{\mathrm{i}\gamma_0\hbar} m_{12}\big)(\dot{m}_{22} + \mathrm{i}\gamma_0\hbar\dot{m}_{21})\Big] \\ &(m_{22} + \mathrm{i}\gamma_0\hbar m_{21})^{-2}\,.\end{aligned} \tag{6.116}$$

Writing out the equations of motion for the elements of the stability matrix (2.263) gives

$$\dot{m}_{1j} = -V'' m_{2j}, \tag{6.117}$$

$$\dot{m}_{2j} = \frac{1}{m} m_{1j}, \tag{6.118}$$

with $j = 1, 2$. Inserting them above leads to

$$\begin{aligned}\dot{\gamma}_t = \Big[&-V''\big(\gamma m_{21} + \frac{1}{\mathrm{i}\hbar} m_{22}\big)(m_{22} + \mathrm{i}\gamma\hbar m_{21}) \\ &-\frac{\gamma}{m}\big(m_{11} + \frac{1}{\mathrm{i}\gamma\hbar} m_{12}\big)(m_{12} + \mathrm{i}\gamma\hbar m_{11})\Big] \\ &(m_{22} + \mathrm{i}\gamma\hbar m_{21})^{-2} \\ = &-\frac{1}{\mathrm{i}\hbar} V'' - \frac{\mathrm{i}\hbar}{m}\gamma_t^2\,,\end{aligned} \tag{6.119}$$

where the definition (6.115) has been used. This is the sought for Riccati equation, analogous to the one given in (2.61) for $\alpha = \gamma/2$.

(b) With the definition $Q = m_{22} + \mathrm{i}\hbar\gamma m_{21}$ the inverse width parameter (6.115) can be rewritten in log-derivative form as

$$\gamma_t = \frac{m}{\mathrm{i}\hbar}\frac{\dot{Q}}{Q}, \tag{6.120}$$

which in turn can be solved for Q by

$$Q = \exp\left\{\int_0^t \mathrm{d}t' \frac{\mathrm{i}\hbar}{m}\gamma_{t'}\right\}. \tag{6.121}$$

The 1D version of the complex conjugate HK-prefactor (2.231) can thus be expressed entirely in terms of γ_t via

$$R^* = \sqrt{\frac{1}{2}(1+\gamma_t/\gamma)}\exp\left\{\frac{1}{2}\int_0^t \mathrm{d}t' \frac{\mathrm{i}\hbar}{m}\gamma_t\right\}$$
$$= \sqrt{\frac{1}{2}(Q+\frac{m}{\mathrm{i}\hbar}\dot{Q})}$$
$$= \sqrt{\frac{1}{2}\left(m_{11}+m_{22}+\mathrm{i}\hbar\gamma m_{21}+\frac{1}{\mathrm{i}\hbar\gamma}m_{12}\right)}. \quad (6.122)$$

With slight generalizations, these calculations can be taken over to the case of many degrees of freedom, involving the block matrices $\mathbf{m}_{ij}$ defined in App. 2.C [9]. The advantage of the reformulation of the prefactor is that it may allow for favorable approximations [10].

6.3 Solutions to Problems in Chap. 3

3.1. This exercise is devoted to the study of classical minimal coupling.

(a) The Euler-Lagrange (vector) equation for the Lagrangian

$$L(\dot{\boldsymbol{r}},\boldsymbol{r},t) = \frac{m}{2}\dot{\boldsymbol{r}}^2 - q\Phi(\boldsymbol{r},t) + q\dot{\boldsymbol{r}}\cdot\boldsymbol{A}(\boldsymbol{r},t) \quad (6.123)$$

reads

$$\frac{\mathrm{d}}{\mathrm{d}t}\frac{\partial L}{\partial\dot{\boldsymbol{r}}} = \frac{\partial L}{\partial\boldsymbol{r}}\,. \quad (6.124)$$

With the total time-derivative

$$\frac{\mathrm{d}}{\mathrm{d}t}\cdots = \frac{\partial}{\partial t}\cdots + (\dot{\boldsymbol{r}}\cdot\frac{\partial}{\partial\boldsymbol{r}})\ldots, \quad (6.125)$$

we get

$$m\ddot{\boldsymbol{r}} + q\frac{\partial\boldsymbol{A}}{\partial t} + q(\dot{\boldsymbol{r}}\cdot\frac{\partial}{\partial\boldsymbol{r}})\boldsymbol{A} = -q\frac{\partial\Phi}{\partial\boldsymbol{r}} + q\frac{\partial}{\partial\boldsymbol{r}}(\dot{\boldsymbol{r}}\cdot\boldsymbol{A})\,. \quad (6.126)$$

Using the vector algebraic rule $\boldsymbol{a}\times(\boldsymbol{b}\times\boldsymbol{c}) = \boldsymbol{b}(\boldsymbol{a}\cdot\boldsymbol{c}) - \boldsymbol{c}(\boldsymbol{a}\cdot\boldsymbol{b})$ this leads to

$$m\ddot{\boldsymbol{r}} = -q\left(\frac{\partial\Phi}{\partial\boldsymbol{r}} + \frac{\partial\boldsymbol{A}}{\partial t}\right) + q\dot{\boldsymbol{r}}\times\left(\frac{\partial}{\partial\boldsymbol{r}}\times\boldsymbol{A}\right)$$
$$= q(\mathcal{E} + \dot{\boldsymbol{r}}\times\mathcal{B})\,, \quad (6.127)$$

where in the last line use has been made of the expressions (3.10) and (3.11) for the fields. This is Newton's equation of motion with the Lorentz force!

(b) The canonical momentum is defined by

$$\boldsymbol{p} = \frac{\partial L}{\partial \dot{\boldsymbol{r}}} = m\dot{\boldsymbol{r}} + q\boldsymbol{A}, \tag{6.128}$$

whereas the mechanical (or kinetic) momentum is given by

$$\boldsymbol{p}_{\mathrm{m}} = m\dot{\boldsymbol{r}} = \boldsymbol{p} - q\boldsymbol{A}. \tag{6.129}$$

(c) To calculate the Hamiltonian we have to replace $\dot{\boldsymbol{r}}$ by $(\boldsymbol{p} - q\boldsymbol{A})/m$ in the Legendre transform according to

$$\begin{aligned} H(\boldsymbol{p}, \boldsymbol{r}, t) &= \dot{\boldsymbol{r}} \cdot \boldsymbol{p} - L(\dot{\boldsymbol{r}}, \boldsymbol{r}, t) \\ &= \frac{\boldsymbol{p} - q\boldsymbol{A}}{m} \cdot \boldsymbol{p} - \frac{m}{2}\left[\frac{\boldsymbol{p} - q\boldsymbol{A}}{m}\right]^2 + q\Phi - q\frac{\boldsymbol{p} - q\boldsymbol{A}}{m} \cdot \boldsymbol{A} \\ &= \frac{1}{2m}(\boldsymbol{p} - q\boldsymbol{A}(\boldsymbol{r}, t))^2 + q\Phi(\boldsymbol{r}, t), \end{aligned} \tag{6.130}$$

where the last line follows by first combining the first with the fourth term of the previous line and subtracting the second term from the result.

3.2. In the case of the minimally coupled Hamiltonian in the TDSE (3.7), for the time-derivative of the probability density we find

$$\begin{aligned} \frac{\partial \rho}{\partial t} &= \frac{\partial \Psi^*}{\partial t}\Psi + \Psi^*\frac{\partial \Psi}{\partial t} \\ &= \frac{\mathrm{i}}{\hbar}\{\Psi(\hat{H}\Psi)^* - \Psi^*(\hat{H}\Psi)\} \\ &= \frac{\mathrm{i}}{\hbar}\left\{\frac{\hbar^2}{2m}[\Psi^*\Delta\Psi - \Psi\Delta\Psi^*]\right. \\ &\quad \left. + \frac{\hbar q}{m\mathrm{i}}[\Psi^*\Psi(\nabla \cdot \boldsymbol{A}) + \Psi^*\boldsymbol{A} \cdot \nabla\Psi + \Psi\boldsymbol{A} \cdot \nabla\Psi^*]\right\} \\ &= -\frac{\partial}{\partial \boldsymbol{r}} \cdot \left\{\frac{1}{2m}\left[\Psi^*\frac{\hbar}{\mathrm{i}}\nabla\Psi - \Psi\frac{\hbar}{\mathrm{i}}\nabla\Psi^*\right] - \frac{q}{m}\boldsymbol{A}\Psi^*\Psi\right\} \\ &= -\nabla \cdot \boldsymbol{j}, \end{aligned} \tag{6.131}$$

from which we can conclude that

$$\boldsymbol{j} = \frac{1}{m}\mathrm{Re}\left\{\Psi^*\hat{\boldsymbol{p}}_{\mathrm{m}}\Psi\right\} \tag{6.132}$$

with the kinetic momentum from (6.129). A local gauge transformation (3.1) applied to the probability current density according to

$$\begin{aligned}
\boldsymbol{j}' &= \frac{1}{m}\mathrm{Re}\left\{\Psi'^*(\hat{\boldsymbol{p}} - q\boldsymbol{A}')\Psi'\right\} \\
&= \frac{1}{m}\mathrm{Re}\left\{\mathrm{e}^{-\mathrm{i}\frac{q}{\hbar}\chi(\boldsymbol{r},t)}\Psi^*\mathrm{e}^{\mathrm{i}\frac{q}{\hbar}\chi(\boldsymbol{r},t)}(\hat{\boldsymbol{p}} - q\boldsymbol{A} - q(\nabla\chi) + q(\nabla\chi))\Psi\right\} \\
&= \frac{1}{m}\mathrm{Re}\left\{\Psi^*(\hat{\boldsymbol{p}} - q\boldsymbol{A})\Psi\right\} \\
&= \boldsymbol{j}
\end{aligned} \tag{6.133}$$

leaves it unchanged.

3.3. In this exercise, we follow [11] on the search for a gauge invariant energy operator.

(a) For the expectation value to be gauge invariant

$$\langle\Psi|\hat{\Theta}(\boldsymbol{A},\Phi)|\Psi\rangle = \langle\Psi'|\hat{\Theta}(\boldsymbol{A}',\Phi')|\Psi'\rangle \tag{6.134}$$

has to hold. In addition, we find by inserting unity twice on the LHS

$$\langle\Psi|\hat{\Theta}(\boldsymbol{A},\Phi)|\Psi\rangle = \langle\Psi'|\mathrm{e}^{\mathrm{i}\frac{q}{\hbar}\chi}\hat{\Theta}(\boldsymbol{A},\Phi)\mathrm{e}^{-\mathrm{i}\frac{q}{\hbar}\chi}|\Psi'\rangle \tag{6.135}$$

and by comparison of the RHSs

$$\hat{\Theta}(\boldsymbol{A}',\Phi') = \mathrm{e}^{\mathrm{i}\frac{q}{\hbar}\chi}\hat{\Theta}(\boldsymbol{A},\Phi)\mathrm{e}^{-\mathrm{i}\frac{q}{\hbar}\chi} \tag{6.136}$$

(b) From

$$\hat{H}(\boldsymbol{A},\Phi) = \frac{1}{2m}(\hat{\boldsymbol{p}} - q\boldsymbol{A})^2 + V + q\Phi, \tag{6.137}$$

we find that

$$\hat{H}(\boldsymbol{A}',\Phi') = \frac{1}{2m}(\hat{\boldsymbol{p}} - q\boldsymbol{A}')^2 + V + q\Phi', \tag{6.138}$$

whereas

$$\mathrm{e}^{\mathrm{i}\frac{q}{\hbar}\chi}\hat{H}(\boldsymbol{A},\Phi)\mathrm{e}^{-\mathrm{i}\frac{q}{\hbar}\chi} = \frac{1}{2m}(\hat{\boldsymbol{p}} - q\boldsymbol{A}')^2 + V + q\Phi, \tag{6.139}$$

because only $\boldsymbol{A}$ is modified (the exponential factors around Φ cancel). An alternative, gauge invariant operator would be given by

$$\hat{H}(\boldsymbol{A},0) = \hat{H}(\boldsymbol{A},\Phi) - q\Phi = \frac{1}{2m}(\hat{\boldsymbol{p}} - q\boldsymbol{A})^2 + V. \tag{6.140}$$

3.4. We follow an idea laid out in the PhD thesis of Maria Göppert-Mayer, published in [12], to derive the length-gauge Lagrangian.

(a) Adding a total time derivative $\frac{\mathrm{d}}{\mathrm{d}t}f(\boldsymbol{r},t)$ to the Lagrangian leads to

$$L' = L + \frac{\mathrm{d}}{\mathrm{d}t}f(\boldsymbol{r},t) \tag{6.141}$$

with

$$\frac{\mathrm{d}}{\mathrm{d}t}f(\boldsymbol{r},t) = \dot{\boldsymbol{r}}\cdot\frac{\partial f}{\partial \boldsymbol{r}} + \frac{\partial f}{\partial t}. \tag{6.142}$$

The Euler-Lagrange equations

$$\begin{aligned}\frac{\mathrm{d}}{\mathrm{d}t}\frac{\partial L'}{\partial \dot{\boldsymbol{r}}} - \frac{\partial L'}{\partial \boldsymbol{r}} &= \frac{\mathrm{d}}{\mathrm{d}t}\frac{\partial L}{\partial \dot{\boldsymbol{r}}} + \frac{\mathrm{d}}{\mathrm{d}t}\frac{\partial}{\partial \dot{\boldsymbol{r}}}\left(\dot{\boldsymbol{r}}\cdot\frac{\partial f}{\partial \boldsymbol{r}} + \frac{\partial f}{\partial t}\right) - \frac{\partial L}{\partial \boldsymbol{r}} - \frac{\partial}{\partial \boldsymbol{r}}\frac{\mathrm{d}f}{\mathrm{d}t}\\ &= \frac{\mathrm{d}}{\mathrm{d}t}\frac{\partial L}{\partial \dot{\boldsymbol{r}}} + \frac{\mathrm{d}}{\mathrm{d}t}\frac{\partial f}{\partial \boldsymbol{r}} - \frac{\partial L}{\partial \boldsymbol{r}} - \frac{\partial}{\partial \boldsymbol{r}}\frac{\mathrm{d}f}{\mathrm{d}t}\\ &= \frac{\mathrm{d}}{\mathrm{d}t}\frac{\partial L}{\partial \dot{\boldsymbol{r}}} - \frac{\partial L}{\partial \boldsymbol{r}} = 0\end{aligned} \tag{6.143}$$

are unaffected by the additional term.
This fact can be explained alternatively by observing that the total time derivative only adds boundary terms to the action S, which are not varied in the derivation of the Euler-Lagrange equations.

(b) For the (velocity-gauge) Lagrangian of (6.123) in dipole approximation and Coulomb gauge, adding $-q\frac{\mathrm{d}}{\mathrm{d}t}(\boldsymbol{r}\cdot\boldsymbol{A})$ leads to

$$\begin{aligned}L_1 &= \frac{m}{2}\dot{\boldsymbol{r}}^2 + q\dot{\boldsymbol{r}}\cdot\boldsymbol{A}(t) - q\frac{\mathrm{d}}{\mathrm{d}t}\left(\boldsymbol{r}\cdot\boldsymbol{A}(t)\right)\\ &= \frac{m}{2}\dot{\boldsymbol{r}}^2 - q\boldsymbol{r}\cdot\frac{\mathrm{d}}{\mathrm{d}t}\boldsymbol{A}(t)\\ &= \frac{m}{2}\dot{\boldsymbol{r}}^2 + q\boldsymbol{r}\cdot\boldsymbol{\mathcal{E}}(t),\end{aligned} \tag{6.144}$$

where

$$\frac{\mathrm{d}}{\mathrm{d}t}\boldsymbol{A}(t) = \frac{\partial}{\partial t}\boldsymbol{A}(t) = -\boldsymbol{\mathcal{E}}(t) \tag{6.145}$$

has been used.

3.5. With the two unitary transformations (3.19), (3.20) and in dipole approximation, we get

$$\mathrm{i}\hbar\dot{\hat{U}}_1^{-1} = \frac{q^2}{2m}\boldsymbol{A}^2(t)\hat{U}_1^{-1}, \tag{6.146}$$

$$\mathrm{i}\hbar\dot{\hat{U}}_2^{-1} = \frac{\mathrm{i}\hbar q}{m}\boldsymbol{A}(t)\cdot\nabla\hat{U}_2^{-1}. \tag{6.147}$$

Under the assumptions in the main text, the TDSE for the velocity gauge wavefunction $\Psi_{\rm v} = \hat{U}_1^{-1}\hat{U}_2^{-1}\Psi_{\rm a}$ is

$$\begin{aligned}
\mathrm{i}\hbar\dot{\Psi}_{\rm v} &= \mathrm{i}\hbar[\dot{\hat{U}}_1^{-1}\hat{U}_2^{-1}\Psi_{\rm a} + \hat{U}_1^{-1}\dot{\hat{U}}_2^{-1}\Psi_{\rm a} + \hat{U}_1^{-1}\hat{U}_2^{-1}\dot{\Psi}_{\rm a}]\\
&= \frac{q^2}{2m}\boldsymbol{A}^2(t)\hat{U}_1^{-1}\hat{U}_2^{-1}\Psi_{\rm a} + \frac{\mathrm{i}\hbar q}{m}\boldsymbol{A}(t)\cdot\nabla\hat{U}_1^{-1}\hat{U}_2^{-1}\Psi_{\rm a} + \mathrm{i}\hbar\hat{U}_1^{-1}\hat{U}_2^{-1}\dot{\Psi}_{\rm a}\\
&= \left[-\frac{\hbar^2}{2m}\Delta + \frac{\mathrm{i}\hbar q}{m}\boldsymbol{A}\cdot\nabla + \frac{q^2}{2m}\boldsymbol{A}^2 + V\right]\hat{U}_1^{-1}\hat{U}_2^{-1}\Psi_{\rm a}.
\end{aligned} \tag{6.148}$$

The terms containing the vector potential in the last line are canceled by the first and second term in the line above. After multiplication from the left with $\hat{U}_2\hat{U}_1$ this is leading to

$$\mathrm{i}\hbar\dot{\Psi}_{\rm a}(\boldsymbol{r},t) = \left[-\frac{\hbar^2}{2m}\Delta + \hat{U}_2V(\boldsymbol{r})\hat{U}_2^{-1}\right]\Psi_{\rm a}(\boldsymbol{r},t). \tag{6.149}$$

With the help of the Baker-Haussdorff lemma, and the abbreviation $\boldsymbol{\alpha}(t)$, given in (3.22), the unitary transformation of the potential is

$$\begin{aligned}
\hat{U}_2V(\boldsymbol{r})\hat{U}_2^{-1} &= \sum_{n=0}^{\infty}\frac{1}{n!}[\boldsymbol{\alpha}\cdot\nabla, V(\boldsymbol{r})]_n\\
&= \sum_{n=0}^{\infty}\frac{1}{n!}(\boldsymbol{\alpha}\cdot\nabla)^nV(\boldsymbol{r}) = V(\boldsymbol{r}+\boldsymbol{\alpha}),
\end{aligned} \tag{6.150}$$

where the last step is due to the fact that the series is a Taylor expansion.

3.6. We study the "free" electron in 1D under the influence of a laser field in length gauge.

(a) Hamilton's equations are

$$\dot{q}_t = p_t/m_{\rm e}, \tag{6.151}$$

$$\dot{p}_t = -e\mathcal{E}_0\cos(\omega t). \tag{6.152}$$

Under the given initial conditions they lead to the solutions

$$q_t = q_0 + \frac{e\mathcal{E}_0}{m_{\rm e}\omega^2}[\cos(\omega t) - 1], \tag{6.153}$$

$$p_t = -\frac{e\mathcal{E}_0}{\omega}\sin(\omega t), \tag{6.154}$$

which are special cases of (3.26) and (3.27).

The classical kinetic energy[1] is

$$T_{\mathrm{k}} = \frac{p_t^2}{2m_{\mathrm{e}}} = \frac{e^2\mathcal{E}_0^2}{2m_{\mathrm{e}}\omega^2}\sin^2(\omega t) = 2U_{\mathrm{p}}\sin^2(\omega t), \tag{6.155}$$

with the ponderomotive energy from (3.28). Averaging over one period of the external field yields

$$\frac{1}{T}\int_0^T \mathrm{d}t \frac{p_t^2}{2m_{\mathrm{e}}} = \frac{e^2\mathcal{E}_0^2}{2m_{\mathrm{e}}\omega^2 T}\int_0^T \mathrm{d}t \sin^2(\omega t) = U_{\mathrm{p}}\,. \tag{6.156}$$

The derivative of the kinetic energy is

$$\frac{\mathrm{d}}{\mathrm{d}t}\frac{p_t^2}{2m_{\mathrm{e}}} = \frac{p_t \dot{p}_t}{m_{\mathrm{e}}} = \frac{p_t F}{m_{\mathrm{e}}}, \tag{6.157}$$

with the force F (the RHS of (6.152)). The average over one period of this quantity vanishes (due to symmetry reasons), as can be seen by inserting (6.154) and (6.152). This is another confirmation of the fact that the kinetic energy oscillates between 0 and $2U_{\mathrm{p}}$ (see Fig. 6.6).

(b) Because $V'' = 0$, we can use (6.46), leading to

$$\alpha_t = \alpha_0/(1 + 2\mathrm{i}\hbar\alpha_0 t/m_{\mathrm{e}}) \tag{6.158}$$

for the inverse width parameter.

(c) The classical action is

$$\begin{aligned}\int_0^t \mathrm{d}t' L &= \int_0^t \mathrm{d}t'(T-V) \overset{\text{Taylor}}{=} \int_0^t \mathrm{d}t'\left[\frac{p_{t'}^2}{2m_{\mathrm{e}}} - V' q_{t'}\right]\\ &= \int_0^t \mathrm{d}t'\left[\frac{p_{t'}^2}{2m_{\mathrm{e}}} + \dot{p}_{t'} q_{t'}\right]\\ &= -\int_0^t \mathrm{d}t' \frac{p_{t'}^2}{2m_{\mathrm{e}}} + p_{t'} q_{t'}|_0^t\,. \end{aligned} \tag{6.159}$$

In the last step, integration by parts has been used. Using the initial conditions, the phase therefore is given by

$$\delta_t = -\int_0^t \mathrm{d}t' \frac{p_{t'}^2}{2m_{\mathrm{e}}} + p_t q_t + \frac{\mathrm{i}\hbar}{2}\ln\{1 + \frac{2\mathrm{i}\hbar}{m_{\mathrm{e}}}\alpha_0 t\}. \tag{6.160}$$

Inserting everything in the GWD expression (2.48)

[1] Note that the quantum expectation value $\langle\Psi(t)|\frac{\hat{p}^2}{2m_{\mathrm{e}}}|\Psi(t)\rangle$ has an additional contribution to the kinetic energy, due to the time-dependent width factor, which is unaffected by the external field.

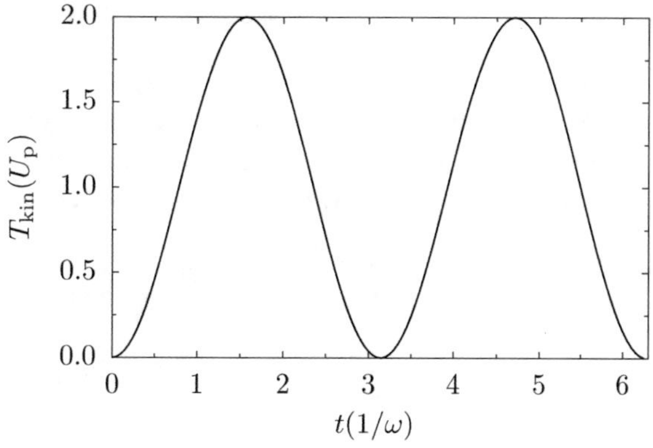

Fig. 6.6 Classical kinetic energy in units of the ponderomotive energy as a function of time

$$\Psi(x,t)=\left(\frac{2\alpha_0}{\pi}\right)^{1/4}\exp\left\{-\alpha_t(x-q_t)^2+\frac{\mathrm{i}}{\hbar}p_t(x-q_t)+\frac{\mathrm{i}}{\hbar}\delta_t\right\} \quad (6.161)$$

and using (6.155) together with the integral

$$\int \mathrm{d}x\,\sin^2(ax)=\frac{1}{2}x-\frac{1}{4a}\sin(2ax), \quad (6.162)$$

after replacing $2\alpha_0$ by γ, leads to

$$\Psi(x,t)=\left(\frac{\gamma}{\pi}\right)^{1/4}\sqrt{\frac{1}{1+\mathrm{i}\gamma\hbar t/m_{\rm e}}}\exp\left\{\frac{\mathrm{i}}{\hbar}\left[\frac{U_{\rm p}}{2\omega}\sin(2\omega t)-U_{\rm p}t+xp(t)\right]\right\}$$
$$\exp\left\{-\frac{\gamma}{2(1+\mathrm{i}\gamma\hbar t/m_{\rm e})}[x-q(t)]^2\right\}, \quad (6.163)$$

given in (3.25). This is an exact solution of the time-dependent Schrödinger equation due to the fact that the potential is linear!

3.7. In the first part of the exercise, the RWA is motivated by an averaging procedure. In the second part, the constants in the solution of the simplified equations are determined from the initial conditions.

(a) Integrating the differential equations (3.62) and (3.63)

$$\mathrm{i}\dot{d}_1=d_2\frac{\boldsymbol{\mu}_{12}\cdot\boldsymbol{\mathcal{E}}_0}{2\hbar}\left\{\exp[\mathrm{i}(\omega-\omega_{21})t]+\exp[-\mathrm{i}(\omega+\omega_{21})t]\right\}, \quad (6.164)$$

$$\mathrm{i}\dot{d}_2=d_1\frac{\boldsymbol{\mu}_{21}\cdot\boldsymbol{\mathcal{E}}_0}{2\hbar}\left\{\exp[-\mathrm{i}(\omega-\omega_{21})t]+\exp[\mathrm{i}(\omega+\omega_{21})t]\right\} \quad (6.165)$$

over a time Δt that fulfills

$$1/\Delta_d \gg \Delta t \gg 1/(\omega + \omega_{21}), \tag{6.166}$$

with $\Delta_d = \omega - \omega_{21}$ and assuming that the coefficients d_1 and d_2 can be taken out of the integral (because they oscillate slowly, see next exercise), the second exponential term averages to zero. This argument does not hold for the first exponential term, which therefore survives, leading to

$$\frac{i}{\Delta t}\int_0^{\Delta t} dt\, \dot{d}_1 = d_2 \frac{\boldsymbol{\mu}_{12}\cdot\boldsymbol{\mathcal{E}}_0}{2\hbar} \exp[i\Delta_d t] \frac{1}{\Delta t}\int_0^{\Delta t} dt, \tag{6.167}$$

$$\frac{i}{\Delta t}\int_0^{\Delta t} dt\, \dot{d}_2 = d_1 \frac{\boldsymbol{\mu}_{21}\cdot\boldsymbol{\mathcal{E}}_0}{2\hbar} \exp[-i\Delta_d t] \frac{1}{\Delta t}\int_0^{\Delta t} dt, \tag{6.168}$$

where we also took out the slowly oscillating exponent from the integration. The left hand sides can be evaluated as a finite difference, which in turn is written as a differentiation, leading to (3.65) and (3.66), i.e., (6.164) and (6.164) without the counter-rotating terms.

(b) For the given initial conditions, the constants in (3.67) and (3.68) are $D = -C = \mu\mathcal{E}_0/(2\hbar\Omega_R)$ yielding

$$d_1(t) = \exp[i\Delta_d t/2]\left[\cos(\Omega_R t/2) - i\frac{\Delta_d}{\Omega_R}\sin(\Omega_R t/2)\right], \tag{6.169}$$

$$d_2(t) = -i\exp[-i\Delta_d t/2]\frac{\mu\mathcal{E}_0}{\hbar\Omega_R}\sin(\Omega_R t/2) \tag{6.170}$$

and thus

$$|d_2(t)|^2 = \left(\frac{\mu\mathcal{E}_0}{\hbar\Omega_R}\right)^2 \sin^2(\Omega_R t/2), \tag{6.171}$$

having the same analytical form as in the constant perturbation case, (3.53), and leading to the same behavior as in Fig. 3.2, just with a slightly different definition of the Rabi frequency (a factor of two is missing).

3.8. A dipole selection rule is established and the conditions for applicability of the RWA are checked for a specific example.

(a) The parity operator $\hat{\mathcal{P}}$ replaces x by $-x$ and is a unitary and Hermitian operator [13] and therefore has real eigenvalues λ. Due to $\hat{\mathcal{P}}^2 = \hat{1}$, for the square of the eigenvalues we get $\lambda^2 = 1$ with the solutions $\lambda = \pm 1$. For a symmetric potential, $V(x) = V(-x)$, the Hamiltonian and the parity operator commute and the two operators have a common system of eigenfunctions, i.e., the eigenfunctions ψ_n of the Hamiltonian are either symmetric or antisymmetric. As the lowest eigenfunction is node-less and indexed by $n = 0$, the even states have even parity and the odd states have odd parity.
For a symmetric potential, a diagonal element of the dipole operator is given by

Table 6.1 (Resonant) Rabi frequencies for different values of the electric field

$\mathcal{E}_0$(V/m)	1	10^6	10^{10}
$\Omega_{\rm R}$(1/s)	0.21×10^6	0.21×10^{12}	0.21×10^{16}

$$\mu_{nn} = e\int \mathrm{d}x|\psi_n|^2 x \tag{6.172}$$

and, due to the antisymmetric nature of the integrand, is always zero. In general, in order for the dipole matrix element not to vanish, for a symmetric potential, the participating eigenfunctions have to have different parity.

(b) The eigenfunctions of the harmonic oscillator are given in (2.157) and the sought for matrix element is (using (1.31))

$$\mu_{01} = \langle\psi_0|e\hat{x}|\psi_1\rangle = e\sigma\sqrt{2/\pi}\int \mathrm{d}x x^2\exp\{-\sigma^2x^2\} = e\sqrt{\frac{\hbar}{2m_{\rm e}\omega}}\,, \tag{6.173}$$

leading to a Rabi frequency of

$$\Omega_{\rm R} = \mu_{01}\mathcal{E}_0/\hbar = \mathcal{E}_0\sqrt{\frac{e^2}{2\hbar m_{\rm e}\omega}}\,. \tag{6.174}$$

Restricting the discussion to the two lowest levels is an artificial oversimplification because in the harmonic oscillator all level separations are equal (see also Exercise 5.14 on the control of the harmonic oscillator in Chap. 5).

For the different field strengths, different values of $\Omega_{\rm R}$ are gathered in Table 6.1, where $\omega = \pi\, 10^{15}\mathrm{s}^{-1}$ has been used. In the last column $\Omega_{\rm R} \approx \omega$ and the assumption that one can take out $d_{1,2}$ of the integral in (6.167) and (6.168), which was necessary for the derivation of the RWA, is not fulfilled.

3.9. We study the analytic solution of the Rosen-Zener model, which is defined by

$$\epsilon = \Delta/2, \qquad \nu(t) = \nu_0\mathrm{sech}(t/T_{\rm p}) \tag{6.175}$$

for the diagonal and off-diagonal elements of the two by two Hamiltonian (3.45).

(a) The coefficients fulfill

$$\mathrm{i}\dot{c}_1 = \epsilon c_1 + \nu c_2, \tag{6.176}$$

$$\mathrm{i}\dot{c}_2 = \nu c_1 - \epsilon c_2, \tag{6.177}$$

which after taking the time derivative of the second equation and elimination of $\dot{c}_1$, using the first equation and elimination of c_1, using the second, leads to

$$\begin{aligned}
\mathrm{i}\ddot{c}_2 &= -\dot{\epsilon}c_2 - \epsilon\dot{c}_2 + \dot{\nu}c_1 + \nu\dot{c}_1 \\
&= -\dot{\epsilon}c_2 - \epsilon\left(\frac{\nu}{\mathrm{i}}c_1 - \frac{\epsilon}{\mathrm{i}}c_2\right) + \dot{\nu}c_1 + \nu\left(\frac{\epsilon}{\mathrm{i}}c_1 + \frac{\nu}{\mathrm{i}}c_2\right) \\
&= \left[-\dot{\epsilon} - \mathrm{i}(\epsilon^2 + \nu^2)\right]c_2 + \dot{\nu}c_1 \\
&= \left[-\dot{\epsilon} - \mathrm{i}\left(\epsilon^2 + \nu^2 + \epsilon\frac{\dot{\nu}}{\nu}\right)\right]c_2 + \mathrm{i}\frac{\dot{\nu}}{\nu}\dot{c}_2.
\end{aligned} \tag{6.178}$$

After multiplication with $-\mathrm{i}$, we get

$$\ddot{c}_2 - \frac{\dot{\nu}}{\nu}\dot{c}_2 + \left(\epsilon^2 + \nu^2 - \mathrm{i}\dot{\epsilon} + \mathrm{i}\epsilon\frac{\dot{\nu}}{\nu}\right)c_2 = 0. \tag{6.179}$$

(b) With the transformation from t to z, we find

$$\dot{c}_2 = \frac{\mathrm{d}c_2}{\mathrm{d}z}\frac{\mathrm{d}z}{\mathrm{d}t} \qquad \ddot{c}_2 = \frac{\mathrm{d}^2c_2}{\mathrm{d}z^2}\left(\frac{\mathrm{d}z}{\mathrm{d}t}\right)^2 + \frac{\mathrm{d}c_2}{\mathrm{d}z}\frac{\mathrm{d}^2z}{\mathrm{d}t^2}, \tag{6.180}$$

with

$$\frac{\mathrm{d}z}{\mathrm{d}t} = \frac{2z(1-z)}{T_\mathrm{p}}, \tag{6.181}$$

$$\frac{\mathrm{d}^2z}{\mathrm{d}t^2} = -\frac{1}{T_\mathrm{p}^2}(2z-1)4z(1-z), \tag{6.182}$$

and in addition

$$\frac{\dot{\nu}}{\nu} = -\frac{1}{T_\mathrm{p}}(2z-1). \tag{6.183}$$

Finally the equation for $c_2(z)$ is

$$\begin{aligned}
&2z(1-z)\frac{\mathrm{d}^2c_2}{\mathrm{d}z^2} - (2z-1)\frac{\mathrm{d}c_2}{\mathrm{d}z} \\
&+ \frac{(\epsilon T_\mathrm{p})^2 + (\nu_0 T_\mathrm{p})^2 4z(1-z) - \mathrm{i}\dot{\epsilon}T_\mathrm{p}^2 - \mathrm{i}\epsilon T_\mathrm{p}(2z-1)}{2z(1-z)}c_2 = 0.
\end{aligned} \tag{6.184}$$

(c) For $\epsilon = 0$, the equation above is a hypergeometric differential equation (see 9.151 in [14]). Its solution is

$$c_2(z) = AF(a, b; c; z) + Bz^{1-c}F(a-c+1, b-c+1; 2-c; z), \tag{6.185}$$

with $a = -b = T_\mathrm{p}\nu_0$ and $c = 1/2$. The initial conditions for $t = -\infty$, i.e. $z = 0$ are:

(i) $c_2(z=0) = 0$,

(ii) $c_1(z = 0) = 1$.

Because $F(\ldots, \ldots, \ldots, 0) = 1$, we get from (i) $A = 0$ and from (ii) using (6.177) and (6.181), we get $B = \frac{2\nu_0 T_\mathrm{p}}{\mathrm{i}}$. The solution at $t = \infty$, i.e., $z = 1$ then is

$$\begin{aligned} c_2(z = 1) &= \frac{2\nu_0 T_\mathrm{p}}{\mathrm{i}} F(a + 1/2, b + 1/2; 3/2; 1) \\ &= \frac{2\nu_0 T_\mathrm{p}}{\mathrm{i}} \frac{\Gamma(3/2)\Gamma(3/2)}{\Gamma(1 - a)\Gamma(1 - b)} . \end{aligned} \tag{6.186}$$

Taking the absolute square, due to $\pi/\sin(x) = \Gamma(1 - x)\Gamma(1 + x)/x$, leads to the resonance limit

$$|c_2(\infty)|^2 = \sin^2(\pi \nu_0 T_\mathrm{p}) \tag{6.187}$$

of the solution (3.77).

6.4 Solutions to Problems in Chap. 4

4.1. As a first application of the Rayleigh-Ritz variational principle, formulated at the end of Sect. 2.1.1, we use the given trial function with the variational parameter α. The expectation value of the energy is then given by (atomic units used)

$$\begin{aligned} E_\mathrm{var} &\equiv \langle \psi_\mathrm{var} | \hat{H} | \psi_\mathrm{var} \rangle \\ &= \langle \psi_\mathrm{var} | \frac{\hat{p}^2}{2} - \frac{1}{r} | \psi_\mathrm{var} \rangle \\ &= \left(\frac{2\alpha}{\pi} \right)^{3/2} \int_0^\infty \mathrm{d}r \int_0^\pi \mathrm{d}\vartheta \int_0^{2\pi} \mathrm{d}\varphi \\ &\quad r^2 \sin\vartheta \exp(-2\alpha r^2) \left(3\alpha - 2\alpha^2 r^2 - \frac{1}{r} \right) . \end{aligned} \tag{6.188}$$

The angular integration yields a factor of 4π and by using the integrals [14]

$$\int_0^\infty \mathrm{d}x\, x^{2n} \mathrm{e}^{-px^2} = \frac{(2n - 1)!!}{2(2p)^n} \sqrt{\pi\, p}, \tag{6.189}$$

$$\int_0^\infty \mathrm{d}x\, x^{2n+1} \mathrm{e}^{-px^2} = \frac{n!}{2p^{n+1}}, \tag{6.190}$$

we arrive at

$$E_\mathrm{var}(\alpha) = 2 \left[\frac{3\alpha}{4} - \left(\frac{2\alpha}{\pi} \right)^{1/2} \right] . \tag{6.191}$$

Setting the derivative of this result with respect to α equal to zero, we get $\alpha_{\text{ex}} = 8/(9\pi)$, which leads to the upper bound

$$E_{\text{var}}(\alpha_{\text{ex}}) = -\frac{4}{3\pi} \approx -0.4244 \tag{6.192}$$

for the ground state energy of hydrogen (whose exact value is -0.5).

4.2. The tunneling length r_f is determined by setting the laser induced potential at the exit point equal to the ground state energy E_{g}, i.e.,

$$-r_f\mathcal{E}_0 = E_{\text{g}} \quad \Rightarrow \quad r_f = -\frac{E_{\text{g}}}{\mathcal{E}_0}. \tag{6.193}$$

The Gamov factor reads

$$\Gamma \sim \exp\left\{-2^{3/2}\int_0^{r_f} \mathrm{d}r\sqrt{V(r) - E_{\text{g}}}\right\}. \tag{6.194}$$

The integral in the exponent yields

$$\int_0^{r_f} \mathrm{d}r\sqrt{-r\mathcal{E}_0 - E_{\text{g}}} = -\frac{2}{3}\frac{(-r\mathcal{E}_0 - E_{\text{g}})^{3/2}}{\mathcal{E}_0}\Bigg|_0^{r_f} = \frac{2}{3}\frac{(-E_{\text{g}})^{3/2}}{\mathcal{E}_0}, \tag{6.195}$$

leading to

$$\Gamma \sim \exp\left\{-\frac{2}{3}\frac{(-2E_{\text{g}})^{3/2}}{\mathcal{E}_0}\right\}, \tag{6.196}$$

which in the case of hydrogen ($E_{\text{g}} = -1/2$) gives the same exponential factor as in (4.30).

4.3. We define

$$I_1(t) = \int_{-\infty}^{t} \mathrm{d}t' \langle \Phi_f(t')|V_{\text{C}}|\Psi_i^+(t')\rangle \tag{6.197}$$

for the integral appearing in the post form of the transition probability amplitude. Applying a time derivative leads to (parameter integral!)

$$\dot{I}_1 = \langle \Phi_f(t)|V_{\text{C}}|\Psi_i^+(t)\rangle. \tag{6.198}$$

Now we apply the time-derivative to

$$f(t) = \langle \Phi_f(t)|\Psi_i^+(t)\rangle. \tag{6.199}$$

Multiplying by i and using the product rule together with the appropriate TDSE, this leads to

$$\begin{aligned}
\mathrm{i}\partial_t\langle\Phi_f(t)|\Psi_i^+(t)\rangle &= \mathrm{i}\left[\langle\dot{\Phi}_f(t)|\Psi_i^+(t)\rangle + \langle\Phi_f(t)|\dot{\Psi}_i^+(t)\rangle\right] \\
&= \langle\Phi_f(t)|(-\hat{H}_f)|\Psi_i^+(t)\rangle + \langle\Phi_f(t)|\hat{H}|\Psi_i^+(t)\rangle \\
&= \langle\Phi_f(t)|V_\mathrm{C}|\Psi_i^+(t)\rangle, \qquad (6.200)
\end{aligned}$$

showing that, up to a factor of i, the two time derivatives are identical.

A similar calculation can be performed for the prior form of the amplitude. The equivalence of the time-derivatives together with the same initial conditions then proves the identity of the integrated form and the integration-less form of the amplitudes.

4.4. In first order Magnus approximation, the solution of the TDSE for the hydrogen atom under a single half cycle pulse is given by

$$\begin{aligned}
|\Psi(t)\rangle &= \exp\left[-\mathrm{i}\left(\int_0^t \mathrm{d}t'\boldsymbol{\mathcal{E}}(t')\right)\cdot\hat{\boldsymbol{r}}\right]\exp[-\mathrm{i}\hat{H}_0 t]|\Psi(0)\rangle \\
&= \exp\left[-\mathrm{i}\boldsymbol{q}\cdot\hat{\boldsymbol{r}}\right]\exp[-\mathrm{i}\hat{H}_0 t]|\Psi(0)\rangle. \qquad (6.201)
\end{aligned}$$

We will determine the probability of excitation of the 2s and the 2p states, starting from the ground state (see Sect. 4.1 for the eigenstates of the 3D Coulomb problem)

$$\langle\boldsymbol{r}|\Psi(0)\rangle = \psi_{100}(\boldsymbol{r}) = \frac{1}{\sqrt{\pi}}\mathrm{e}^{-r} \qquad (6.202)$$

of the hydrogen atom.

First the transition amplitude $\langle\psi_{210}|\Psi(t)\rangle$ to the $2\mathrm{p}_0$-state (see Sect. 4.1.1)

$$\psi_{210}(\boldsymbol{r}) = \frac{1}{\sqrt{32\pi}}r\mathrm{e}^{-r/2}\cos\vartheta \qquad (6.203)$$

will be looked for.[2] With the wavefunction from (6.201), and using spherical coordinates with the z-axis along the laser polarization, we find

$$\begin{aligned}
\langle\psi_{210}|\Psi(t)\rangle &= \langle\psi_{210}|\exp\left[-\mathrm{i}\boldsymbol{q}\cdot\hat{\boldsymbol{r}}\right]\exp[-\mathrm{i}\hat{H}_0 t]|\psi_{100}\rangle \\
&= \mathrm{e}^{-\mathrm{i}E_0 t}\langle\psi_{210}|\exp\left[-\mathrm{i}qr\cos(\vartheta)\right]|\psi_{100}\rangle \\
&= \sqrt{\frac{1}{32\pi^2}}\mathrm{e}^{-\mathrm{i}E_0 t}\int \mathrm{d}^3r\, r\,\mathrm{e}^{-3r/2}\cos\vartheta\,\mathrm{e}^{-\mathrm{i}qr\cos\vartheta} \\
&= \frac{1}{\sqrt{8}}\mathrm{e}^{-\mathrm{i}E_0 t}\int_0^\infty \mathrm{d}r\, r^3\,\mathrm{e}^{-3r/2}\int_{-1}^{1}\mathrm{d}\zeta\,\zeta\mathrm{e}^{-\mathrm{i}qr\zeta}, \qquad (6.204)
\end{aligned}$$

[2]The transition amplitudes to the $2\mathrm{p}_{\pm1}$ states are zero (why?).

where we have used the substitution $\zeta = \cos\vartheta$. Now [14]

$$\int \mathrm{d}x\, x\mathrm{e}^{ax} = \mathrm{e}^{ax}(x/a - 1/a^2) \tag{6.205}$$

and thus

$$\langle\psi_{210}|\Psi(t)\rangle = \frac{\mathrm{i}}{\sqrt{2}}\mathrm{e}^{-\mathrm{i}E_0 t}\int_0^\infty \mathrm{d}r\left[\frac{r^2}{q}\cos(qr) - \frac{r}{q^2}\sin(qr)\right]\mathrm{e}^{-3r/2}. \tag{6.206}$$

We again consult [14], this time for the definite integrals

$$\int_0^\infty \mathrm{d}x\, x^n\mathrm{e}^{-\beta x}\cos(bx) = (-1)^n\frac{\partial^n}{\partial\beta^n}\left(\frac{\beta}{b^2+\beta^2}\right), \tag{6.207}$$

$$\int_0^\infty \mathrm{d}x\, x^n\mathrm{e}^{-\beta x}\sin(bx) = (-1)^n\frac{\partial^n}{\partial\beta^n}\left(\frac{b}{b^2+\beta^2}\right), \tag{6.208}$$

leading to

$$\langle\psi_{210}|\Psi(t)\rangle = -\mathrm{i}\,3\sqrt{8}\mathrm{e}^{-\mathrm{i}E_0 t}\frac{q}{\left[q^2+(3/2)^2\right]^3} \tag{6.209}$$

for the amplitude. The corresponding probability is

$$|\langle\psi_{210}|\Psi(t)\rangle|^2 = \frac{72\,q^2}{\left[q^2+(3/2)^2\right]^6}. \tag{6.210}$$

An analogous calculation for the probability to excite the 2s state

$$\psi_{200}(\boldsymbol{r}) = \frac{1}{\sqrt{4\pi}}\frac{1}{2^{3/2}}(2-r)\mathrm{e}^{-r/2} \tag{6.211}$$

gives

$$|\langle\psi_{200}|\Psi(t)\rangle|^2 = \frac{32\,q^4}{\left[q^2+(3/2)^2\right]^6}, \tag{6.212}$$

which is much smaller than (6.210) for $q \ll 1$.

4.5. This exercise follows closely the work in [15], where additional information can be found.

(a) Analogous to the interaction picture in Sect. 2.2.5, we find in second order Magnus expansion in the Schrödinger picture (in atomic units)

$$\hat{U}^{(2)}(T_\mathrm{p},0) = \exp\left\{-\mathrm{i}\int_0^{T_\mathrm{p}} \mathrm{d}t'\,\hat{H}(t') + \frac{1}{2}\int_0^{T_\mathrm{p}} \mathrm{d}t'' \int_0^{t''} \mathrm{d}t'[\hat{H}(t'),\hat{H}(t'')]\right\} \tag{6.213}$$

for the time-evolution operator. The Hamiltonian in velocity gauge is

$$\hat{H}_\mathrm{v}(t) = \frac{[\hat{\boldsymbol{p}} + \boldsymbol{A}(t)]^2}{2} + V(\boldsymbol{r}) = \hat{H}_0 + \hat{\boldsymbol{p}}\cdot\boldsymbol{A}(t) + \frac{1}{2}\boldsymbol{A}^2(t) \tag{6.214}$$

and thus

$$\begin{aligned} [\hat{H}_\mathrm{v}(t'),\hat{H}_\mathrm{v}(t'')] &= [\hat{H}_0,\hat{\boldsymbol{p}}\cdot\boldsymbol{A}(t'')] + [\hat{\boldsymbol{p}}\cdot\boldsymbol{A}(t'),\hat{H}_0] \\ &= [V,\hat{\boldsymbol{p}}\cdot\boldsymbol{A}(t'')] + [\hat{\boldsymbol{p}}\cdot\boldsymbol{A}(t'),V] \\ &= \mathrm{i}\boldsymbol{\nabla}V\cdot\boldsymbol{A}(t'') - \mathrm{i}\boldsymbol{\nabla}V\cdot\boldsymbol{A}(t'). \end{aligned} \tag{6.215}$$

For an integral over a triangular region, we find

$$\begin{aligned} \int_0^{T_\mathrm{p}} \mathrm{d}t'' \int_0^{t''} \mathrm{d}t' \ldots &= \int_0^{T_\mathrm{p}} \mathrm{d}t'' \int_0^{T_\mathrm{p}} \mathrm{d}t' \cdots - \int_0^{T_\mathrm{p}} \mathrm{d}t'' \int_{t''}^{T_\mathrm{p}} \mathrm{d}t' \ldots \\ &= \int_0^{T_\mathrm{p}} \mathrm{d}t'' \int_0^{T_\mathrm{p}} \mathrm{d}t' \cdots - \int_0^{T_\mathrm{p}} \mathrm{d}t' \int_0^{t'} \mathrm{d}t'' \ldots \end{aligned} \tag{6.216}$$

and with the definitions

$$\boldsymbol{\alpha}(T_\mathrm{p}) = \int_0^{T_\mathrm{p}} \mathrm{d}t'\boldsymbol{A}(t'), \tag{6.217}$$

$$\bar{\boldsymbol{\alpha}}(T_\mathrm{p}) = \frac{1}{T_\mathrm{p}}\int_0^{T_\mathrm{p}} \mathrm{d}t'' \int_0^{t''} \mathrm{d}t'\boldsymbol{A}(t'), \tag{6.218}$$

$$\Phi(T_\mathrm{p}) = \frac{1}{2}\int_0^{T_\mathrm{p}} \mathrm{d}t'\boldsymbol{A}^2(t'), \tag{6.219}$$

where the first one is (3.22) in atomic units, we can write

$$\begin{aligned} \hat{U}_\mathrm{v}^{(2)}(T_\mathrm{p},0) = \exp\Big\{&-\mathrm{i}\hat{H}_0T_\mathrm{p} - \mathrm{i}\hat{\boldsymbol{p}}\cdot\boldsymbol{\alpha} - \mathrm{i}\Phi \\ &+\mathrm{i}\boldsymbol{\nabla}VT_\mathrm{p}\cdot(\boldsymbol{\alpha}/2 - \bar{\boldsymbol{\alpha}})\Big\}. \end{aligned} \tag{6.220}$$

$\boldsymbol{\alpha}$ and $\bar{\boldsymbol{\alpha}}$ are of second order and Φ is of third order in the pulse length T_p [15]. The final result is correct up to third order in T_p. We now use the Zassenhaus formula

$$\mathrm{e}^{\hat{x}+\hat{y}} = \mathrm{e}^{\hat{x}}\mathrm{e}^{\hat{y}}\mathrm{e}^{-1/2[\hat{x},\hat{y}]}\ldots \tag{6.221}$$

to disentangle the exponents and get in the velocity gauge

$$\hat{U}_{\mathrm{v}}^{(2)}(T_{\mathrm{p}},0)=\exp\{-\mathrm{i}\Phi\}\exp\left\{-\mathrm{i}\hat{\boldsymbol{p}}\cdot\boldsymbol{\alpha}\right\}$$
$$\exp\left\{-\mathrm{i}\nabla V T_{\mathrm{p}}\cdot\bar{\boldsymbol{\alpha}}\right\}\exp\left\{-\mathrm{i}\hat{H}_0 T_{\mathrm{p}}\right\}, \tag{6.222}$$

again up to third order in T_{p} (the $\boldsymbol{\alpha}/2$ term cancels due to the commutator $[\hat{H}_0,\hat{\boldsymbol{p}}\cdot\boldsymbol{\alpha}]$).

(b) In first order Magnus approximation

$$\hat{U}^{(1)}(T_{\mathrm{p}},0)=\exp\left\{-\mathrm{i}\int_0^{T_{\mathrm{p}}}\mathrm{d}t'\,\hat{H}(t')\right\} \tag{6.223}$$

and with the Kramers-Henneberger frame Hamiltonian

$$\hat{H}_{\mathrm{a}}=\frac{\hat{\boldsymbol{p}}^2}{2}+V[\boldsymbol{r}+\boldsymbol{\alpha}(t)] \tag{6.224}$$

after expansion of the potential up to second order in T_{p}

$$V[\boldsymbol{r}+\boldsymbol{\alpha}(t)]=V(\boldsymbol{r}(t))+\nabla V\cdot\boldsymbol{\alpha}+O(T_{\mathrm{p}}^3), \tag{6.225}$$

we get

$$\hat{U}_{\mathrm{a}}^{(1)}(T_{\mathrm{p}},0)=\exp\left\{-\mathrm{i}\hat{H}_0 T_{\mathrm{p}}-\mathrm{i}\nabla V T_{\mathrm{p}}\cdot\bar{\boldsymbol{\alpha}}\right\} \tag{6.226}$$

up to third order in T_{p}. Using the Zassenhaus formula, we get

$$\hat{U}_{\mathrm{a}}^{(1)}(T_{\mathrm{p}},0)=\exp\left\{-\mathrm{i}\nabla V T_{\mathrm{p}}\cdot\bar{\boldsymbol{\alpha}}\right\}\exp\left\{-\mathrm{i}\hat{H}_0 T_{\mathrm{p}}\right\}, \tag{6.227}$$

again up to third order in the pulse length.

(c) The difference between the velocity gauge and Kramers-Henneberger frame evolution operators is compensated by the two unitary transformations in (3.18)

$$\hat{U}_1=\exp\{\mathrm{i}\Phi\}, \tag{6.228}$$
$$\hat{U}_2=\exp\{\mathrm{i}\hat{\boldsymbol{p}}\cdot\boldsymbol{\alpha}\}, \tag{6.229}$$

such that

$$|\Psi_{\mathrm{a}}^{(1)}(T_{\mathrm{p}})\rangle=\hat{U}_2\hat{U}_1|\Psi_{\mathrm{v}}^{(2)}(T_{\mathrm{p}})\rangle=\hat{U}_2\hat{U}_1\hat{U}_{\mathrm{v}}^{(2)}|\Psi(0)\rangle=\hat{U}_{\mathrm{a}}^{(1)}|\Psi(0)\rangle \tag{6.230}$$

to the same (3rd) order in the pulse length.

For the excitation probability from the (hydrogen) ground state, a similar reasoning leads to

$$
\begin{aligned}
P(T_\mathrm{p}) &= |\langle \Psi_\mathrm{v}(T_\mathrm{p})|\hat{U}_\mathrm{v}^{(2)}(T_\mathrm{p})|\psi_{100}\rangle|^2 \\
&= |\langle \Psi_\mathrm{v}(T_\mathrm{p})| \exp\{-\mathrm{i}\Phi\} \exp\left\{-\mathrm{i}\hat{\boldsymbol{p}}\cdot\boldsymbol{\alpha}\right\} \\
&\quad \exp\left\{-\mathrm{i}\nabla V T_\mathrm{p}\cdot\bar{\boldsymbol{\alpha}}\right\} \exp\left\{-\mathrm{i}\hat{H}_0 T_\mathrm{p}\right\} |\psi_{100}\rangle|^2 \\
&= |\langle \Psi_\mathrm{a}(T_\mathrm{p})| \exp\left\{-\mathrm{i}\nabla V T_\mathrm{p}\cdot\bar{\boldsymbol{\alpha}}\right\} \exp\left\{-\mathrm{i}\hat{H}_0 T_\mathrm{p}\right\} |\psi_{100}\rangle|^2 \\
&= |\langle \Psi_\mathrm{a}(T_\mathrm{p})|\hat{U}_\mathrm{a}^{(1)}(T_\mathrm{p})|\psi_{100}\rangle|^2 .
\end{aligned}
\tag{6.231}
$$

The Kramers-Henneberger frame is the natural frame to describe the driven system in the Magnus approximation [15].

4.6. The dipole matrix element

$$
d(t) = \langle \psi_n(t)|\hat{x}|\psi_n(t)\rangle \tag{6.232}
$$

is a periodic function of time (due to the periodic nature of the Floquet functions). Its Fourier components can be written as

$$
\begin{aligned}
d_m &= \frac{1}{T}\int_0^T \mathrm{d}t\, \mathrm{e}^{\mathrm{i}m\omega t} d(t) \\
&= \frac{1}{T}\left[\int_0^{T/2} \mathrm{d}t\, \mathrm{e}^{\mathrm{i}m\omega t} d(t) + \int_0^{T/2} \mathrm{d}t\, \mathrm{e}^{\mathrm{i}m\omega(t+T/2)} d(t+T/2)\right] \\
&= \frac{1}{T}\int_0^{T/2} \mathrm{d}t\, \mathrm{e}^{\mathrm{i}m\omega t} [d(t) \pm d(t+T/2)] ,
\end{aligned}
\tag{6.233}
$$

where the plus sign holds for even m and the minus sign holds for odd values of m. Now due to invariance under the generalized parity transformation, discussed in Appendix 3.A,

$$
\begin{aligned}
d(t+T/2) &= \langle \psi_n(t+T/2)|\hat{x}|\psi_n(t+T/2)\rangle \\
&= \int \mathrm{d}x\, \psi_n^*(x, t+T/2) x \psi_n(x, t+T/2) \\
&= \int_{-\infty}^{\infty} \mathrm{d}x\, \psi_n^*(-x, t) x \psi_n(-x, t) \\
&\overset{\tilde{x}=-x}{=} \int_{\infty}^{-\infty} \mathrm{d}\tilde{x}\, \psi_n^*(\tilde{x}, t)\tilde{x}\psi_n(\tilde{x}, t) \\
&= -\int_{-\infty}^{\infty} \mathrm{d}x\, \psi_n^*(x, t) x \psi_n(x, t) \\
&= -d(t)
\end{aligned}
\tag{6.234}
$$

holds. Inserting this result into (6.233) reveals that for even m the Fourier components do vanish while for odd m in general they do not. Performing a second time-derivative, this statement can be taken over to the dipole acceleration.

6.5 Solutions to Problems in Chap. 5

5.1. The overlap integral is given by

$$S(R) = \frac{1}{\pi}\int \mathrm{d}V\,\mathrm{e}^{-(r_a+r_b)}$$
$$= \frac{1}{\pi}\int \mathrm{d}V\,\mathrm{e}^{-\mu R}, \tag{6.235}$$

which, with the volume element $\mathrm{d}V = \frac{1}{8}R^3(\mu^2 - \nu^2)\mathrm{d}\mu\mathrm{d}\nu\mathrm{d}\varphi$ and due to the independence on the angular variable φ gives

$$S(R) = \frac{R^3}{4}\int_{-1}^{1}\mathrm{d}\nu\int_1^{\infty}\mathrm{d}\mu(\mu^2 - \nu^2)\mathrm{e}^{-\mu R}$$
$$= \frac{R^3}{4}\left(\int_{-1}^{1}\mathrm{d}\nu\int_1^{\infty}\mathrm{d}\mu\,\mu^2\mathrm{e}^{-\mu R} - \int_{-1}^{1}\mathrm{d}\nu\,\nu^2\int_1^{\infty}\mathrm{d}\mu\,\mathrm{e}^{-\mu R}\right). \tag{6.236}$$

We now use the general formula (2-11) in [16]

$$\int \mathrm{d}x x^n \mathrm{e}^{-ax} = -\frac{n!\mathrm{e}^{-ax}}{a^{n+1}}\left[1 + ax + \frac{1}{2!}(ax)^2 + \cdots + \frac{1}{n!}(ax)^n\right], \tag{6.237}$$

leading to the final result (5.8)

$$S(R) = \mathrm{e}^{-R}(1 + R + R^2/3), \tag{6.238}$$

repeated here for convenience.

The Coulomb integral is given by

$$C(R) = -\frac{1}{\pi}\int \mathrm{d}V\mathrm{e}^{-2r_a}1/r_b$$
$$= -\frac{1}{\pi}\int \mathrm{d}V\mathrm{e}^{-(\mu+\nu)R}\,2/[R(\mu - \nu)]. \tag{6.239}$$

Again inserting the volume element, using $u^2 - v^2 = (u + v)(u - v)$ to cancel $(u - v)$ in the denominator, and integrating over φ leads to

$$
\begin{aligned}
C(R) &= -\frac{R^2}{2}\int_{-1}^{1}\mathrm{d}\nu\int_{1}^{\infty}\mathrm{d}\mu(\mu+\nu)\mathrm{e}^{-(\mu+\nu)R} \\
&= -\frac{R^2}{2}\left(\int_{-1}^{1}\mathrm{d}\nu\,\mathrm{e}^{-\nu R}\int_{1}^{\infty}\mathrm{d}\mu\,\mu\mathrm{e}^{-\mu R}\right. \\
&\quad \left.+\int_{-1}^{1}\mathrm{d}\nu\,\nu\mathrm{e}^{-\nu R}\int_{1}^{\infty}\mathrm{d}\mu\,\mathrm{e}^{-\mu R}\right).
\end{aligned} \tag{6.240}
$$

Using the general formula (6.237) (for both, the integral over μ, as well as that over ν) and gathering terms leads to

$$
C(R) = -\left[1-(1+R)\mathrm{e}^{-2R}\right]/R. \tag{6.241}
$$

This result could have also been gained more easily by using spherical polar coordinates [16].

The exchange integral is finally given by

$$
\begin{aligned}
D(R) &= -\frac{1}{\pi}\int \mathrm{d}V\,\mathrm{e}^{-(r_a+r_b)}1/r_a \\
&= -\frac{1}{\pi}\int \mathrm{d}V\,\mathrm{e}^{-\mu R}2/[R(\mu+\nu)] \\
&= -\frac{R^2}{2}\int_{-1}^{1}\mathrm{d}\nu\int_{1}^{\infty}\mathrm{d}\mu\,(\mu-\nu)\mathrm{e}^{-\mu R} \\
&= -\frac{R^2}{2}\left(\int_{-1}^{1}\mathrm{d}\nu\int_{1}^{\infty}\mathrm{d}\mu\,\mu\mathrm{e}^{-\mu R}-\int_{-1}^{1}\mathrm{d}\nu\,\nu\int_{1}^{\infty}\mathrm{d}\mu\,\mathrm{e}^{-\mu R}\right) \\
&= -(1+R)\mathrm{e}^{-R}.
\end{aligned} \tag{6.242}
$$

Here the second integral in the next to last line vanishes due to symmetry reasons and we have again used $u^2-v^2=(u+v)(u-v)$ as well as (6.237).

5.2. To improve the LCAO energy, we use the variational principle of Rayleigh and Ritz (see Sect. 2.1.1). First, we take the expectation value of the Hamiltonian on the LHS of (5.3) with the variational wavefunction. From

$$
\Delta\mathrm{e}^{-\alpha r_a} = \left(\frac{\partial^2}{\partial r_a^2}+\frac{2}{r_a}\frac{\partial}{\partial r_a}\right)\mathrm{e}^{-\alpha r_a} = \left(\alpha^2-\frac{2\alpha}{r_a}\right)\mathrm{e}^{-\alpha r_a} \tag{6.243}
$$

and using integrations analogous to the ones in the previous exercise (some of which can be more easily done in spherical coordinates [16]), one finds for the expectation value of the kinetic energy

$$
\int \mathrm{d}V\,\psi_\mathrm{e}\left(-\frac{1}{2}\Delta\right)\psi_\mathrm{e} = \alpha^2 F_1(\alpha R), \tag{6.244}
$$

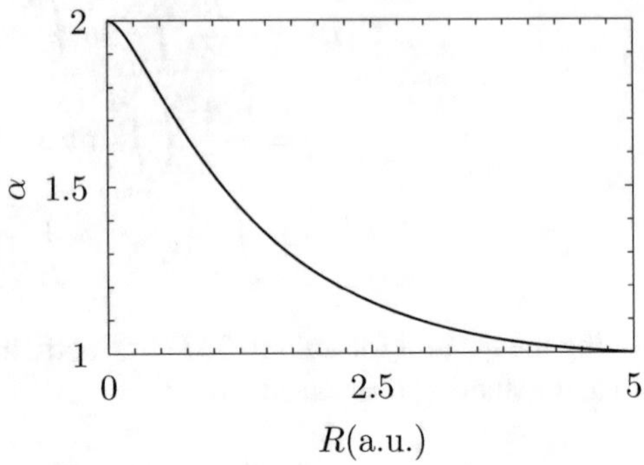

Fig. 6.7 The variational parameter α as a function of the internuclear distance R interpolates between the helium ($R = 0$) and the hydrogen ($R = \infty$) limit

and for the expectation value of the potential energy

$$\int \mathrm{d}V \psi_\mathrm{e} \left(-\frac{1}{r_a} - \frac{1}{r_b} \right) \psi_\mathrm{e} = \alpha F_2(\alpha R), \tag{6.245}$$

with

$$F_1(w) = \frac{1 + \mathrm{e}^{-w}(1 + w - w^2/3)}{2 + 2\mathrm{e}^{-w}(1 + w + w^2/3)}, \tag{6.246}$$

$$F_2(w) = -\frac{1 + 2\mathrm{e}^{-w}(1 + w) - 1/w - (1/w + 1)\mathrm{e}^{-2w}}{1 + \mathrm{e}^{-w}(1 + w + w^2/3)}, \tag{6.247}$$

where $w = \alpha R$. Minimizing the total energy expression with respect to α at constant R leads to

$$\begin{aligned} 0 &= 2\alpha F_1 + \alpha^2 R \frac{\mathrm{d}F_1}{\mathrm{d}w} + F_2 + \alpha R \frac{\mathrm{d}F_2}{\mathrm{d}w} \\ &= \alpha \left(2F_1 + w \frac{\mathrm{d}F_1}{\mathrm{d}w} \right) + \left(F_2 + w \frac{\mathrm{d}F_2}{\mathrm{d}w} \right), \end{aligned} \tag{6.248}$$

which can be resolved for α, yielding

$$\alpha = -\frac{F_2 + w \frac{\mathrm{d}F_2}{\mathrm{d}w}}{2F_1 + w \frac{\mathrm{d}F_1}{\mathrm{d}w}}. \tag{6.249}$$

Calculating α as a function of w and then using $R = w/\alpha$, the variational parameter can be plotted a function of R, see Fig. 6.7 and [16].

5.3. We have to apply the radial part of the kinetic energy (5.24) to the function v_n of (5.26). Looking at the series expansion for the Bessel function (5.27) of zeroth order (using $\Gamma(j + 1) = j!$)

$$J_0(x) = \sum_{j=0}^{\infty} \frac{(-1)^j}{j!j!} \left(\frac{x}{2}\right)^{2j} , \tag{6.250}$$

we have to calculate the following derivatives (for $j \geq 1$)

$$\frac{\partial}{\partial \rho} \left(\frac{x_n}{2L}\rho\right)^{2j} = j\frac{x_n}{L} \left(\frac{x_n}{2L}\rho\right)^{2j-1} , \tag{6.251}$$

$$\frac{\partial^2}{\partial \rho^2} \left(\frac{x_n}{2L}\rho\right)^{2j} = j\frac{2j-1}{2} \left(\frac{x_n}{L}\right)^2 \left(\frac{x_n}{2L}\rho\right)^{2j-2} , \tag{6.252}$$

leading to

$$\begin{aligned} \hat{T}_\rho \left(\frac{x_n}{2L}\rho\right)^{2j} &= \left(-\frac{1}{2}\frac{\partial^2}{\partial \rho^2} - \frac{1}{2\rho}\frac{\partial}{\partial \rho}\right) \left(\frac{x_n}{2L}\rho\right)^{2j} \\ &= -\frac{1}{2} j^2 \left(\frac{x_n}{L}\right)^2 \left(\frac{x_n}{2L}\rho\right)^{2j-2} . \end{aligned} \tag{6.253}$$

In J_0, the factor $j!$ appears twice in the denominator and cancellation of the j^2 term from (6.253) leads to the double appearance of $(j-1)!$ in the denominator, if $\hat{T}_\rho$ is applied to the function

$$v_n(\rho) = \frac{\sqrt{2}}{L J_1(x_n)} J_0(x_n \rho / L). \tag{6.254}$$

$\hat{T}_\rho$ thus reproduces the full series representation of the Bessel function (6.250), starting from $j = 0$, and (5.29) holds.

5.4. Using the LCAO solution from (5.14) and (5.15) for large R, we find with the normalized 1s functions $|\psi_{a,b}\rangle = |1\mathrm{s}_{\mathrm{a,b}}\rangle$

$$|1\sigma_g\rangle \approx \frac{1}{\sqrt{2}}(|1\mathrm{s}_\mathrm{a}\rangle + |1\mathrm{s}_\mathrm{b}\rangle) \qquad |1\sigma_\mathrm{u}\rangle \approx \frac{1}{\sqrt{2}}(|1\mathrm{s}_\mathrm{a}\rangle - |1\mathrm{s}_\mathrm{b}\rangle) , \tag{6.255}$$

due to the fact that the overlap integral $S \approx 0$, see (5.8).

The dipole matrix element then is

$$\begin{aligned} \langle 1\sigma_g | z | 1\sigma_u \rangle &= \langle 1\sigma_g | z + R/2 - R/2 | 1\sigma_u \rangle \\ &\approx \frac{1}{2} \langle (1\mathrm{s}_\mathrm{a} + 1\mathrm{s}_\mathrm{b}) | \mathrm{z} + \mathrm{R}/2 - \mathrm{R}/2 | (1\mathrm{s}_\mathrm{a} - 1\mathrm{s}_\mathrm{b}) \rangle , \end{aligned} \tag{6.256}$$

where we have inserted a "nutritious zero" to make use of the fact that $1\mathrm{s}_\mathrm{a}$ is localized around $z = -R/2$ and $1\mathrm{s}_\mathrm{b}$ around $z = R/2$. In this way, we find

$$
\begin{aligned}
\langle 1\sigma_g|z|1\sigma_u\rangle \approx\ & \frac{1}{2}\langle 1\mathrm{s_a}|\mathrm{z}+\mathrm{R}/2|1\mathrm{s_a}\rangle \\
& -\frac{1}{2}\langle 1\mathrm{s_b}|\mathrm{z}-\mathrm{R}/2|1\mathrm{s_b}\rangle \\
& +\frac{1}{2}\langle 1\mathrm{s_a}|-\mathrm{R}/2|1\mathrm{s_a}\rangle \\
& -\frac{1}{2}\langle 1\mathrm{s_b}|\mathrm{R}/2|1\mathrm{s_b}\rangle \\
=\ & \frac{1}{2}(-R/2-R/2) = -\frac{1}{2}R\,. \qquad (6.257)
\end{aligned}
$$

Here, we have again neglected the overlap, and the first two terms are zero due to symmetry. Furthermore, in the last two terms, we have used the normalization of the 1s functions at the same location.

5.5. The full molecular problem satisfies the TISE

$$\hat{H}_{\mathrm{mol}}\psi(\boldsymbol{x},\boldsymbol{X}) = [\hat{T}_{\mathrm{e}}+\hat{T}_{\mathrm{N}}+V(\boldsymbol{x},\boldsymbol{X})]E\psi(\boldsymbol{x},\boldsymbol{X}), \qquad (6.258)$$

whereas for the factorized BO solution

$$\psi^{\mathrm{BO}}(\boldsymbol{x},\boldsymbol{X}) = \phi(\boldsymbol{x},\boldsymbol{X})\chi(\boldsymbol{X}), \qquad (6.259)$$

the equations

$$\left[\hat{T}_{\mathrm{e}}+V(\boldsymbol{x},\boldsymbol{X})\right]\phi = E(\boldsymbol{X})\phi, \qquad (6.260)$$

$$\left[\hat{T}_{\mathrm{N}}+E(\boldsymbol{X})\right]\chi = \epsilon\chi \qquad (6.261)$$

do hold.

The exact energy is

$$E\langle\psi|\psi\rangle_{\boldsymbol{x},\boldsymbol{X}} = \langle\psi|\hat{H}|\psi\rangle_{\boldsymbol{x},\boldsymbol{X}}, \qquad (6.262)$$

where the subscripts indicate the integration over both sets of variables and summation over spin in the scalar product. Now for fixed $\boldsymbol{X}$, the exact wavefunction is only an approximate one for the Hamiltonian without the nuclear kinetic energy, and we get

$$\langle\psi|\hat{T}_{\mathrm{e}}+V(\boldsymbol{x},\boldsymbol{X})|\psi\rangle_{\boldsymbol{x}} \geq E(\boldsymbol{X})\langle\psi|\psi\rangle_{\boldsymbol{x}}, \qquad (6.263)$$

due to the Rayleigh-Ritz variational principle for the electronic Schrödinger equation (6.260). Adding the nuclear kinetic energy and using (6.262), we get after integration over the nuclear variables

$$E\langle\psi|\psi\rangle_{x,X} \geq \langle\psi|\hat{T}_{\mathrm{N}} + E(X)|\psi\rangle_{x,X}. \tag{6.264}$$

For fixed electronic coordinates, by analogous reasoning, we find

$$\langle\psi|\hat{T}_{\mathrm{N}} + E(X)|\psi\rangle_X \geq \epsilon\langle\psi|\psi\rangle_X \tag{6.265}$$

and after integration over those electronic coordinates and comparing with the equation above, we find

$$E \geq \epsilon. \tag{6.266}$$

This proof has been given by Epstein [17], commenting on previous work by Brattsev [18].

5.6. The Hellman-Feynman theorem and the derivative coupling term are studied in this exercise.

(a) For the derivative of the energy eigenvalue, we find with the product rule

$$\frac{\mathrm{d}E_\nu(R)}{\mathrm{d}R} = \left\langle\frac{\mathrm{d}\phi_\nu(R)}{\mathrm{d}R}\middle|\hat{H}_{\mathrm{e}}(R)\middle|\phi_\nu(R)\right\rangle + \left\langle\phi_\nu(R)\middle|\hat{H}_{\mathrm{e}}(R)\middle|\frac{\mathrm{d}\phi_\nu(R)}{\mathrm{d}R}\right\rangle + \left\langle\phi_\nu(R)\middle|\frac{\mathrm{d}\hat{H}_{\mathrm{e}}(R)}{\mathrm{d}R}\middle|\phi_\nu(R)\right\rangle. \tag{6.267}$$

Using the electronic eigenvalue equation twice and the product rule in reverse order yields

$$\begin{aligned}\frac{\mathrm{d}E_\nu(R)}{\mathrm{d}R} &= E_\nu(R)\left\langle\frac{\mathrm{d}\phi_\nu(R)}{\mathrm{d}R}\middle|\phi_\nu(R)\right\rangle + E_\nu(R)\left\langle\phi_\nu(R)\middle|\frac{\mathrm{d}\phi_\nu(R)}{\mathrm{d}R}\right\rangle \\ &\quad + \left\langle\phi_\nu(R)\middle|\frac{\mathrm{d}\hat{H}_{\mathrm{e}}(R)}{\mathrm{d}R}\middle|\phi_\nu(R)\right\rangle \\ &= E_\nu(R)\frac{\mathrm{d}}{\mathrm{d}R}\langle\phi_\nu(R)|\phi_\nu(R)\rangle + \left\langle\phi_\nu(R)\middle|\frac{\mathrm{d}\hat{H}_{\mathrm{e}}(R)}{\mathrm{d}R}\middle|\phi_\nu(R)\right\rangle \\ &= \left\langle\phi_\nu(R)\middle|\frac{\mathrm{d}\hat{H}_{\mathrm{e}}(R)}{\mathrm{d}R}\middle|\phi_\nu(R)\right\rangle,\end{aligned} \tag{6.268}$$

where the normalization condition

$$\langle\phi_\nu(R)|\phi_\nu(R)\rangle = 1 \tag{6.269}$$

has been used in the last step.

(b) Now we use that

$$\langle\phi_\mu(R)|\hat{H}_e(R)|\phi_\nu(R)\rangle = 0 \quad \text{if} \quad \mu \neq \nu \tag{6.270}$$

and take the derivative to get

$$\begin{aligned}
0 &= \left\langle \phi_\mu(R) \left| \frac{\mathrm{d}\hat{H}_e(R)}{\mathrm{d}R} \right| \phi_\nu(R) \right\rangle \\
&+ \left\langle \frac{\mathrm{d}\phi_\mu(R)}{\mathrm{d}R} \left| \hat{H}_e(R) \right| \phi_\nu(R) \right\rangle + \left\langle \phi_\mu(R) \left| \hat{H}_e(R) \right| \frac{\mathrm{d}\phi_\nu(R)}{\mathrm{d}R} \right\rangle \\
&= \left\langle \phi_\mu(R) \left| \frac{\mathrm{d}\hat{H}_e(R)}{\mathrm{d}R} \right| \phi_\nu(R) \right\rangle \\
&+ E_\nu \left\langle \frac{\mathrm{d}\phi_\mu(R)}{\mathrm{d}R} \middle| \phi_\nu(R) \right\rangle + E_\mu \left\langle \phi_\mu(R) \middle| \frac{\mathrm{d}\phi_\nu(R)}{\mathrm{d}R} \right\rangle .
\end{aligned} \tag{6.271}$$

This can be expressed with the help of the derivative coupling matrix element from (5.57), yielding

$$0 = \left\langle \phi_\mu(R) \left| \frac{\mathrm{d}\hat{H}_e(R)}{\mathrm{d}R} \right| \phi_\nu(R) \right\rangle + \frac{M_r}{\hbar^2}(E_\nu - E_\mu)Q_{\mu\nu}, \tag{6.272}$$

where we have used anti-hermiticity, $Q_{\mu\nu} = -Q^*_{\nu\mu}$, and have chosen the electronic eigenfunctions to be real-valued. Finally, we get

$$Q_{\mu\nu} = \frac{\hbar^2}{M_r} \frac{\left\langle \phi_\mu(R) \left| \frac{\mathrm{d}\hat{H}_e(R)}{\mathrm{d}R} \right| \phi_\nu(R) \right\rangle}{E_\nu - E_\mu}, \tag{6.273}$$

increasing if E_μ and E_ν approach each other. We note that for an exact crossing (conical intersection) of two adiabatic surfaces, at least two parameters have to be varied.

5.7. In the case of three atoms on a line, we can define one center of mass and two relative (one dimensional) coordinates according to

$$X_S = \frac{M_1X_1 + M_2X_2 + M_3X_3}{M_S}, \tag{6.274}$$

$$X_{R_1} = X_2 - X_1, \tag{6.275}$$

$$X_{R_2} = X_3 - X_2, \tag{6.276}$$

with the total mass $M_S = M_1 + M_2 + M_3$. The inverse transformation is

$$X_1 = X_S - \frac{M_2 + M_3}{M_S} X_{R_1} - \frac{M_3}{M_S} X_{R_2}, \tag{6.277}$$

$$X_2 = X_S + \frac{M_1}{M_S} X_{R_1} - \frac{M_3}{M_S} X_{R_2}, \tag{6.278}$$

$$X_3 = X_S + \frac{M_1}{M_S} X_{R_1} + \frac{M_1 + M_2}{M_S} X_{R_2}. \tag{6.279}$$

The (classical) kinetic energy of the nuclei therefore can be written as

$$\begin{aligned} T_{\mathrm{N}} &= \frac{M_1}{2}\dot{X}_1^2 + \frac{M_2}{2}\dot{X}_2^2 + \frac{M_3}{2}\dot{X}_3^2, \\ &= \frac{M_S}{2}\dot{X}_S^2 + \frac{P_{R_1}^2}{2M_{R_1}} - \frac{P_{R_1} P_{R_2}}{M_2} + \frac{P_{R_2}^2}{2M_{R_2}}, \end{aligned} \tag{6.280}$$

with the canonically conjugate relative momenta

$$P_{R_1} = \frac{\partial L}{\partial \dot{X}_{R_1}} = \frac{M_1(M_2 + M_3)}{M_S} \dot{X}_{R_1} + \frac{M_1 M_2}{M_S} \dot{X}_{R_2}, \tag{6.281}$$

$$P_{R_2} = \frac{\partial L}{\partial \dot{X}_{R_2}} = \frac{M_1 M_3}{M_S} \dot{X}_{R_1} + \frac{(M_1 + M_2) M_3}{M_S} \dot{X}_{R_2}, \tag{6.282}$$

and the corresponding reduced masses

$$M_{R_1} = \frac{M_1 M_2}{M_1 + M_2}, \tag{6.283}$$

$$M_{R_2} = \frac{M_2 M_3}{M_2 + M_3}. \tag{6.284}$$

5.8. From (5.14) and (5.15), we find with the normalized 1s functions $|\psi_{a,b}\rangle = |1\mathrm{s}_{\mathrm{a,b}}\rangle$

$$|1\sigma_g\rangle \approx \frac{1}{\sqrt{2(1 + S(R))}} (|1\mathrm{s_a}\rangle + |1\mathrm{s_b}\rangle) \,. \tag{6.285}$$

The dipole matrix element in the electronic ground state is

$$\begin{aligned} \boldsymbol{\mu}_{gg} &= \langle 1\sigma_g | \boldsymbol{R}_a + \boldsymbol{R}_b - \boldsymbol{r} | 1\sigma_g \rangle \\ &= \frac{1}{2(1 + S(R))} \langle (1\mathrm{s_a} + 1\mathrm{s_b}) | \boldsymbol{R}_\mathrm{a} + \boldsymbol{R}_\mathrm{b} - \boldsymbol{r} | (1\mathrm{s_a} + 1\mathrm{s_b}) \rangle. \end{aligned} \tag{6.286}$$

The two terms with the nuclear coordinates, due to the homonuclear nature of H_2^+, are proportional to the center of mass of the two nuclei and can be neglected, because we are interested only in the coupling of a laser field to the relative motion.

The remaining term is

$$\begin{aligned}\boldsymbol{\mu}_{gg} &= \frac{1}{2(1+S(R))}\langle(1\mathrm{s}_\mathrm{a}+1\mathrm{s}_\mathrm{b})|-\boldsymbol{r}|(1\mathrm{s}_\mathrm{a}+1\mathrm{s}_\mathrm{b})\rangle \\ &= 0,\end{aligned} \tag{6.287}$$

where the last step follows from the fact that the electronic ground state is of even parity.

5.9. We wish to calculate the exponentiated matrix

$$\exp\{-\mathrm{i}\Delta t\mathbf{V}\} = \exp\left\{-\mathrm{i}\Delta t\begin{pmatrix} V_{11} & V_{12} \\ V_{12} & V_{22}\end{pmatrix}\right\}. \tag{6.288}$$

This task shall be performed by diagonalization, using

$$\exp\{-\mathrm{i}\Delta t\mathbf{V}\} = \mathbf{S}^{-1}\exp\{-\mathrm{i}\Delta t\mathbf{S}\mathbf{V}\mathbf{S}^{-1}\}\mathbf{S}, \tag{6.289}$$

with the orthogonal matrix $\mathbf{S}^{-1} = \mathbf{S}^\mathrm{T}$.

The eigenvalues of the potential matrix are

$$\chi_{1,2} = \frac{V_{11}+V_{22}}{2} \pm \frac{\phi}{\Delta t}, \tag{6.290}$$

with

$$\phi = \Delta t\sqrt{V_{12}^2+\lambda^2}, \tag{6.291}$$

and half the potential energy difference

$$\lambda = \frac{V_{11}-V_{22}}{2}. \tag{6.292}$$

The eigenvectors are given by

$$\boldsymbol{\xi}_1 = \begin{pmatrix}\cos(\Theta) \\ \sin(\Theta)\end{pmatrix} \quad \boldsymbol{\xi}_2 = \begin{pmatrix}-\sin(\Theta) \\ \cos(\Theta)\end{pmatrix}, \tag{6.293}$$

with

$$\cos(\Theta) = \left(V_{12}^2 + \left(\lambda - \frac{\phi}{\Delta t}\right)^2\right)^{-1/2} V_{12}, \tag{6.294}$$

$$\sin(\Theta) = \left(V_{12}^2 + \left(\lambda - \frac{\phi}{\Delta t}\right)^2\right)^{-1/2}\left(\lambda - \frac{\phi}{\Delta t}\right). \tag{6.295}$$

They constitute the orthogonal matrix via

$$\mathbf{S} = \begin{pmatrix} \cos(\Theta) & -\sin(\Theta) \\ \sin(\Theta) & \cos(\Theta) \end{pmatrix}. \tag{6.296}$$

Using the eigenvalues, we find that

$$\exp\{-\mathrm{i}\Delta t \mathbf{SVS}^{-1}\} = \exp\left\{-\mathrm{i}\Delta t \left(\frac{V_{11}+V_{22}}{2}\right)\right\} \begin{pmatrix} \mathrm{e}^{-\mathrm{i}\phi} & 0 \\ 0 & \mathrm{e}^{\mathrm{i}\phi} \end{pmatrix}. \tag{6.297}$$

Orthogonal transformation of this intermediate result with the matrix **S**, after some algebra, leads to

$$\exp\{-\mathrm{i}\Delta t \mathbf{V}\} = \exp\left\{-\mathrm{i}\Delta t \left(\frac{V_{11}+V_{22}}{2}\right)\right\} \begin{pmatrix} A & B \\ B & A^* \end{pmatrix}, \tag{6.298}$$

with the expressions

$$A = \cos\phi - \mathrm{i}\Delta t \lambda \frac{\sin\phi}{\phi}, \tag{6.299}$$

$$B = -\mathrm{i}\Delta t V_{12} \frac{\sin\phi}{\phi}, \tag{6.300}$$

given in (5.69).

5.10. The transition dipole μ_{01} in case of the harmonic oscillator was calculated already in Exercise 3.8, and we have found (generalizing to arbitrary mass m and charge q in atomic units)

$$\mu_{01} = q\sqrt{\frac{1}{2m\omega}} = \frac{q}{\sigma\sqrt{2}}, \tag{6.301}$$

with $\sigma = \sqrt{m\omega}$.

For the transition dipole from state one to two, we get (using the eigenfunctions (2.157) and the Gaussian integration formula (1.31))

$$\begin{aligned}\mu_{12} &= q\sqrt{\frac{\sigma^2}{2!2^2\pi\, 1!2!}} \int \mathrm{d}x\, H_1(\sigma x) H_2(\sigma x) x \exp\{-\sigma^2 x^2\} \\ &= q\frac{\sigma}{4\sqrt{\pi}} \int \mathrm{d}x\, 2\sigma x[4(\sigma x)^2 - 2] x \exp\{-\sigma^2 x^2\} \\ &= \frac{q}{4\sigma\sqrt{\pi}} \left(\frac{8\cdot 3\sqrt{\pi}}{4} - \frac{4\cdot\sqrt{\pi}}{2}\right) \\ &= \frac{q}{\sigma}.\end{aligned} \tag{6.302}$$

This leads to the scaling relation $\mu_{12}^2 = 2\mu_{01}^2$.

5.11. The discretized solution (5.119) can be written explicitly for $n = 1$ as

$$\begin{aligned}\chi_1(R, \Delta t) &= \frac{1}{\mathrm{i}}\Delta t \hat{U}_1(\Delta t)\mu_{10}\mathcal{E}(0)\hat{U}_0(0)\chi_0(R, 0)\\ &+ \frac{1}{\mathrm{i}}\Delta t \mu_{10}\mathcal{E}(\Delta t)\hat{U}_0(\Delta t)\chi_0(R, 0),\end{aligned} \tag{6.303}$$

where the first line is zero, due to the vanishing of the electric field. For $n = 2$ we find

$$\begin{aligned}\chi_1(R, 2\Delta t) &= \frac{1}{\mathrm{i}}\Delta t \hat{U}_1(2\Delta t)\mu_{10}\mathcal{E}(0)\hat{U}_0(0)\chi_0(R, 0)\\ &+ \frac{1}{\mathrm{i}}\Delta t \hat{U}_1(\Delta t)\mu_{10}\mathcal{E}(\Delta t)\hat{U}_0(\Delta t)\chi_0(R, 0)\\ &+ \frac{1}{\mathrm{i}}\Delta t \mu_{10}\mathcal{E}(2\Delta t)\hat{U}_0(2\Delta t)\chi_0(R, 0)\\ &= \hat{U}_1(\Delta t)\chi_1(\Delta t) + \frac{1}{\mathrm{i}}\Delta t \mu_{10}\mathcal{E}(2\Delta t)\hat{U}_0(2\Delta t)\chi_0(R, 0).\end{aligned} \tag{6.304}$$

Here $\hat{U}_1(2\Delta t) = \hat{U}_1(\Delta t)\hat{U}_1(\Delta t)$ and the result for $\chi_1(R, \Delta t)$ have been used. Finally for $n = 3$

$$\begin{aligned}\chi_1(R, 3\Delta t) &= \frac{1}{\mathrm{i}}\Delta t \hat{U}_1(3\Delta t)\mu_{10}\mathcal{E}(0)\hat{U}_0(0)\chi_0(R, 0)\\ &+ \frac{1}{\mathrm{i}}\Delta t \hat{U}_1(2\Delta t)\mu_{10}\mathcal{E}(\Delta t)\hat{U}_0(\Delta t)\chi_0(R, 0)\\ &+ \frac{1}{\mathrm{i}}\Delta t \hat{U}_1(\Delta t)\mu_{10}\mathcal{E}(2\Delta t)\hat{U}_0(2\Delta t)\chi_0(R, 0)\\ &+ \frac{1}{\mathrm{i}}\Delta t \mu_{10}\mathcal{E}(3\Delta t)\hat{U}_0(3\Delta t)\chi_0(R, 0)\\ &= \hat{U}_1(\Delta t)\chi_1(2\Delta t) + \frac{1}{\mathrm{i}}\Delta t \mu_{10}\mathcal{E}(3\Delta t)\hat{U}_0(3\Delta t)\chi_0(R, 0)\end{aligned} \tag{6.305}$$

is found, where again the previous result has been used and which is a special case of the given iteration formula.

The proof of the iteration prescription for arbitrary n can be performed with the help of mathematical induction.

5.12. The characteristic polynomial of the STIRAP matrix is

$$\begin{vmatrix} -\omega & -\frac{1}{2}\Omega_{\mathrm{P}} & 0 \\ -\frac{1}{2}\Omega_{\mathrm{P}} & -\omega & -\frac{1}{2}\Omega_{\mathrm{S}} \\ 0 & -\frac{1}{2}\Omega_{\mathrm{S}} & -\omega \end{vmatrix} = \omega\left(-\omega^2 + \frac{1}{4}(\Omega_{\mathrm{S}}^2 + \Omega_{\mathrm{P}}^2)\right) = 0, \tag{6.306}$$

with the solutions

$$\omega_{0,\pm}(t) = 0, \pm\frac{\Omega(t)}{2}, \tag{6.307}$$

given in (5.157) and where

$$\Omega(t) = \sqrt{\Omega_S^2(t) + \Omega_P^2(t)}. \tag{6.308}$$

The eigenvector corresponding to the zero eigenvalue is obtained from the solution of

$$\begin{pmatrix} 0 & -\frac{1}{2}\Omega_P & 0 \\ -\frac{1}{2}\Omega_P & 0 & -\frac{1}{2}\Omega_S \\ 0 & -\frac{1}{2}\Omega_S & 0 \end{pmatrix} \begin{pmatrix} f_1^0 \\ f_2^0 \\ f_3^0 \end{pmatrix} = \begin{pmatrix} 0 \\ 0 \\ 0 \end{pmatrix}. \tag{6.309}$$

It follows that $f_2^0 = 0$ and the remaining two entries of the eigenvector are obtained as in the two-level case (compare to the calculations in Sect. 3.2.2) from a (mixing) angle Θ, via $f_1^0 = \cos\Theta$, $f_3^0 = -\sin\Theta$, leading to

$$\frac{1}{2}(\Omega_P \cos\Theta - \Omega_S \sin\Theta) = 0 \tag{6.310}$$

and thus

$$\Theta(t) \equiv \arctan\left(\frac{\Omega_P(t)}{\Omega_S(t)}\right). \tag{6.311}$$

Exemplarily, the eigenvector corresponding to the positive eigenvalue is obtained from the solution of

$$\begin{pmatrix} \sqrt{\Omega_P^2 + \Omega_S^2} & \Omega_P & 0 \\ \Omega_P & \sqrt{\Omega_P^2 + \Omega_S^2} & \Omega_S \\ 0 & \Omega_S & \sqrt{\Omega_P^2 + \Omega_S^2} \end{pmatrix} \begin{pmatrix} f_1^+ \\ f_2^+ \\ f_3^+ \end{pmatrix} = \begin{pmatrix} 0 \\ 0 \\ 0 \end{pmatrix}. \tag{6.312}$$

The second equation can be solved by again using the same mixing angle as above, via the Ansatz $f_1^+ = \sin\Theta$, $f_2^+ = -1$, $f_3^+ = \cos\Theta$, leading to

$$\Omega_P \sin\Theta - \sqrt{\Omega_S^2 + \Omega_P^2} + \Omega_S \cos\Theta = 0. \tag{6.313}$$

A proof of this equation is possible with the relations

$$\arctan(x) = \arccos(1/\sqrt{1+x^2}) = \arcsin(x/\sqrt{1+x^2}). \tag{6.314}$$

In Dirac notation, the normalized eigenvector is thus given by

$$|g_+\rangle(t) = \frac{1}{\sqrt{2}}\left(\sin[\Theta(t)]|1\rangle - |2\rangle + \cos[\Theta(t)]|3\rangle\right). \tag{6.315}$$

It is orthogonal to the eigenvector of the zero eigenvalue.

The eigenvector corresponding to the negative eigenvalue is almost identical to the one in (6.315), but has a plus sign in front of $|2\rangle$, and can be shown to be orthogonal to the other two eigenvectors.

5.13. The variation of $J_{\mathcal{E}}$ given in (5.165) yields

$$\frac{\delta J_{\mathcal{E}}}{\delta \mathcal{E}^*} = \lambda \mathcal{E}(t), \tag{6.316}$$

whereas the $\mathcal{E}^*$-dependent part of J_H can be read off from

$$\begin{aligned} J_H &= \int_0^{T_t} \mathrm{d}t \left[\langle \boldsymbol{\xi}(t)|\frac{\hat{\mathbf{H}}}{\mathrm{i}}|\boldsymbol{\chi}(t)\rangle + \langle \boldsymbol{\chi}(t)|\frac{\hat{\mathbf{H}}}{-\mathrm{i}}|\boldsymbol{\xi}(t)\rangle \right] + \dots \\ &= \int_0^{T_t} \mathrm{d}t \, \frac{\mathcal{E}^*}{\mathrm{i}} \left[\langle \xi_\mathrm{g}|\mu|\chi_\mathrm{e}\rangle - \langle \chi_\mathrm{g}|\mu|\xi_\mathrm{e}\rangle \right] + \dots \, . \end{aligned} \tag{6.317}$$

The variation yields

$$\frac{\delta J_H}{\delta \mathcal{E}^*} = \left[\frac{1}{\mathrm{i}} \langle \xi_\mathrm{g}|\mu|\chi_\mathrm{e}\rangle - \langle \chi_\mathrm{g}|\mu|\xi_\mathrm{e}\rangle \right]. \tag{6.318}$$

Setting $-\frac{\delta J_{\mathcal{E}}}{\delta \mathcal{E}^*} + \frac{\delta J_H}{\delta \mathcal{E}^*} = 0$ leads to the expression (5.177) for the field.

5.14. To study the controllability of the population of harmonic oscillator eigenstates, we need the solution of the TDSE in the externally forced case.

(a) The solution of the TDSE of the driven harmonic oscillator

$$\hat{H} = \frac{\hat{p}^2}{2} + \frac{1}{2}x^2 - \mathcal{E}(t)x \tag{6.319}$$

can either be gained by using the appropriate propagator [6] or by using the explicit expression given by Husimi [19]

$$\begin{aligned} \chi_m(x,t) = \psi_m(x - \xi(t)) \\ \exp\left\{ \mathrm{i} \left[\int_0^t \mathrm{d}\tau L + \dot{\xi}(x - \xi) - E_m t \right] \right\}, \end{aligned} \tag{6.320}$$

with $\psi_m(x)$ being an eigenfunction of the undriven oscillator, see, e. g., (2.157) and with the Lagrangian

$$L = \frac{1}{2}\dot{\xi}^2 - \frac{1}{2}\xi^2 + \mathcal{E}(t)\xi. \tag{6.321}$$

This Lagrangian leads to the Newton equation

$$\ddot{\xi} + \xi - \mathcal{E}(t) = 0, \tag{6.322}$$

which is fulfilled by the special solution

$$\xi(t) = \int_0^t \mathrm{d}\tau \mathcal{E}(\tau) \sin(t - \tau). \tag{6.323}$$

You may want to check the solution given in (6.320) by inserting it into the TDSE!

(b) We now specialize to the case $m = 0$ for the initial state, and calculate the overlap with the time-independent final state $\chi_n(x) = \psi_n(x)$ with quantum number n, using $\int \mathrm{d}x\, \mathrm{e}^{-(x-y)^2} H_n(x) = \pi^{1/2}(2y)^n$ [14]. As a result we get

$$\begin{aligned} a_{n0}(t) &= \int \mathrm{d}x \psi_n^*(x)\psi_0(x - \xi(t)) \exp\left\{\mathrm{i}\left[\left[\int_0^t \mathrm{d}\tau L + \dot{\xi}\,(x - \xi) - E_0 t\right]\right\}\right. \\ &= \frac{1}{\sqrt{n! 2^n \pi}} \int \mathrm{d}x\, H_n(x) \exp\left\{-\frac{x^2}{2} - \frac{(x - \xi)^2}{2}\right. \\ &\left. + \mathrm{i}\left[\left[\int_0^t \mathrm{d}\tau L + \dot{\xi}(x - \xi) - E_0 t\right]\right]\right\} \\ &= \frac{1}{\sqrt{n!}}\left[\frac{1}{\sqrt{2}}(\xi + \mathrm{i}\dot{\xi})\right]^n \exp\left\{-\left(\frac{\xi^2}{4} + \frac{\dot{\xi}^2}{4}\right)\right. \\ &\left. + \mathrm{i}\left(\int_0^t \mathrm{d}\tau L - \frac{\dot{\xi}\xi}{2} - E_0 t\right)\right\}. \end{aligned} \tag{6.324}$$

(c) For the probability, we find

$$|a_{n0}(t)|^2 = \frac{1}{n!}\left[\frac{1}{2}(\xi^2 + \dot{\xi}^2)\right]^n \exp\left\{-\frac{1}{2}(\xi^2 + \dot{\xi}^2)\right\}. \tag{6.325}$$

This result is depending on the "parameter" $y = \frac{1}{2}(\xi^2 + \dot{\xi}^2)$ and in order to determine the maximum of $|a_{n0}(t)|^2$, one has to differentiate

$$\frac{\mathrm{d}}{\mathrm{d}y} y^n \mathrm{e}^{-y}|_{y_0} = n y^{n-1}\mathrm{e}^{-y} - y^n \mathrm{e}^{-y}|_{y_0} \stackrel{!}{=} 0, \tag{6.326}$$

leading to $y_0 = n$, implying the condition

$$\left[\int_0^t \mathrm{d}\tau \mathcal{E}(\tau) \sin(t - \tau)\right]^2 + \left[\int_0^t \mathrm{d}\tau \mathcal{E}(\tau) \cos(t - \tau)\right]^2 = 2n \tag{6.327}$$

for the field. The maximal probability is then given by

$$\mathrm{Max}\left(|a_{n0}|^2\right) = \frac{n^n \mathrm{e}^{-n}}{n!}. \tag{6.328}$$

For $n = 1$ this leads to around 37%. The simplest field that can achieve the optimum in this case is the impulsive field $\mathcal{E}(t) = \sqrt{2}\delta(t)$, as can be seen by inserting it into (6.327) [20].

References

1. K. Shimoda, *Introduction to Laser Physics* (Springer, Berlin, 1984)
2. W.P. Schleich, *Quantum Optics in Phase Space* (Wiley-VCH, Berlin, 2001)
3. E. Kamke, *Differentialgleichungen, Lösungsmethoden und Lösungen: I. Gewöhnliche Differentialgleichungen*, 7th edn. (Akademische Verlagsanstalt, Leipzig, 1961)
4. F. Grossmann, M. Werther, L. Chen, Y. Zhao, Chem. Phys. **481**, 99 (2016)
5. D.J. Tannor, *Introduction to Quantum Mechanics: A Time-Dependent Perspective* (University Science Books, Sausalito, 2007)
6. R.P. Feynman, A.R. Hibbs, *Quantum Mechanics and Path Integrals*, emended edn. (Dover, Mineola, 2010)
7. C. Leforestier, R.H. Bisseling, C. Cerjan, M.D. Feit, R. Friesner, A. Guldberg, A. Hammerich, G. Jolicard, W. Karrlein, H.D. Meyer, N. Lipkin, O. Roncero, R. Kosloff, J. Comp. Phys. **94**, 59 (1991)
8. W.H. Louisell, *Quantum Statistical Properties of Radiation* (Wiley, New York, 1973)
9. R. Gelabert, X. Giménez, M. Thoss, H. Wang, W.H. Miller, J. Phys. Chem. A **104**, 10321 (2000)
10. G. Di Liberto, M. Ceotto, J. Chem. Phys. **145**, 144107 (2016)
11. D.H. Kobe, E.C.T. Wen, J. Phys. A **15**, 787 (1982)
12. M. Göppert-Mayer, Ann. Phys. (Leipzig) **9**, 273 (1931)
13. J.J. Sakurai, *Modern Quantum Mechanics* (Addison-Wesley, Reading, 1994)
14. I.S. Gradshteyn, I.M. Ryzhik, *Table of Integrals Series and Products*, 5th edn. (Academic Press, New York, 1994)
15. M. Klaiber, D. Dimitrovski, J.S. Briggs, J. Phys. B **41**, 175002 (2008)
16. J.C. Slater, *Quantum Theory of Molecules and Solids*, vol. 1 (McGraw-Hill, New York, 1963)
17. S.T. Epstein, J. Chem. Phys. **44**, 836 (1966)
18. V.F. Brattsev, Dokl. Akad. Nauk SSSR **160**, 570 (1965)
19. K. Husimi, Prog. Theor. Phys. **9**, 381 (1953)
20. A.G. Butkovskiy, Y.I. Samoilenko, *Control of Quantum Mechanical Processes and Systems* (Kluwer, Dordrecht, 1990)

Erratum to: Time-Dependent Quantum Theory

Erratum to:
Chapter 2 in: F. Grossmann, *Theoretical Femtosecond Physics*, Graduate Texts in Physics, https://doi.org/10.1007/978-3-319-74542-8_2

In the original version of the book, the belated correction to update Fig. 2.7 has to be incorporated in Chapter 2. The erratum chapter and the book have been updated with the change.

The updated online version of this chapter can be found at
https://doi.org/10.1007/978-3-319-74542-8_2

F. Grossmann, *Theoretical Femtosecond Physics*, Graduate Texts in Physics,
https://doi.org/10.1007/978-3-319-74542-8_7

Index

F. Grossmann, *Theoretical Femtosecond Physics*, Graduate Texts in Physics,
https://doi.org/10.1007/978-3-319-74542-8

Printed in the United States
By Bookmasters